Layered Deposits: Stratigraphic Studies

Layered Deposits: Stratigraphic Studies

Edited by **Joe Carry**

R CALLISTO REFERENCE

New York

Published by Callisto Reference,
106 Park Avenue, Suite 200,
New York, NY 10016, USA
www.callistoreference.com

Layered Deposits: Stratigraphic Studies
Edited by Joe Carry

International Standard Book Number: 978-1-63239-447-7 (Hardback)

Printed in the United States of America.

Contents

Preface   VII

Section 1   Application of Geophysical Techniques
in Stratigraphic Investigations   1

Chapter 1   Seismic Stratigraphy and Marine Magnetics
of the Naples Bay (Southern Tyrrhenian Sea, Italy):
The Onset of New Technologies in Marine
Data Acquisition, Processing and Interpretation   3
Gemma Aiello, Laura Giordano,
Ennio Marsella and Salvatore Passaro

Chapter 2   Medium to Shallow Depth Stratigraphic
Assessment Based on the Application
of Geophysical Techniques   43
Roberto Balia

Chapter 3   Orbital Control on Carbonate-Lignite Cycles
in the Ptolemais Basin, Northern Greece
– An Integrated Stratigraphic Approach   61
M.E. Weber, N. Tougiannidis, W. Ricken,
C. Rolf, I. Oikonomopoulos and P. Antoniadis

Chapter 4   Ground Penetrating Radar:
A Useful Tool for Shallow
Subsurface Stratigraphy Characterization   79
Giovanni Leucci

Section 2   Biostratigraphy   105

Chapter 5   The Muhi Quarry:
A Fossil-Lagerstätte from the Mid-Cretaceous
(Albian-Cenomanian) of Hidalgo, Central México   107
Victor Manuel Bravo Cuevas, Katia A. González Rodríguez,
Rocío Baños Rodríguez and Citlalli Hernández Guerrero

Chapter 6   The Paleogene Dinoflagellate Cyst
             and Nannoplankton Biostratigraphy
             of the Caspian Depression   123
             Olga Vasilyeva and Vladimir Musatov

Chapter 7   Pliocene Mediterranean Foraminiferal Biostratigraphy:
             A Synthesis and Application to the Paleoenvironmental
             Evolution of Northwestern Italy   157
             Donata Violanti

Chapter 8   Late Silurian-Middle Devonian Miospores   195
             Adnan M. Hassan Kermandji

Section 3   Sequence Stratigraphy   225

Chapter 9   Paleocene Stratigraphy in Aqra
             and Bekhme Areas, Northern Iraq   227
             Nabil Y. Al-Banna, Majid M. Al-Mutwali and Zaid A. Malak

Section 4   Tectonostratigraphy   253

Chapter 10   Tektono-Stratigraphy as a Reflection
             of Accretion Tectonics Processes
             (on an Example of the Nadankhada-Bikin
             Terrane of the Sikhote-Alin Jurassic
             Accretionary Prism, Russia Far East)   255
             Igor V. Kemkin

Chapter 11   Sedimentary Tectonics and Stratigraphy:
             The Early Mesozoic Record in Central
             to Northeastern Mexico   275
             José Rafael Barboza-Gudiño

             Permissions

             List of Contributors

Preface

The world is advancing at a fast pace like never before. Therefore, the need is to keep up with the latest developments. This book was an idea that came to fruition when the specialists in the area realized the need to coordinate together and document essential themes in the subject. That's when I was requested to be the editor. Editing this book has been an honour as it brings together diverse authors researching on different streams of the field. The book collates essential materials contributed by veterans in the area which can be utilized by students and researchers alike.

Stratigraphic analyses of layered deposits are described in this all-inclusive book. Stratigraphy is a geological sub-specialty which is defined as the science of describing the lateral and vertical relationships of distinct rock formations formed through time to comprehend the earth history. These interrelations might be the result of variety of factors including fossil content, magnetic properties, reflection seismology or archaeological deposits. Their study and analysis comprise of various sub-specialties of stratigraphy including lithostratigraphy, biostratigraphy, magnetostratigraphy, chemostratigraphy, seismic stratigraphy, chronostratigraphy and archaeological stratigraphy. During the late 1700s, James Hutton said -"the present is the key to the past", a concept further developed by Charles Lyell in the early 1800s. This has evolved as the basis on which stratigraphy has worked ever since. This book lays specific emphasis on applications of geophysical techniques in stratigraphic investigations and analysis of layered basin deposits from distinct geologic settings across continents ranging from Mexico region (North America) through the Alpine belt including Greece, Italy, and Iraq to Russia (Northern Asia).

Each chapter is a sole-standing publication that reflects each author´s interpretation. Thus, the book displays a multi-facetted picture of our current understanding of application, resources and aspects of the field. I would like to thank the contributors of this book and my family for their endless support.

Editor

Section 1

Application of Geophysical Techniques in Stratigraphic Investigations

Seismic Stratigraphy and Marine Magnetics of the Naples Bay (Southern Tyrrhenian Sea, Italy): The Onset of New Technologies in Marine Data Acquisition, Processing and Interpretation

Gemma Aiello, Laura Giordano, Ennio Marsella and Salvatore Passaro
Istituto per l'Ambiente Marino Costiero (IAMC),
Consiglio Nazionale delle Ricerche (CNR),
Napoli,
Italy

1. Introduction

Seismic stratigraphy and marine magnetics in the case histories of the Somma-Vesuvius volcanic complex, Phlegrean Fields offshore and Ischia and Procida islands offshore (Naples Bay, Southern Tyrrhenian sea) are here discussed. Detailed geo-volcanologic setting of these areas is presented to give a better framework of the presented data. Seismo-stratigraphic techniques and methodologies are discussed, focussing, in particular, on the Naples area, where the Quaternary volcanic activity prevented the application of classical stratigraphic concepts, due to the occurrence of interlayered sedimentary sequences and intervening volcanic bodies (volcanites and volcaniclastites). The onset of new technologies in marine data acquisition, processing and interpretation is also discussed taking into account some historical aspects.

2. Seismo-stratigraphic techniques and methodologies

The applied stratigraphic subdivision derives from the type of data utilized in marine geology (reflection seismics) and by the methods of seismic interpretation (high resolution sequence stratigraphy). The geological structures recognized through the seismic interpretation are the acoustically-transparent volcanic units, representing the rocky acoustic basement and the systems tracts of the Late Quaternary depositional sequence (Fabbri et al., 2002). The widespread volcanic activity, which controlled the stratigraphic architecture of the Naples Bay during the Late Quaternary, has disallowed the application of a classical stratigraphic approach, due to the occurrence of interlayered sedimentary sequences and intervening volcanic bodies (volcanites and volcaniclastites).

In the Late Quaternary Depositional Sequence (SDTQ) the seismo-stratigraphic analysis has allowed to characterize depositional systems respectively referred to the sea level fall (FST; Helland Hansen & Gjelberg, 1994), to the sea level lowstand (LST) and related internal subdivisions (Posamentier et al., 1991), to the transgressive phase (TST; Posamentier &

Allen, 1993; Trincardi et al., 1994) and to the highstand phase of sea level (HST; Posamentier & Vail, 1988). A sketch stratigraphic diagram of the SDTQ and its components has been constructed in order to clarify the stratigraphic relationships between systems tracts (Fig. 1).

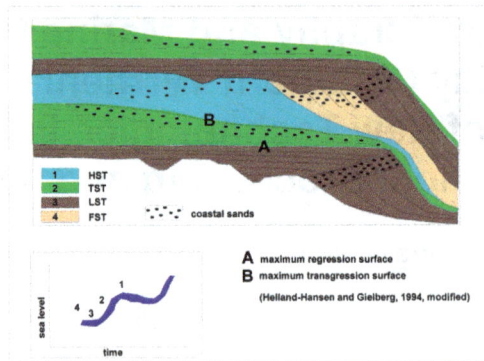

Fig. 1. Sketch stratigraphic diagram as a function of both depth (upper diagram) and time (lower diagram) showing the geometric relationships between systems tracts and distribution of siliciclastic facies in unconformity-bounded depositional sequences (modified after Vail et al., 1977; Christie-Blick, 1991; Helland Hansen & Gjelberg, 1994).

2.1 Main components of the Late Quaternary depositional sequence

The main characteristics of the system tracts related to main phases of the last sea level glacio-eustatic cycle are here discussed, focussing, in particular, on the Naples Bay. The stratigraphic meaning and the map representation of each system tract are also resumed.

2.1.1 Highstand deposits (HST)

The highstand deposits are younger than the phase of maximum marine ingression happened at the end of the sea-level rise (about 4-5 ky B.P.) and show their maximum thickness in the inner shelf, next to the main deltas (i.e. Po, Tiber, Arno, etc.) along the Italian coast, while reduce to a few meters on the outer shelf (Fabbri et al., 2002). Highstand deposits of the Naples Bay have been intensively studied in the frame of research projects of marine geological mapping of the Campania Region (Aiello et al., 2001; D'Argenio et al., 2004; Sacchi et al., 2005; Insinga et al., 2008; Molisso et al., 2010; Fig. 2).

2.1.2 Transgressive deposits (TST)

The transgressive deposits, originated in continental, coastal paralic or marine environment during the phases successive to the Late Quaternary sea level rise generally appear reduced in thickness and studied with very high resolution seismic profiles and piston cores. The Italian continental margins document the variability of facies, internal geometry, sedimentologic expression and marker horizons (Trincardi et al., 1994). In the Naples Bay the TST was deposited during the rising of sea level (18-6 ky). It has been widely documented by Milia & Torrente (2000; 2003) and consists of three minor stratigraphic units. The second of them corresponds to a thick progradational unit overlying the Neapolitain Yellow Tuff (18 ky B.P.), the Penta Palummo Bank and the Miseno Bank.

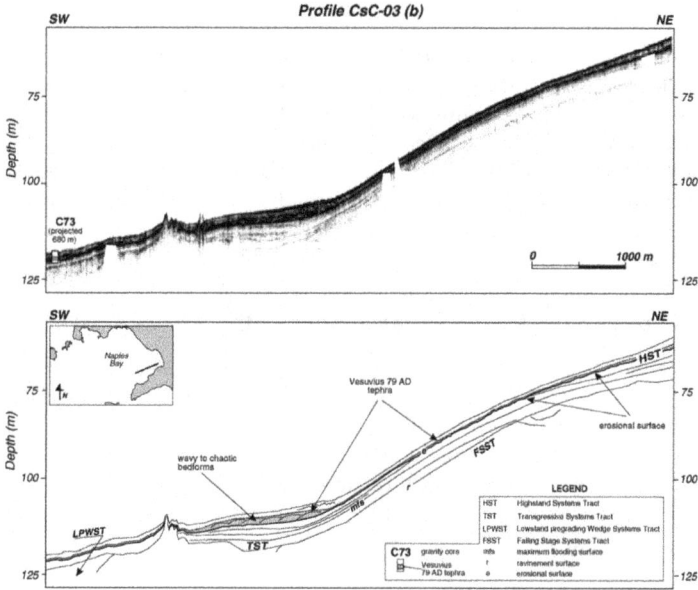

Fig. 2. Subbottom Chirp profile CsC-03 (b) across the Sarno prodelta highstand deposits (eastern Naples Bay) and corresponding geological interpretation (slightly modified after Sacchi et al., 2005).

2.1.3 Lowstand system tract deposits (LST)

The deposits originated in sea level lowstand during the last Quaternary glacial episode (isotopic stage 2; Martinson et al., 1987) may be separated in mass transport deposits, base of slope turbiditic systems and shelf margin progradational wedges. Each sector of continental margin does not include all the three types of deposits, but only one or two. The development of each of the three types of lowstand deposits is a function of the morphological setting and regime of clastic supply. The mass transport deposits usually have a great lateral extension and are characterized by chaotic reflections or acoustic transparence, erosional base and thickness from several meters (Marani et al., 1986; Mongardi et al., 1995; Trincardi & Normark, 1988; Trincardi et al., 1994).

2.1.4 Falling sea level system tract deposits (FST)

The Mediterranean continental margins show several examples of falling sea level deposits, which can be characterized by different geometries, thickness, areal extension and lithology (Tesson et al., 1993; Trincardi & Field, 1991; Hernandez Molina et al., 1994). They consist of progradational wedges emplaced through a mechanism of erosional or forced regression recognized through the progressive seaward and downward shifting of the coastal onlap. In the Naples Bay progradational units (FST and LST), the tops of which are located at depths ranging from – 130 m and – 150 m, occur at the seaward termination of the wide subaerial erosional surface that affects the volcanic banks (Milia & Torrente, 2000; 2003; Aiello et al., 2005).

2.1.5 Sequence boundaries (SB)

Two sequence boundaries have been defined as a function of the ratio between the rate of sea level fall and the rate of subsidence at the shelf margin (Vail et al., 1984; Posamentier & Allen, 1993). Type 1 sequence boundaries form when the rate of eustatic sea level fall exceeds the rate of subsidence; as a consequence, the subaerial exposure of the whole continental shelf occurs. Type 2 sequence boundaries characterize the continental margins in which the rate of subsidence of the outer shelf is higher than the rate of sea level fall; more or less extended parts of the continental shelf rest submerged or subject to deposition. Type 1 sequence boundaries are characterized by more extended phenomena of fluvial incision.

2.2 Facies analysis and schematic representation of the depositional environments

The system tracts of the Late Quaternary depositional sequence include deposits characterized by facies related to continental, paralic coastal, shelf and deep sea depositional environments. The marker horizons constituting the base and the top of the system tracts may be represented by variable sedimentological expressions due to the differences between the facies and the occurrence and entity of related erosional phenomena.

2.2.1 Continental deposits

The continental deposits may occur in shallow areas controlled by subaerial exposure during Quaternary glacial periods. They consist of alluvial plain deposits in which extended fluvial systems have been recognized, characterized by channel deposits with incised thalwegs and levees. The inter-channel zones are characterized by soil formation. The filling of the fluvial incisions may be characterized by sediments highly varying in grain-size and by filling geometries related to meanders or braided streams.

2.2.2 Paralic and coastal deposits

The coastal and paralic depositional systems greatly vary both in morphology and depositional style. This variability reflects different budgets between the available sediments (type and quantity) and the oceanographic regime (wave-dominated, tide-dominated or mixed). As a general rule, the Mediterranean is characterized by a microtidal regime; the most of the coastal systems on the Italian margins is dominated by the waves. Coastal and paralic deposits may theoretically form in each phase of a relative sea level fluctuation cycle, but are characterized by different facies; regressive systems form in condition of sea level fall (forced regressions; Posamentier et al., 1992) or when the siliciclastic supply counterbalances the relative sea level rate.

2.2.3 Continental shelf deposits

The sediments of the actual continental shelves may be summarized into three main types (Fabbri et al., 2002):

- Sediments deposited in a phase during which the shoreline was seaward advanced with respect to the present-day location and successively drowned (relic sediments).
- Sediments deposited in a phase during which the shoreline was seaward advanced with respect to the present-day location and successively drowned, but then reworked due to currents, storm waves or tides (palinsest sediments).

- Sediments related to the Late Quaternary highstand in equilibrium with the present-day depositional processes.

The continental shelf of the Naples Bay has a variable width, ranging between 2.5 km (offshore the western sector of Capri island) and 10-15 km (offshore the Sorrento coast). Such a submarine topography is controlled by the interactions between subaerial and submarine volcanism, strongly involving the Gulf during the Late Pleistocene and the linear erosion and sediment drainage along main axis of Dohrn and Magnaghi canyons (Aiello et al., 2005; Di Fiore et al., 2011). The eruption centres occurring on the islands of Procida, Vivara and Ischia range in age between 150 kyr and historical times (Rosi & Sbrana, 1987; Vezzoli, 1988).

2.2.4 Deep sea deposits

Slopes, basins or submarine highs are less influenced by the sea level fluctuations during the Late Quaternary. Based on the piston core data acquired during the last 30 ky in all the Mediterranean sea it is clear that the last sea level rise and the successive sea level highstand are represented by drapes of clayey sediments (Holocene drapes). Under the Holocene drapes four main types of deposits occur (Fabbri et al., 2002):

- Turbidite deposits having variable nature (referred to specific depositional elements as channel-levee systems, lobes and distal, not channellised deposits).
- Mass gravity transport deposits.
- Deposits originated by bottom currents and related erosional or condensed surfaces.
- Pelagic drapes.

Different types of turbiditic deposits, mass gravity transport deposits and pelagic drapes have been widely recognized on the sea bottom of the Naples Bay, in the frame of research programmes of submarine geological mapping (Aiello et al., 2001; 2008; 2009b; 2009c).

2.2.5 Mass gravity transport deposits

Mass gravity transport deposits, varying in nature, internal organization and areal extension have been recognized in the Late Quaternary successions of the Italian Peninsula. Their emplacement may happen under conditions of lowstand, relative sea level rise and highstand of the sea level (Galloway et al., 1991; Correggiari et al., 1992; Trincardi & Field, 1991; Trincardi et al., 2003; Aiello et al., 2009c; Di Fiore et al., 2011). Sketched tables of the geological interpretation of selected Subbottom Chirp profiles have been constructed in order to show significant instability processes occurring in the Naples Bay (Aiello et al., 2009c; Fig. 3).

3. Geo-volcanologic setting

The Campania Tyrrhenian margin is characterized by the occurrence of marine areas, strongly subsident during the Plio-Quaternary, sites of thick sedimentation, as the Volturno Basin, the Naples Bay, the Salerno Valley and the Sapri and Paola basins (peri-tyrrhenian basins; Argnani & Trincardi, 1990; Boccaletti et al., 1990; Tramontana et al., 1995; Gabbianelli et al., 1996; Aiello et al., 2000). Under the Plio-Quaternary sedimentary cover, the Campania continental margin is characterized by the occurrence of tectonic units of Apenninic chain, resulting from the seaward prolongation of corresponding units cropping out on the coastal belt of Southern Apennines (D'Argenio et al., 1973; Bigi et al., 1992; Fig. 4).

DESCRIPTION	GEOLOGIC INTERPRETATION
A Chaotic reflections with interrupted lateral continuity	Volcanoclastic and/or pyroclastic deposits (mud flow, lahars, piroclastic flows), genetically linked to the deposition in submarine environment of the pyroclastic fluxes of the recent vesuvian activity(< 2 ky; see also Milia et al., 2008)
B Acoustically-transparent bodies strongly convex buried and/or cropping out at the sea bottom alternated to parallel reflectors wavy or concave.	Pyroclastic mounds constituted by alternating coarse volcanogenic sands and pumice levels, with fillings or covers of shales. The mounds are genetically linked to the "Tufo Giallo Napoletano" Auct. (12 ka) and/or to the "Pomici Principali" Auct. (10Ka), since they are located in the upper part of a seismic unit interpreted as the"Tufo Giallo Napoletano" Auct., developed starting from the coastal cliff off Posillipo (Naples town) up to the outer shelf off the Naples town (see also Aiello et al., 2001).
C Convex reflectors with high lateral continuity (periodically interrupted) overlying a strong seismic reflector bounding their base.	Holocene marine sediments involved into creeping (probably due to high contents of organic matter) overlying a sharp surface of separation corresponding to the maximum flooding surface of the last glacio-eustatic cycle (see also Aiello et al., 2001).
D Parallel to slightly inclined seismic reflectors,grading upwards to an acoustically- transparent seismic facies, overlain by a thin drape of parallel and continuous seismic reflectors	Outcrops of relic sands deposited during the last phase of sea level lowstand (18-20 ky) of the last glacio-eustatic cycle, covered by a thin drape of Holocene sediments (highstand drape).

Fig. 3. Sketched table showing significant acoustic facies related to submarine gravity instability on the continental shelf of the Naples Bay (reported after Aiello et al., 2009c).

The main structural trends of the Campania margin are NW-SE and NNW-SSE (Apenninic) and are characterized, on the continental slope and in the bathyal plain, by the occurrence of intra-slope basins and structural highs, showing hints of intense synsedimentary tectonics (Aiello et al., 2009a). Two main NE-SW (counter-Apenninic) trending lineaments, i.e. the Phlegrean Fields-Ischia fault and the Capri-Sorrento Peninsula fault, control the structural setting of the Naples Bay. These lineaments have controlled the emplacement of main morpho-structures on the continental slope and in the bathyal plain. In a time interval spanning from the Middle to the Late Pleistocene the synsedimentary tectonics has played a major role in triggering submarine gravity instabilities.

3.1 Somma-Vesuvius volcanic complex

The Vesuvius volcano has been intensively studied, mainly with respect to the eruptive events, the recent seismicity, the geochemistry and the ground movements of the volcano and the related volcanic hazard (Cassano & La Torre, 1987; Santacroce et al., 1987; Castellano et al., 2002; Esposti Ongaro et al., 2002; Mastrolorenzo et al., 2002; Saccorotti et al., 2002; Scarpa et al., 2002; Todesco et al., 2002).

The eruption of the Campanian Ignimbrite pyroclastic flow deposits (37 ky B.P.; Rosi & Sbrana, 1987) covered the whole Campania Region and part of the adjacent offshore with grey tuff deposits, upon which the Somma edifice started to grow. The eruptive activity ranges from the "Pomici di Base" (18 ky) and the "Pomici Verdoline" plinian eruptions, enabling the collapse of the Somma edifice and the consequent calderization, with the formation of a new volcanic edifice, i.e. the Vesuvius. The period from 8000 B.C. to 79 A.D. was characterized by

Seismic Stratigraphy and Marine Magnetics of the Naples Bay (Southern Tyrrhenian Sea, Italy): The Onset of
New Technologies in Marine Data Acquisition, Processing and Interpretation

9

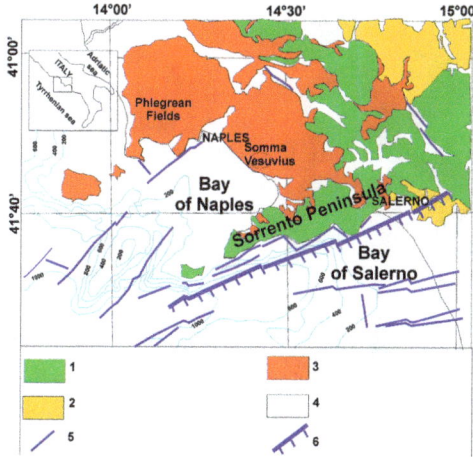

Fig. 4. Tectonic sketch map of western Campania Apennines (modified after D'Argenio et al., 2004). Key. 1: Shallow water carbonate and deep basinal units (Mesozoic). 2: Piggy-back siliciclastic units (Tertiary). 3: Pyroclastic deposits and lavas (Quaternary). 4: Continental and marine deposits (Quaternary). 5: Normal fault. 6: Detachment faults (barbs indicate downthrown side).

three main plinian eruptions: "Mercato" (7900 B.C.), "Avellino" (3800 B.C.) and the Pompei eruptions (79 A.D.). The activity continued with the Pollena eruption (472 A.D.) and the 1631 A.D. eruption and then with several small energy, effusive and explosive, giving rise to lava flows along the western and southern slopes of the volcano (Sheridan et al., 1981; 1982; Sigurdsson et al., 1982; 1984; Santacroce, 1987). The variability in the eruptive behaviour of the Vesuvius volcano has been explained by volcanologists with an alternation between periods of open conduits and periods of closed conduits, the latter being characterized by a relative quiescence followed by Plinian eruptions. The periods with open conduits were characterized by a permanent strombolian activity, frequent lava flows and mixed eruptions, both effusive and explosive (Rosi & Santacroce, 1983; Rosi et al., 1983; Arnò et al., 1987). Constraints on the sedimentary basement overlying the volcano and its stratigraphic relationships with the Phlegrean volcanic products can be explained by the deep geothermal well Trecase 1, drilled by the AGIP-ENEL joint venture on the south-eastern slopes of the volcano (Balducci et al., 1985; Brocchini et al., 2001). Sketch stratigraphy of the Trecase 1 exploration well is reported in Fig. 5.

The total magnetic field map of of Somma-Vesuvius volcano shows interpretative elements that have an indicated value for the trend of volcanites in the volcanic complex's peripheral areas (Cassano & La Torre, 1987; fig. 6). From the main sub-circulary anomaly centred on the volcano, two positive appendages diverge towards SE and SW. They might correspond to a great thickness of lava products, possibly in pre-existing depressions of the sedimentary basement of the graben of the Campania Plain. This assumption might explain an elongated magnetized body, which tends to move towards Naples Bay from the Vesuvius volcano through Torre del Greco; an alternative explanation would be the presence of a strip of eruptive vents, settled on a system of NE-SW normal faults (Bernabini et al., 1971; Finetti & Morelli, 1973).

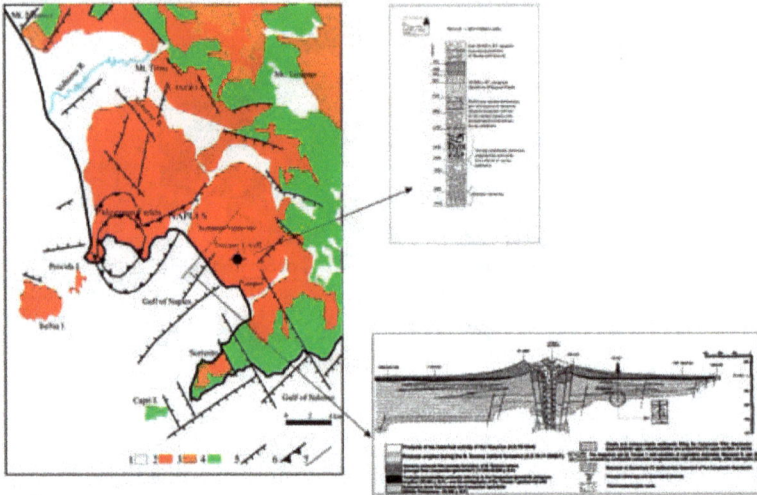

Fig. 5. Geological and structural sketch map of the Southern Campania Plain (after Aiello et al., 2010) . 1: Quaternary siliciclastic sediments. 2: Somma-Vesuvius volcanic deposits; Neapolitain-Phlegrean, Procida and Ischia volcanic deposits. 3: Pliocene and Miocene siliciclastic sediments. 4: Meso-Cenozoic carbonatic units. 5: faults. 6: caldera rims. 7: geological cross-sections. In the inset on the right: geological cross-section of the Somma-Vesuvius volcanic complex (from Principe et al., 1987). In the other inset: stratigraphy of the Trecase 1 exploration well drilled on the Somma-Vesuvius volcanic complex.

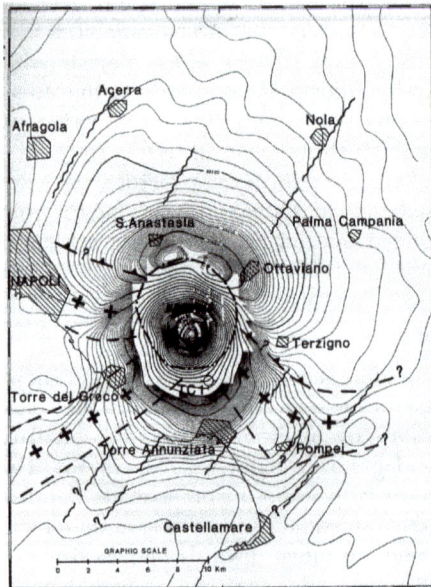

Fig. 6. Total magnetic field map of the Somma-Vesuvius area with sketch structural interpretation (slightly modified after Cassano & La Torre, 1987)

A new aeromagnetic map of the Vesuvian area has been recently produced (Paoletti et al., 2005; fig. 7). It is dominated by a large dipolar anomaly related to the Somma-Vesuvius volcanic complex, having an elliptical shape elongated towards south. Main geological structures of the area are a narrow anomaly on the western flank of the edifice (A in fig. 7) and an irregular shape of the anomaly on the south-eastern slope of the volcano, where small anomalies have been observed (B, C and D). A double minimum at the top of the volcano is articulated in a bigger one placed north of Mt. Somma and a larger one next to Valle dell'Inferno. High frequency anomalies occur in the area surrounding the edifice, related to the high cultural noise of this densely inhabited area.

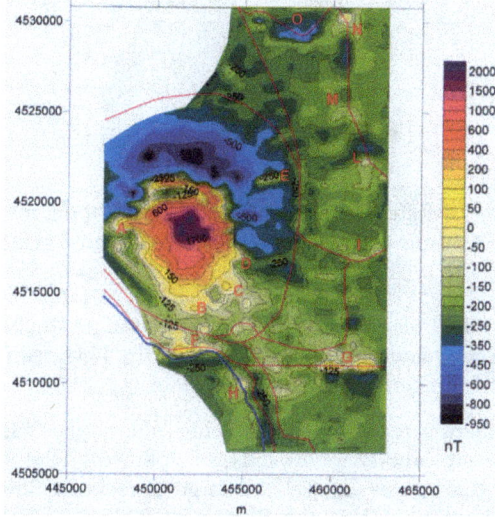

Fig. 7. Aeromagnetic map of the Vesuvian area (after Paoletti et al., 2005). The red lines show the railway lines, the blue line shows the coastline (see the text for the description of the magnetic anomalies).

3.2 Phlegrean Fields volcanic complex

The Phlegrean Fields are a volcanic district surrounding the western part of the Gulf of Naples, where volcanism has been active since at least 50 kyr (Rosi & Sbrana, 1987). They correspond to a resurgent caldera (Rosi & Sbrana, 1987; Orsi et al., 2002) with a diameter of 12 km (Phlegrean caldera) and resulting from the volcano-tectonic collapse induced from the eruption of pyroclastic flow deposits of the Campanian Ignimbrite (37 ky B.P.). Coastal sediments ranging in age from 10.000 to 5300 years crop out at 50 m altitude on the sea level in the marine terrace of "La Starza" (Gulf of Pozzuoli), indicating a volcano-tectonic uplift of the caldera center (Rosi & Sbrana, 1987; Dvorak & Mastrolorenzo, 1991).

Monogenic volcanic edifices, probably representing the offshore rim of the caldera center (Banco di Pentapalummo, Banco di Miseno, Banco di Nisida) are well known from a geological and volcanological point of view (Latmiral et al., 1971; Pescatore et al., 1984; Fusi et al., 1991; Milia, 1996; Aiello et al., 2001; 2005; Fig. 8).

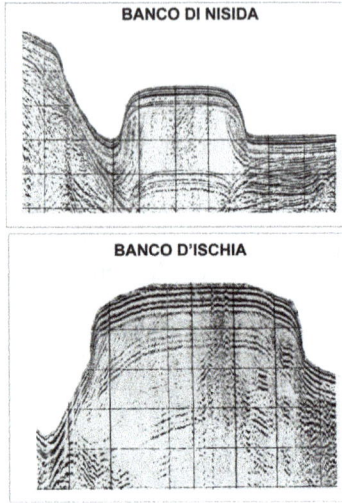

Fig. 8. Sparker seismic profiles showing the Banco di Nisida (upper inset) and the Banco di Ischia (lower inset) relic volcanic edifices (modified after Latmiral et al., 1971). The vertical scale is of 250 msec (twt).

The geological setting of the Phlegrean Fields and their stratigraphy have been discussed by Rosi & Sbrana (1987). The Quaternary volcanic area of the Phlegrean Fields is located in a central position within the graben of the Campania Plain.

The main structural element is represented by a wide caldera (the Phlegrean caldera; fig. 9), individuated after the volcano-tectonic collapse following the emplacement of the Campanian Ignimbrite (Barberi et al., 1991), a large pyroclastic flow, which covered the whole plain about 37 ky ago. Within the Phlegrean caldera and along its margins, the volcanic activity continued since to historical times (Rosi & Sbrana, 1987).

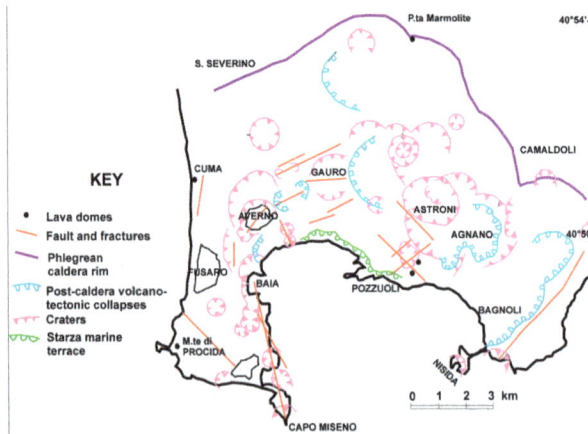

Fig. 9. Volcano-tectonic sketch map of the Phlegrean Fields volcanic complex (modified after Rosi & Sbrana, 1987).

Main volcano-tectonic structures are the Starza marine terrace, the Miseno-Baia area., the
Mofete area and the central zone of Pozzuoli-Solfatara-Agnano (Rosi & Sbrana, 1987; fig. 9).
The Starza terrace is a marine erosional terrace placed near the centre of the caldera, composed
of littoral deposits overlain by thin subaerial pyroclastic deposits. It is articulated in two levels
separated by a step; the upper one, on which the town of Pozzuoli is superimposed, reaches a
height of 50-54 m, while the lower one develops at heights of 40 m. The Miseno-Baia area is
characterized by an active fault at the Bacoli harbour, downthrowing the tuffs of the Bacoli
volcano. Contemporaneous eruption in different sectors of the caldera have been suggested
(Isaia et al., 2009).

Gravimetric and magnetometric informations available for Phlegrean Fields have been
summarized (Cassano and La Torre, 1987) focussing on volcanological and structural
reconstruction of the area. From north to south, the most important gravimetric elements are
the positive anomaly related to the carbonatic horst of Massico Mt., the negative anomaly of
the Volturno graben, the positive gravimetric anomalies of Villa Literno and Parete, a marked
gravimetric gradient with a counter-Apenninic trend, crossing the Somma-Vesuvius volcanic
complex, and, to the south, separating the Acerra graben from the Pompei graben and finally,
the gravimetric gradient corresponding to the Sorrento Peninsula. The total magnetic field
map (Cassano & La Torre, 1987; fig. 10) has evidenced a strong positive anomaly in the area of
Monte di Procida, related to the weaker anomalies of the Procida Channel, Procida and Ischia.
It may be related to considerable volumes of lavas, confirmed by the presence of trachybasaltic
and latitic eruptive centres at Procida. Another large magnetic anomaly characterizes the
Astroni-Agnano volcanic area, probably the result of the overlapping of several lava bodies.
Positive anomalies have been found at Camaldoli, probably related to pre-calderic lavas. The
absence of magnetic anomalies in the Bagnoli-Posillipo area may be due to several factors,
such as the limited presence of buried lavas or hydrothermal phenomena.

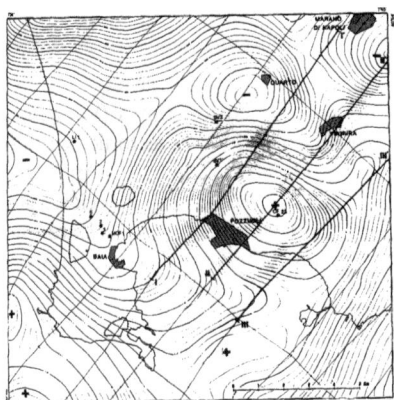

Fig. 10. Total magnetic field map of the Phlegrean Fields area (after Cassano and La Torre,
1987).

A new aeromagnetic map supplement of the northern sector of the Phlegrean Fields allows
for a better geological interpretation of the structural patterns and morpho-structural
features of the Volturno Plain and the Gulf of Pozzuoli and its offshore areas (fig. 11). Main
magneto-structural features are the caldera rims of the Neapolitain Yellow Tuff (fig. 11; A)

and the Torregaveta anomaly (fig. 11; B). A small anomaly corresponds to an isolated volcanic body (fig. 11; C). The Patria Lake anomaly (fig. 11; D) has a sub-circular shape and a diameter of about 10 km. A complex pattern of magnetic anomalies (fig. 11; E) coincides with the Parete volcanic complex (Aiello et al., 2011), while another isolated anomaly (fig. 11; F) corresponds to the Volturno river.

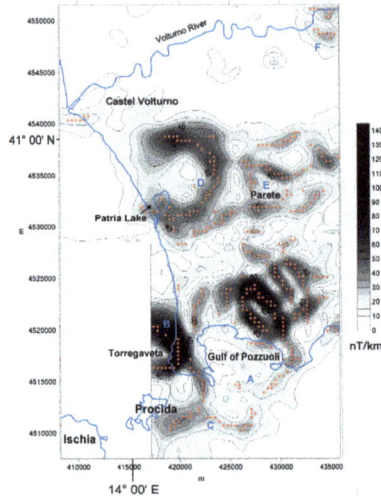

Fig. 11: Map of the horizontal derivative plotted in the gray scale of the southern Volturno Plain (reported from Paoletti et al., 2004). The letters A-F indicate the main magnetic anomalies recognized in the area.

3.3 Ischia and Procida volcanic complexes

The Ischia island represents an alkali-trachytic volcanic complex, whose eruptive activity lasted from the Late Pleistocene up to historical times (Vezzoli, 1988). A resurgent caldera, about 10 km wide, where the eruptive activity and tectonics gave rise to the uplift along faults of the Mount Epomeo block (Orsi et al., 1991). The main eruptive events of the Ischia-Procida-Phlegrean Fields system suggest at least five eruptive cycles, ranging in age from 135 ky to prehistorical and historical times. On the Ischia island volcanic deposits, resulting from both effusive and eruptive eruptions, extensively crop out and have constructed volcanic edifices; some of them are already well preserved, other ones are completely dismantled or buried (Forcella et al., 1981; Gillot et al., 1982; Luongo et al., 1987; Vezzoli, 1988). On the island landslide deposits, derived from the accumulation and fragmentation of pre-existing volcanic rocks, extensively crop out (Guadagno & Mele, 1995; Mele & Del Prete, 1998; Calcaterra et al., 2003; De Vita et al., 2006; 2007; Di Maio et al., 2007; Di Nocera et al., 2007).

Many geo-volcanologic studies have been carried out on the island, starting from the syntheses of Rittmann (1930; 1948) and then on the specific aspects of the eruptive activity of the island and related geological processes (Forcella et al., 1981; Gillot et al., 1982; Chiesa et al., 1985; 1987; Poli et al., 1987; 1989; Civetta et al., 1991; Orsi et al., 1991; Luongo et al., 1997). Particular meaning is covered by the aspects concerning the geochronology of the volcanic deposits in the island (Orsi et al., 1996) and the time evolution of the magmatic system (Luongo et al., 1997).

The geologic and volcanologic history of the Ischia island has been characterized by a main event, represented by the eruption of the Green Tuff of the Epomeo Mt., which verified 55 ky B.P. ago, allowing for the downthrowing of the central sector of the island consequent to a caldera formation (Orsi et al., 1991; Acocella et al., 1997; Acocella & Funiciello, 1999). Consequently, the volcanic activity of the island has been conditioned by a complex phenomenon of calderic resurgence, started from 30 ky B.P., allowing for the gradual uplift and emersion of the rocks deposited in the caldera, initially submerged under the sea level. The rate of uplift, indicating the caldera resurgence, has been evaluated in about 800-1100 m (Barra et al., 1992).

The tectonic activity is characterized by systems of extensional faults, NW-SE and NE-SW trending, Plio-Quaternary in age (Acocella & Funiciello, 1999; Acocella et al., 2004). NW-SE and NE-SW systems of extensional fractures predominate in all the island and around the resurgent caldera block, suggesting a relationship with regional extensional structures. N-S and E-W trending normal faults have been found along the rims of the Epomeo block and interpreted as controlled by the caldera resurgence. The process of resurgence has locally substituted the volcanic activity during the last 33 ky, since the most of the pyroclastic products coeval with the resurgence has been erupted out of the uplifted area.

Marine geological studies already showed that the Ischia island lies on a E-W trending volcanic ridge (Bruno et al. 2002; Passaro, 2005; de Alteriis et al., 2005). A Digital Elevation Model (DEM) of the Ischia island, based on Multibeam bathymetric surveys and integrated by onshore topography is shown in fig. 12.

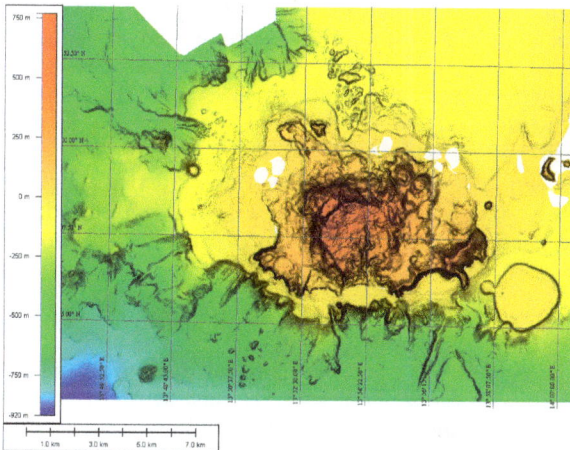

Fig. 12. DEM of the Ischia island resulting from the merging of different datasets of Multibeam bathymetry. The marine DEM has been merged with a Digital Terrain Model of the coastal area derived from topographic maps.

The continental slope off south-western Ischia island is incised by a dense network of canyons and tributary channels, starting from a retreating shelf break, parallel to the coastline and located at varying depths. Large scars characterize the platform margin off south-western Ischia island, in particular the scar of the southern flank of the island, corresponding onshore to the Mount Epomeo block and probably at the origin of the Ischia

Debris Avalanche (Chiocci & de Alteriis, 2006).Volcanic banks, having irregular morphologies, have been identified on the south-western flank of the island, as the "Banco di Capo Grosso" and the banks "G. Buchner" and "P. Buchner" (Passaro, 2005; de Alteriis et al., 2006). A large field of hummocky deposits, named the Ischia Debris Avalanche has put in evidence by swath bathymetric surveys coupled with Sidescan Sonar imagery and seismic profiles. Detailed piston coring and tephrostratigraphy suggested that the volcano-tectonic collapse originating the avalanche occurred during prehistorical times (Chiocci & de Alteriis, 2006). A stratigraphic framework for the last 23 ky marine record in the southern Ischia offshore has been recently constructed based on AMS ^{14}C dating and tephrostratigraphic analysis (de Alteriis et al., 2010).

Previous studies on the stratigraphic sequences cropping out in the Procida island have been carried out (Rosi et al., 1988a; 1988b), improving the geological knowledge of the volcanic district (Di Girolamo & Stanzione, 1973; Pescatore & Rolandi, 1981; Di Girolamo et al., 1984). Five monogenic volcanoes (Vivara, Terra Murata, Pozzo Vecchio, Fiumicello and Solchiaro) have been active over the last 80 ky, producing pyroclastic deposits and lava domes. New stratigraphic data on Procida based on geochemistry of major and trace element of volcanic deposits older than 14 ky have been recently presented (De Astis et al., 2004).

4. Marine seismic reflection and magnetic data in the Naples Bay: from old to new technologies

Seismic exploration is commonly performed by means of sources that can generate elastic waves from a rapid expansion of underwater gas bubbles. This can generate many pulses that take the form of double exponential spikes of gradually decreasing amplitude (Cole, 1965). Several technologies can be used in order to produce an acoustic pressure wave into water such as free-falling weights, chemical explosives, piezoelectric or magneto-resistive sources, sparkers, boomers, airguns and water-guns. Each of these sources has a precise signature and wave frequency that can be considered optimal in function of depth, resolution, etc. The main characteristic of a seismic source is to produce a single high-energy spike that is detectable, despite the presence of noise, after crossing the portion of the seabed that we wish to study. A broad range of frequencies can be reproduced, as well as a broad range of waveforms can be generated in function of frequency-dependent absorption of elastic waves and nearby boundaries presence.

Seismic sources for offshore investigation may be impulsive, providing a short-lived burst of elastic wave energy and swept-frequency, producing a low-amplitude sinusoidal signal. Impulsive sources such as explosives can cause damages to marine flora and fauna; for this reason towed sources activated for only few seconds must be preferred. The type of source should be chosen depending on the required resolution and signal penetration. Vibration of piezoelectric and magnetic materials, electric pulses, or pressured fluid discharge, often organised into arrays, can be considered good seismic sources whose signature, spectra and energy output can vary considerably. Sparkers (Knott & Hersey, 1956) and Boomers (Edgerton & Hayward, 1964) systems are based respectively on an electrode array powered by high voltage capacitor bank and on an electromagnetic source. Sparkers and boomers can generate seismic energy to explore continental margin when there are near surface or deep-towed (10-50 m off the sea), moreover boomers with pulse length of 0.1-0.2 ms can be used to explore very shallow waters. Sparker system can produce low-frequency acoustic wave (the maximum frequency contained in the spectrum of acoustic signal is approximately 2000 Hz) that can penetrate several hundred meters of sediment.

The Multispot Extended Array Sparker (M.E.A.S.) is a seismic source consisting of sparker electrodes disposed on a square metal cage. This kind of system, patented by Institute of Oceanology of Istituto Universitario Navale of Naples (Italy) allows obtaining a good signal penetration and high resolution seismic data with relative small energy use. The M.E.A.S. signal is a short impulse with a large frequency spectrum content (fig. 13).

Mirabile et al. (1991) tested the acquisition geometry in order to reduce a superimposing of source signal with return echoes that respect the "far field" condition and demonstrated the utility of some techniques for signal de-convolution in order to produce the so-called seismic profiles "deghosting". Seismic reflection data require a complex series of numerical treatments to increase the signal/ noise ratio of a single profile as well as obtaining a high resolution seismic section to improve the geological interpretation.

A more recent technology is the Sparker source SAM that is characterized by a varying number of electrodes that can be disposed as "dual-in line" (SAM96) and "planar array" multi electrode electro-acoustic source (SAM 400/800; fig. 14).

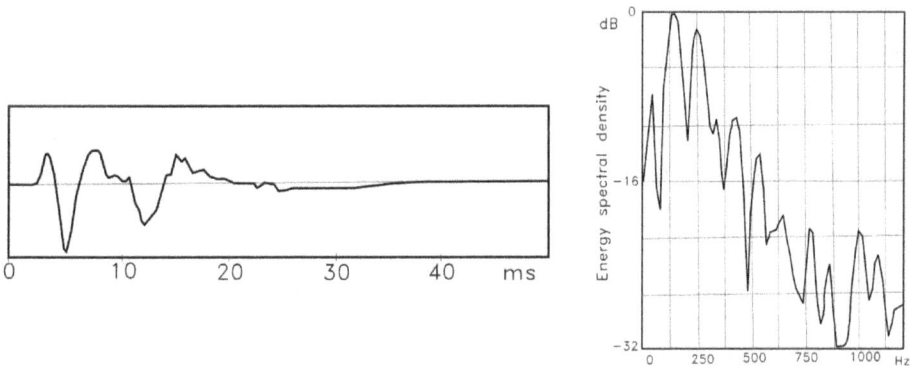

Fig. 13. Signature and spectrum of Multispot Exended Array System (modified after Mirabile et al., 1991).

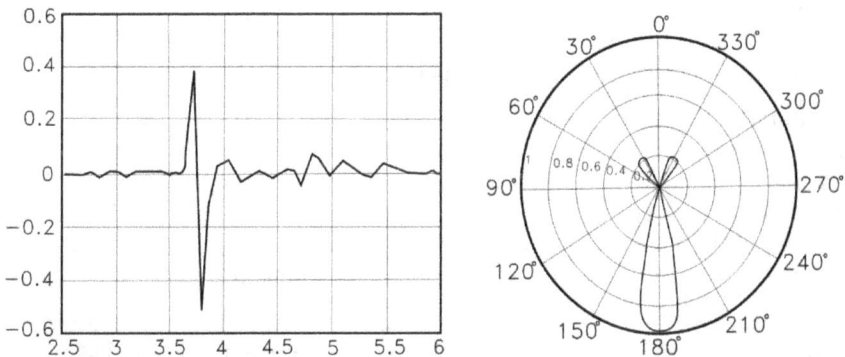

Fig. 14. Left: Signal (received by a sub-surface hydrophone) generated by SAM system at firing energy 200 J duration of primary impulse 0.3 ms (from Corradi et al., 2009); Rigth: Square radiation diagram that shows the high system directivity.

Other seismic sources are the Airguns and Waterguns, recently used in the Naples Bay in submarine geological mapping and basin studies (D'Argenio et al., 2004; Aiello et al., 2005; 2011). The former one produces high-energy seismic pulses short in duration by means of a discharge of compressed air into water (fig. 15), while the latter one produces the sudden collapse of a cavitation volume into water that is proportional to kinetic energy of the water plug. Airguns produce a wide range of pulse shapes and source spectra.

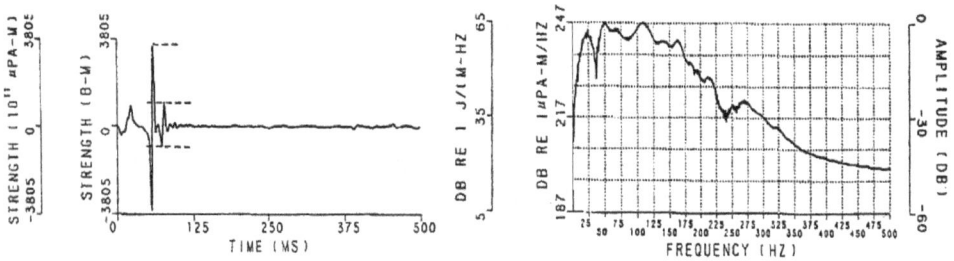

Fig. 15. Example of water-gun signature in far field and the frequency spectum (modified after Ranieri & Mirabile, 1991).

The seismic exploration of Naples Bay has been performed mainly through Sparker systems and Watergun sources. The evolution of sources capability in terms of technological advances together with processing techniques refinement allowed high resolution studies of main intermediate and deep geological structures in the Bay of Naples. Historically some of the first surveys were conducted by R/V Atlantis II (Woods Hole Ocean. Cruse 59) using an Airgun System Subsequently, in 1970 R/V Dectra owned by Istituto Universitario Navale of Naples (Italy) obtained a densely-spaced seismic survey through SPARKER E.G.G. (8 kjoules) and BOOMER systems in the Naples and Pozzuoli Bays (Latmiral et al., 1971; Bernabini et al., 1973).

Since the 70's until now many attempts have been carried out in order to improve seismic technologies performance, data acquisition, and processing. In the practice of seismic prospecting, Sparker systems technologies were widely analyzed using different acquisition systems. Some ones consist of a single electrode hotter than a mass electrode, other ones of more electrodes over distributed mass (eg. Sparker Teledyne and Sparker EGG).

De Vita et al. (1979) tried to identify, also based on experimental data, which one is more appropriate than the two configurations (single electrode or multi-electrode) based on the fundamental equations for the design of an "array". Sparker signals are the base band signals, transitory and continuous spectrum. Based on these measurements it has been demonstrated that energy should never exceed 400 joules/electrode to achieve the best compromise between resolution and electro-acoustic performance.

Ranieri and Mirabile (1991) reported technical and scientific results obtained through the geophysical survey of the deep geological structure of the Phlegrean Fields volcanic complex. It was aimed at improving the knowledge on technologies and sources that are more appropriated for the investigation of the continental margins, particularly in complex volcanic areas like the Gulf of Naples (Fusi et al., 1991).

Among the sources tested in studies of the Gulf of Naples there are the explosives (Mirabile et al., 1989), the Sparker and the Watergun, while the details to study geomorphological data were analyzed through the Surfboom and the Side Scan Sonar. MEAS (Multispot Extended Array Sparker; Mirabile et al., 1991) seismic source (12 and 16 KJ), consists of an array of 36 (6x6) electrodes placed inside a metal cage in a square size 4.5x4.5 m, spaced 0.75 m and fed in phase. The energy used by the MEAS has a pulse of short duration, the order of 10 milliseconds and a significant spectral content up to 1000 Hz, with maximum energy output around 150 - 200Hz. Each echo corresponds to an acoustic discontinuity (impedance contrast) that can generally be interpreted in geological terms.

MEAS system has been largely used in order to acquire a large database of single channel reflection seismics in the Bay of Naples (Mirabile, 1969; Latmiral et al., 1971; Mirabile et al., 1991).

Recently, by means of Multi- tip SAM 96 (0.1-1kJ), SAM400 (1-4KJ) transducer it was possible to record high resolution seismic data in the Bay of Naples both in coastal and deep sea research (Corradi et al., 2009). Some evidences on magnetic field anomalies in the Gulf of Pozzuoli come from the magnetic map of Galdi et al. (1988) who reported a NE-SW interruption of main regional trend where some circular local anomalies are related to products of post-calderic volcanic activity (Rosi & Sbrana, 1987).

Significant correlations between geophysical data come from the comparative analysis of seismic and magnetometric datasets. A magnetometer usually measures the strength or direction of the Earth's magnetic field. This last can vary both temporally and spatially for various reasons, including discontinuities between rocks and interaction among charged particles from the Sun and the magnetosphere. Most technological advances dedicated to measure the Earth's magnetic field have taken place during World War II. Presently, the most common are: the fluxgate, the proton precession, Zeeman-effect, sensor suspended-magnet, and satellite magnetometers. The fluxgate and the proton precession are effectively the most used for marine surveys, they are both cable drawn. The fluxgate magnetometer was the first ship-towed instrument, and it can measure vector components of the magnetic field. Its sensor consists of two magnetic alloy cores that are mounted in parallel configuration with the windings in opposition. The proton precession magnetometer consists of a sensor containing a liquid rich in protons surrounded by a coil conductor, the sensor is towed from the vessel through an armoured coaxial cable whose length depends on vessel length and seafloor depth. Circulating current within the coil generates a magnetic field of approximately two orders of magnitude the Earth's field, in this way 1 proton each 10 will follow the coil positioning. Stopping the induced magnetic field, the protons will align according to the Earth's magnetic field through a movement of precession.

The proton precession magnetometer is one of the most used for offshore surveys and it records the strength of the total field by determining the precessional frequency (f) of protons spinning about the total field vector (F) as follows:

$$f = \gamma_p F / 2\pi \tag{1}$$

where γ_p is the gyromagnetic ratio of the proton uncorrected for the diamagnetic effect, so that knowing its from laboratory measurements, the total field in nanotesla can be calculated as:

$$F=23.4866 \times f \hspace{6cm} (2)$$

The total field calculated by means of equation (2) is stored by magnetometer into a string of data containing position data that is displayed as an x,y chart. The signal frequency is measured on a time span of 0.5 seconds when the signal-noise ratio is highest. To ensure a maximum value of initial value of proton precession the angle between the axis of the coil and the Earth's field it is necessary to use two orthogonal coils. The measured field must be corrected with respect to the regional field in order to evaluate the anomalies.

The proton precession magnetometer was largely used to explore magnetic anomalies in the Bay of Naples. Interesting examples of magnetic data acquisition related in the Bay of Pozzuoli and Naples is reported in Galdi et al. (1988) and Aiello et al. (2004). As shown in Fig. 16 (modified after Galdi et al., 1988) both positive and negative anomalies were detected, using a magnetometer model Geometrics G-856, globally the area shows an interruption of the regional trend from NE-SW where circular anomalies are probably connected to a post-calderic activity of the Phlegrean Fields. Moreover, the internal area of the Pozzuoli Bay is characterized by a negative anomaly that increases towards the south. Conversely, in the external area there is mainly an alternance of positive and negative anomalies with a dominance of positive values near the area of Bagnoli. For a detailed analysis of the magnetic anomaly field of the volcanic district of the bay of Naples see Secomandi et al. (2003). Recently, Aiello et al. (2004) presented a high resolution map of the Bay of Naples based on data acquired during oceanographic cruise GMS2000-05 performed in October-November 2001 on board of the R/V Urania using the EG&G Geometrics proton magnetometer G-811.

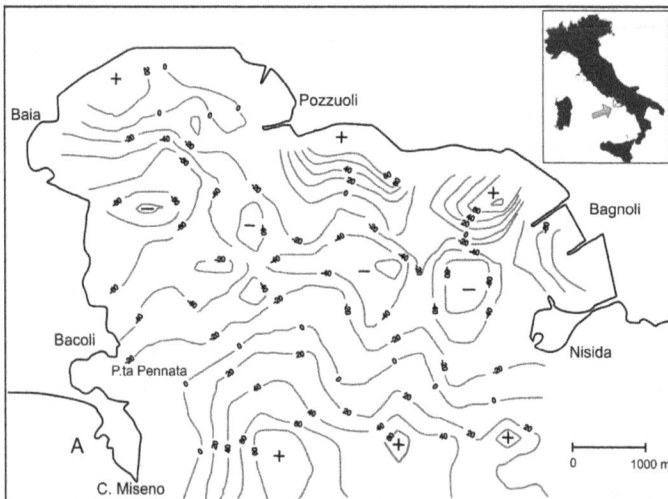

Fig. 16. Magnetic field anomalies in the Bay of Pozzuoli (modified after Galdi et al., 1988).

5. Results

Main results on seismic stratigraphy and marine magnetics of selected areas in the Naples Bay, i.e. Somma-Vesuvius volcanic complex offshore, Naples Gulf and Phlegrean Fields volcanic complex offshore and Ischia and Procida volcanic complexes offshore are illustrated in the following paragraphs based on seismic and magnetic datasets.

5.1 Seismic stratigraphy and marine magnetics of the Somma-Vesuvius volcanic complex offshore

A three-dimensional reconstruction of a large volcanic structure located offshore the Somma-Vesuvius volcano, next to the town of Torre del Greco (Naples, Italy) has been recently carried out (Aiello et al., 2010). It represents the seaward prolongation of the Vesuvius volcano and has been carried out using integrated geological interpretation of densely spaced Watergun seismic profiles and magnetic data recorded on the same navigation lines. Magneto-seismic modelling makes available new information on the geological structure of the Vesuvius volcano, relatively to its offshore.

In the study area the magnetic properties allow one to categorize the volcanic nature of seismo-stratigraphic units recognized through seismic interpretation. A semi-quantitative integrated interpretation of bathymetric and seismic data has been obtained resulting in a 3D topographic and seismic reconstruction of the Torre del Greco volcanic structure.

Significant results on the shallow crustal structure of the Vesuvius volcano and the relationships between seismic velocities and rock lithologies in volcanic environment have been recently obtained based on seismic passive tomography of the volcano (Zollo et al., 1996; 1998; 2003; Capuano et al., 2003). Onshore seismic reflection data on the volcano indicated a SW lateral collapse, which probably occurred between 35 and 11 ky ago (Bruno and Rapolla, 1999). Buried seismic units with reflection free interiors have been interpreted as volcanic deposits erupted during and since the formation of the breached crater of the Monte Somma volcano, preceding the growth of Vesuvius (Milia et al., 1998). Other features include the warping of lowstand marine deposits by undersea cryptodomes, normal faults indicating a seaward collapse of a volcano and a small undersea slump produced by Vesuvius eruption of 1631. The AD 79 Plinian eruption of Vesuvius that buried Pompei and Herculaneum began with pumice falls followed by pyroclastic currents (Milia et al., 2008). These currents reached Herculaneum and entered the sea, forming a fan.

A belt of sharp magnetic anomalies has been already highlighted offshore the Vesuvius volcano (Aiello et al., 2004), suggesting the occurrence of a NNW-SSE structural alignment of magnetic anomalies and related seismic structures. This has not been mentioned by previous authors, who had suggested NE-SW trending normal faults (Bernabini et al., 1973; Finetti & Morelli, 1973; Cassano & La Torre, 1987). Slight magnetic anomalies, located offshore the town of Torre Annunziata, probably correspond to the seaward prolongation of the Vesuvian lavas.

Seismic interpretation already enabled the identification of acoustically transparent, mound-shaped volcanic structures. These correspond to sharp and delineated magnetic anomalies, overlying the top of a seismic unit, interpreted as Campanian Ignimbrite pyroclastic flow deposits (CI; 35 ky B.P.; fig. 17). The volcanic domes represent submerged or buried parasitic vents, genetically related to the activity of the Somma-Vesuvius volcano during recent times (Aiello et al., 2004; 2005).

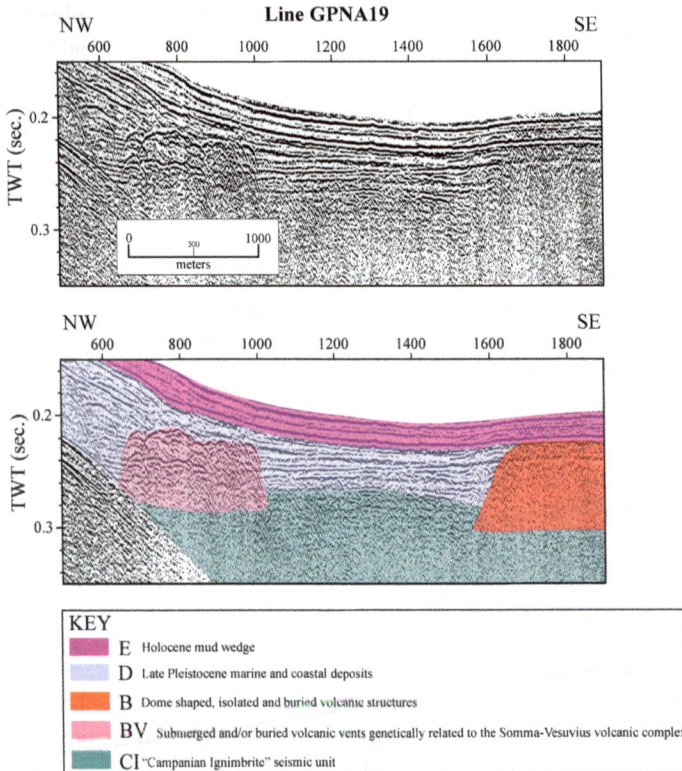

Fig. 17. Multichannel seismic profile GPNA19 located offshore the Vesuvius and corresponding geologic interpretation (reported after Aiello et al., 2010). Two isolated buried volcanic mounds occur near the top of the Campanian Ignimbrite volcanic unit (in grey-blue in the profile). If we consider the CI unit as a stratigraphic marker (35 ky B.P.) the age of the establishment of the volcanic domes on the Naples Bay continental shelf is probably post 35 ky B.P.

Several seismic units and related unconformities have been recognized (D, CI, BV, B, E units in the figures). The deepest one (D unit) is represented by the upper part of a Middle-Late Pleistocene prograding wedge, supplied by the Sarno river mouth, characterized by low angle dipping reflectors, indicating a NW-SE progradation. Its top is truncated by an erosional unconformity marking also the base of the CI unit.

The CI represents an important seismic unit occurring in the eastern Naples Bay (Fig. 17). The CI pyroclastic flow deposits carpeted the whole Campania Plain during a major eruption related to the Phlegrean Fields about 35 ky B.P. (Rosi & Sbrana, 1987). This unit underlies both the Torre del Greco volcanic structure and buried and isolated mounds. Since the CI represents an important stratigraphic marker in the Naples Bay, it can be assessed that both the isolated domes and the Torre del Greco volcanic structure are younger than 35 ky B.P. (age dating of the CI deposits). Buried and isolated volcanic mounds, genetically related to the Vesuvius activity, have been distinguished through seismic interpretation (fig. 17).

The Torre del Greco volcanic structure extends for about 7.5 km offshore the Vesuvius and corresponds to a main magnetic anomaly (fig. 18) reaching intensity of 400 nT. It shows an acoustically transparent seismic facies and three main elevated peaks.

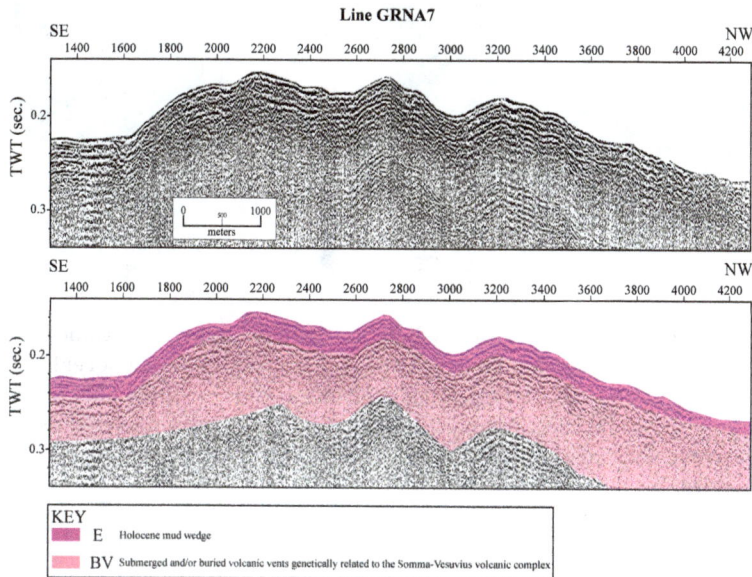

Fig. 18. Multichannel seismic profile GRNA07 located offshore the Vesuvius and corresponding geologic interpretation (reported after Aiello et al., 2010). The Torre del Greco volcanic structure is depicted by the profile.

The total magnetic field offshore the Somma-Vesuvius volcanic complex shows that the shape of the anomalies is dipolar; there is no apparent effect caused by the occurrence of remnant magnetization with a different direction to that of the present-day main field (fig. 19). Three main anomalies are evident following a NW-SE direction, the most southern of which being the least intense, in a relatively magnetically quiet area. The volcanic structures recognized on seismic profiles are located in a complex magnetic anomaly area, which is made up of several anomalies, reaching a maximum intensity of 400 nT. This is one of the highest values detected in the whole Naples Bay. These volcanic bodies represent the seaward prolongation of the Vesuvius volcano. They are interpreted as a strip of volcanic vents, which have been settled on a system of NNW-SSE normal faults, as confirmed by the integrated interpretation of seismic, magnetic and bathymetric data.

5.2 Seismic stratigraphy and marine magnetics of the Naples Bay and Phlegrean Fields volcanic complex offshore

A grid of Sparker Multitip seismic profiles recorded in the Gulf of Pozzuoli in the frame of research projects of submarine geologic cartography has been interpreted to give new insights on seismic stratigraphy of Pozzuoli, i.e. the submarine prolongation of the Phlegrean Fields volcanic complex. The navigation map of the interpreted sections in the Gulf of Pozzuoli is shown in fig. 20.

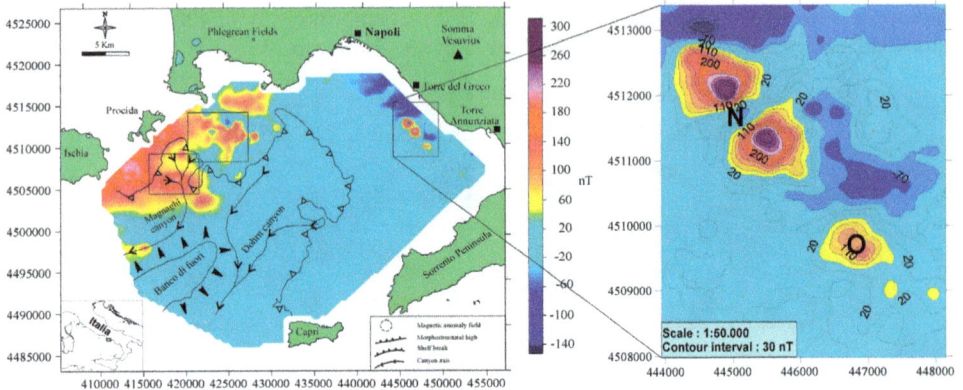

Fig. 19. Map of the total magnetic field offshore the Somma-Vesuvius volcanic complex. In the inset on the left map of the total magnetic field of the Naples Bay (reported after Aiello et al.., 2010). In the inset on the right detailed map of the magnetic anomalies off the volcano. Three intense and dipolar magnetic anomalies are aligned along a direction parallel to the Tyrrhenian coast, having settled along a system of NNW-SSE trending normal faults offshore the volcano.

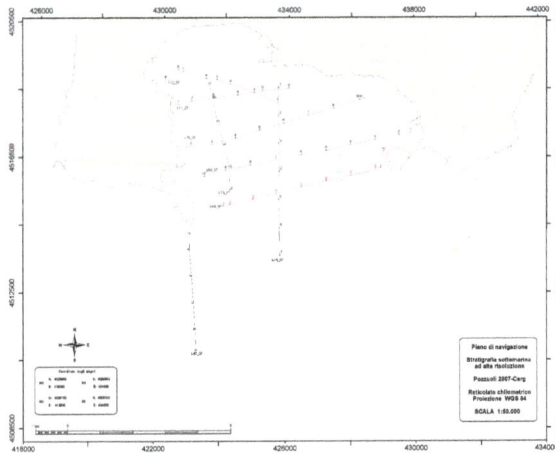

Fig. 20. Navigation map of interpreted seismic profiles in the Pozzuoli Bay

The seismic profile L68_07 (fig. 21) running from the western continental shelf of the Pozzuoli Gulf and the Nisida island has been interpreted to show the main stratigraphic and structural features of the Pozzuoli Gulf, reported in the geological interpretation (in the low inset of fig. 21).

A sketch stratigraphic table (Fig. 22) represents the key to the geological section of fig. 21 and describes the main characteristics and possible chronostratigraphic attribution of the seismic units (Milia, 1998). Large compressional features have been individuated on the seismic section, i.e. the Punta Pennata anticline, the central syncline of the Pozzuoli Gulf and the

Nisida anticline. These features involve intensively in deformation the volcano-sedimentary unit V3 (fig. 21) and have individuated during compressional events genetically related to main tectonic and magmatic events involving the Pozzuoli area during the Late Quaternary.

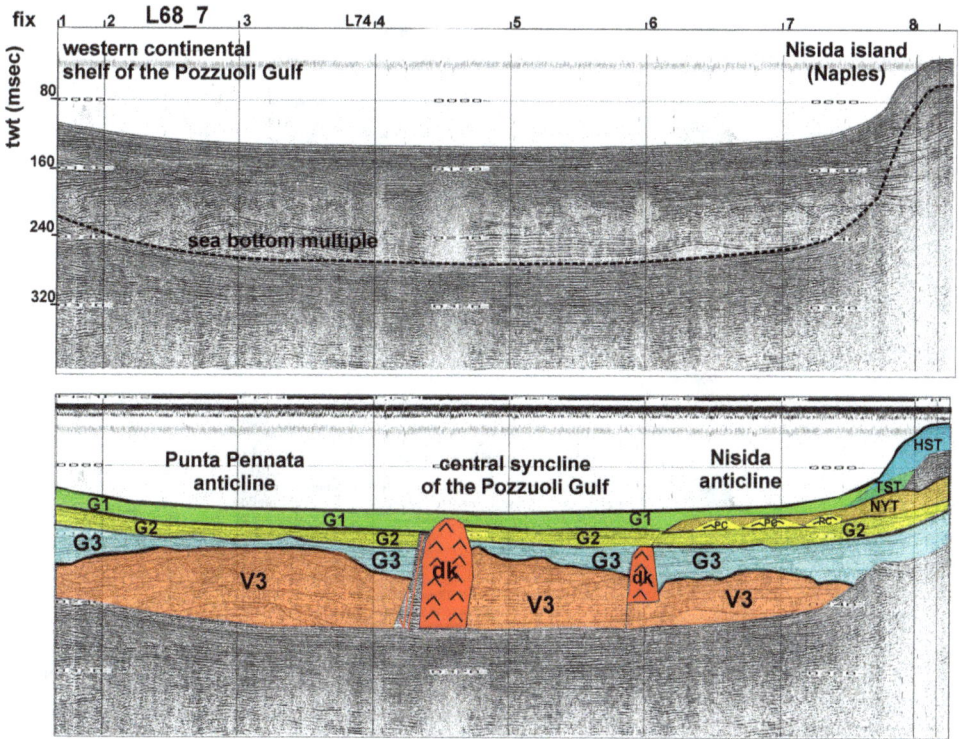

Fig. 21. Seismic profile L68_07 in the Pozzuoli Gulf and corresponding geological interpretation

Kilometer-scale folding deformed the Pozzuoli sequences during an important compressional event. In fact, the uplift of the marine terrace of "La Starza", on which the Pozzuoli town is located (Colantoni et al., 1972; Dvorak & Mastrolorenzo, 1991; Barra, 1992) and the formation of an erosional platform on the inner Pozzuoli continental shelf are linked to an anticlinal crest, while the present basin depocenter is located on a syncline (fig. 21). These folds formed during the deposition of the seismic sequence G3 (fig. 21), characterized by wedging geometries thinning towards the hinge of the anticline.

The seismo-stratigraphic analysis has allowed to distinguish eight main seismic units (figs. 27 and 28). The oldest one (V3 figs. 21 and 22) is a volcano-sedimentary unit related to the northern margin of the Pentapalummo Bank, characterized by discontinuous seismic reflectors. The unit is intensively deformed in correspondence to Punta Pennata and Nisida anticlines, separated by the central syncline of the Pozzuoli Gulf. The overlying unit (G3 figs. 21 and 22) is composed of clastic deposits, characterized by discontinuous to parallel seismic reflectors. It has deposited in the whole Pozzuoli Gulf and is characterized by wedging and growth due to synsedimentary deformation contemporaneous to the individuation of folds.

KEY

HST	Highstand system tract of the Late Quaternary depositional sequence, characterized by progradational seismic reflectors at the foot of the Nisida island
TST	Transgressive system tract of the Late Quaternary depositional sequence, characterized by retrogradational seismic reflectors at the foot of the Nisida island
G3	Sedimentary unit probably composed of clastic deposits (4 ky B.P.) characterized by parallel seismic reflectors, deposited in whole Pozzuoli Gulf
PC **NYT**	NYT: Wedge shaped unit composed of the pyroclastic deposits of the Neapolitain Yellow Tuff (12 kyB.P.) deposited in the Naples offshore; PC: tuff cones of the Nisida complex, genetically related to the Nisida Bank and to the Nisida island, in facies hetherop with the NYT unit.
G2	Sedimentary unit probably composed of clastic deposits (8-4 ky B.P.) characterized by parallel seismic reflectors, deposited in whole Pozzuoli Gulf
dk	Volcanic dykes, characterized by sub-vertical volcanic bodies, acoustically transparent and locally bounded by normal faults
G3	Sedimentary unit probably composed of clastic deposits (12-8 ky B.P.) characterized by discontinuous to parallel seismic reflectors, deposited in whole Pozzuoli Gulf; characterized by wedging and growth due to synsedimentary deformation contemporaneous to the individuation of anticlines and synclines
V3	Volcano-sedimentary unit related to the northern margin of the Pentapalummo bank (150<V3<18 ky B.P.), characterized by discontinuous seismic reflectors; intensively deformed in correspondence to anticlines (Punta Pennata anticline and Nisida anticline) and synclines (central syncline of the Pozzuoli Gulf), individuated due to compressional deformation genetically related to main magmatic events

Fig. 22. Sketch table of the seismic units recognized in the stratigraphic sketch diagram of fig. 21 (Pozzuoli Gulf).

The dk unit distinguishes volcanic dykes, characterized by acoustically transparent sub-vertical volcanic bodies., locally bounded by normal faults. The G2 unit is composed of clastic deposits and is characterized by parallel seismic reflectors in the whole Pozzuoli Gulf.

From the central Pozzuoli Gulf to Nisida a wedge-shaped seismic unit, genetically related to the Neapolitain Yellow Tuff (NYT; 12 ky B.P.; Scarpati et al., 1993) has been identified (fig. 21). It interstratifies with the tuff cones of the Nisida complex, genetically related to the Nisida bank and the Nisida island (PC fig. 21). The NYT/PC unit is overlain by the G3 unit, the most recent one in the sedimentary filling of the Pozzuoli area (fig. 21). TST and HST deposits of the Late have also been identified off the Nisida island.

The interpreted map of the magnetic anomalies in the Gulf of Pozzuoli is shown in fig. 23 (modified after Galdi et al., 1988). It has allowed to distinguish both areas characterized by positive anomalies (represented in yellow) and areas characterized by negative anomalies (represented in light yellow). The inner continental shelf of the Gulf of Pozzuoli is regarded as negative magnetic anomalies. In particular, the area surrounding the Pozzuoli harbour (from the Caligola pier to the Pirelli jetty) does not show significant magnetic anomalies. On the contrary the area adjacent the resort Lucrino-Punta Pennata owns a negative anomaly increasing southwards up to the magnetic minimum at 600-700 m in correspondence to the

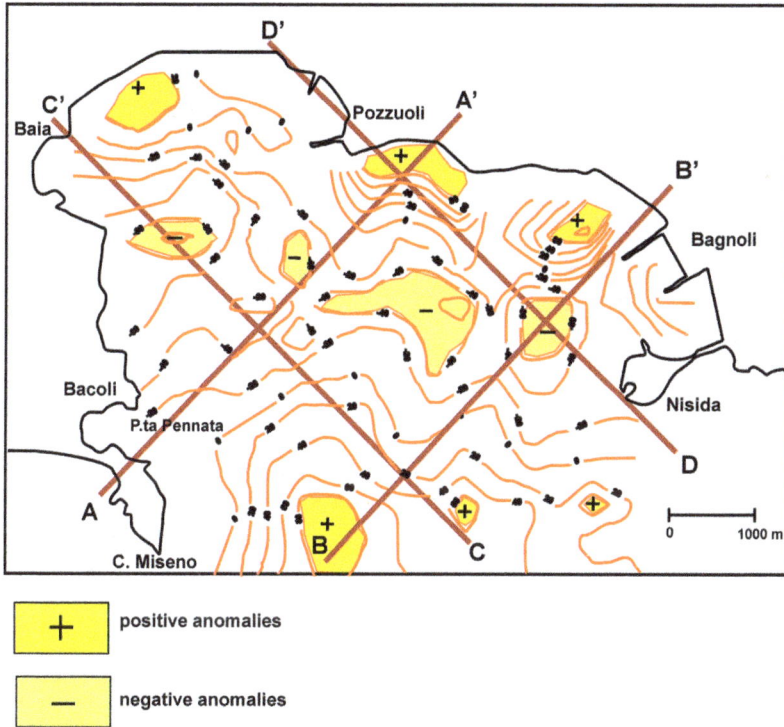

+ positive anomalies

− negative anomalies

Fig. 23. Interpreted map of the magnetic anomalies of the Gulf of Pozzuoli (modified after
Galdi et al., 1988). The positive anomalies are represented in yellow and the negative
anomalies in light yellow.

Baia Castle (- 100 nT). On the outer shelf of the Gulf of Pozzuoli it is possible to observe
alternating magnetic maxima and minima. In particular, an area of magnetic maximum is
located on a belt long about 1.7 km, NE-SW oriented. At the same time, the inner continental
shelf of the Gulf of Pozzuoli, from Bagnoli to the Rione Terra of Pozzuoli shows two strong
magnetic anomalies, separated by a thin belt having a normal magnetic value. Proceeding
seawards, in the offshore surrounding Bagnoli, two magnetic minima (- 40 nT and – 60 nT)
are positioned, which result slightly E-W elongated, culminating with the absolute magnetic
minimum (-100 nT) in correspondence to the Baia Castle. Four magnetic sections,
respectively NE-SW and NW-SE oriented have also been constructed (fig. 24; modified after
Galdi et al., 1988). On the vertical axis the magnetic anomalies (nT) and the depths (m) have
been reported on the same scale, while on the horizontal axis the distances, expressed in
meters have been reported (fig. 24).

The magnetic section A-A' (in the upper inset of fig. 24) runs from Punta Pennata to the
Pozzuoli town (Via Napoli). The total magnetic intensity shows a trending with a magnetic
minimum of – 80 nT in the central area (corresponding to a depth of the sea bottom of – 90
m) and a magnetic maximum of 70 nT in correspondence to the Pozzuoli shoreline. The
magnetic section B-B', translated of 2.4 km towards south-east, shows, starting from southwest a monotonous magnetic trend up to the offshore surrounding Nisida, where a strong

increase of the gradient occurs. The magnetic highs occurring nearshore appear to be related not to the geology, but to the occurrence of the industrial systems of Bagnoli.

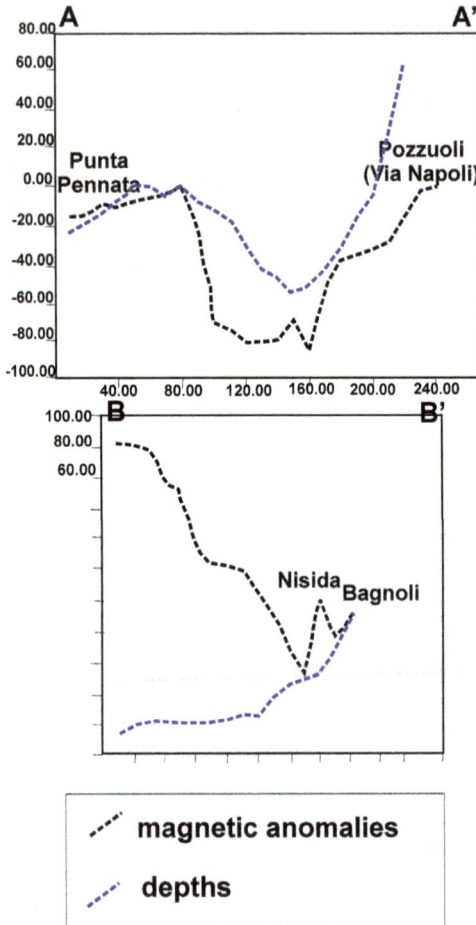

Fig. 24. Magnetic sections of the iso-anomalies NE-SW oriented in the Pozzuoli Gulf (modified after Galdi et al., 1988).

5.3 Seismic stratigraphy and marine magnetics of Ischia and Procida volcanic complexes offshore

Marine geophysical data, in particular Multibeam bathymetry and reflection seismics have allowed to study the submerged sectors of the Ischia island (Naples Bay; Aiello et al., 2009c; Passaro, 2005). They are the site of submarine instability processes, having both catastrophic (instantaneous) and continuous characteristics (accelerated erosion along submarine canyons or channels, debris fluxes along channels and creeping). The geological interpretation of the marine DEM of Ischia island, already shown in fig. 12 has put in evidence an articulated topography of the sea bottom. A complex stratigraphic architecture, with intercalations between volcanic and sedimentary units is revealed by the interpretation of high resolution seismic reflection profiles.

A sketch stratigraphic scheme has been constructed in the northern Ischia offshore (fig. 25). Forced regression prograding wedges (FST), pertaining to the Late Quaternary depositional sequence, appear on the continental shelf off the northern Ischia island. Debris avalanche deposits, having a wedge-shaped external geometry and chaotic facies are arranged in two distinct, superimposed bodies (H1 and H2). The wedges are characterized by facies hetheropy with the upper seismic unit of the basin filling. The lower seismic unit has parallel reflectors and shows bidirectional onlaps in correspondence to depressions eroding the top of the underlying seismic unit. The intermediate unit is characterized by parallel to sub-parallel seismic reflectors. It shows a strong wedging in correspondence to a normal

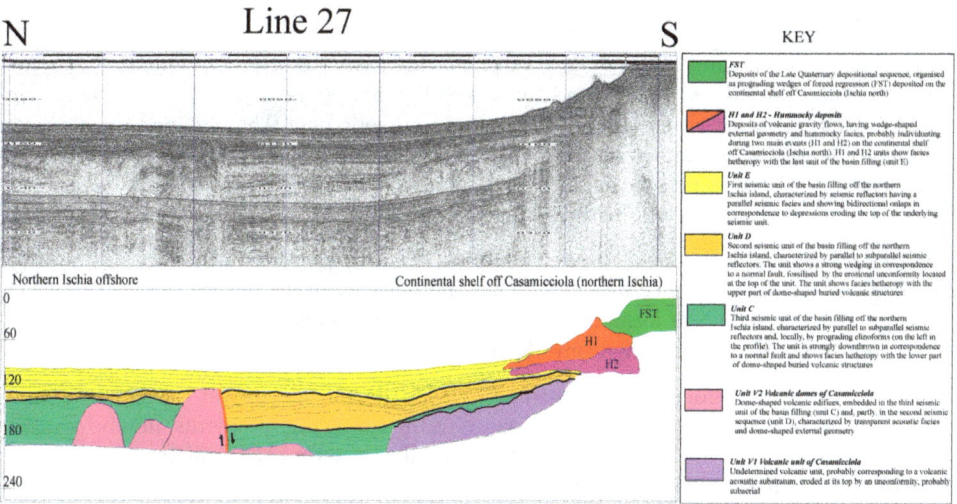

Fig. 25. Seismic profile L27 offshore northern Ischia and corresponding geological interpretation (modified after Aiello et al., 2009).

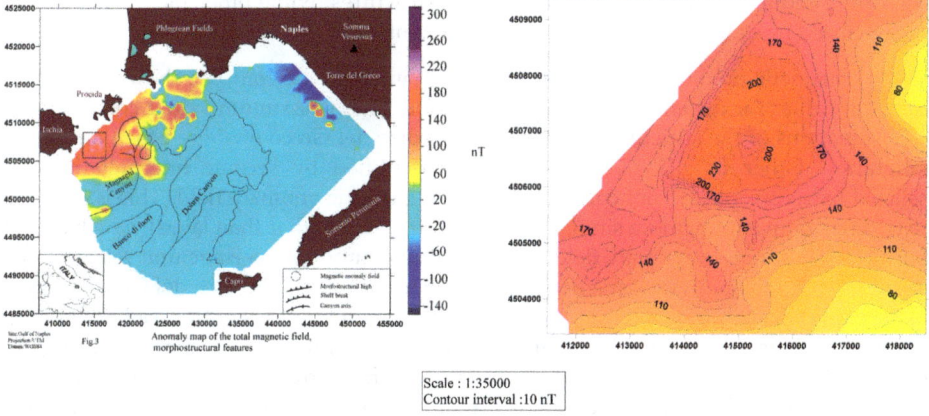

Fig. 26. Total magnetic field map of the Ischia Bank (south-eastern Ischia offshore)

fault (fossilized by an erosional unconformity located at the top of the unit) and stratigraphic relationships of facies hetheropy with the upper part of dome-shaped, buried volcanic structures. The upper unit is characterized by parallel to sub-parallel seismic reflectors and locally by prograding clinoforms. It appears to strongly downthrown in correspondence to a normal fault and shows facies hetheropy with the lower part of dome-shaped buried volcanic structures. Mounded volcanic edifices are in lateral contact with the lower seismic unit of the basin filling and partly, with the second one and are truncated by an erosional unconformity located at the top of the unit 3. An undetermined volcanic unit, having facies hetheropy with the unit 3 is eroded at the top by a subaerial unconformity and interpreted as volcanic acoustic basement. A total magnetic field map of the Ischia Bank is shown in fig. 26.

6. Conclusions

Seismic stratigraphy and marine magnetics in the case histories of Somma Vesuvius offshore, Phlegrean Fields offshore and Ischia and Procida offshore (Naples Bay) have been studied through the interpretation of seismic and magnetic data. The obtained results have improved the geological knowledge in an active volcanic area such as the Naples Bay, since they have disclosed and located several main volcanic bodies based on seismic interpretation, well constrained by the occurrence of significant magnetic anomalies.

Offshore the Somma-Vesuvius volcanic complex, the integrated interpretation of seismic and magnetic data suggests a correlation of the anomalies with three main elevated peaks of the large volcanic structure located offshore the Torre del Greco town, along a NNW-SSE direction in water depths ranging from – 80 m to – 110 m.

Seismo-stratigraphic evidence is represented by acoustically-transparent seismic facies and high contrasts of acoustic impedance compared to the overlying sediments, mound-shaped external geometry and average dimensions measurable in terms of kilometres. The base of the volcanic bodies is not acoustically evident, because they overlie the seismic unit correlated with the Campanian Ignimbrite pyroclastic flow deposits. The top of the structures is irregularly eroded and can show several culminations, as in the case of the Torre del Greco volcanic structure. The thickness of the overlying Holocene sediments is significantly reduced in correspondence to most structures, while other mounds appear to have fossilized by Late Pleistocene-Holocene sediments.

The total magnetic field offshore the Somma-Vesuvius volcanic complex shows three main maximum values for the anomalies, dipolar in shape. These maximum values correspond to the culminations of the structure observed in the Torre del Greco offshore based on seismic stratigraphy. Other minor volcanic structures, identified by seismic interpretation and fossilized by sediments, do not correspond to any magnetic anomaly field. This is probably due to their composition, more similar to that of tuff cones (rather than lavas), which are not related to any magnetic anomaly field, similarly to the important seismic unit related to the Campanian Ignimbrite pyroclastic flow deposits. The rising of the Torre del Greco volcanic structure corresponds to the occurrence of topographic undulations of the sea bottom of up to ten metres. This is confirmed by the interpretation of seismic profiles, showing three main vertical culminations of the volcanic structure, where overlying sediment drape is significantly reduced. These culminations are linked to magnetic anomaly extremes, with values ranging between 250-350 nT.

Offshore the Phlegrean Fields volcanic complex significant magnetic anomalies are located
in a belt of submarine volcanic banks located in the outer shelf of the Gulf of Pozzuoli (fig.
27). Box 2 in fig. 27 shows that the Phlegrean Fields offshore represents a relatively complex
magnetic anomaly area. Two dipolar anomalies, characterized by a maximum-minimum
couple, have been identified. The first anomaly, E-W oriented and located in the northern
area shows a minimum of – 200 nT, associated to a maximum of + 185 nT. These values may
be associated with volcanic bodies not cropping out at the sea bottom, but buried by
sediments. The second anomaly, NW-SE oriented and located in the eastern area, shows a
maximum-minimum couple with a similar intensity. Other anomalies, not dipolar and of
lower intensity, ranging between 40 and 135 nT are due to the occurrence of small volcanic
edifices (fig. 27). A significant magnetic anomaly, in the order of 150 nT occurs at the
Magnaghi canyon head (Box 3 in fig. 27), deeply eroding the volcanic deposits of the
continental slope of Procida island. This confirms that the Magnaghi canyon is incised in
volcanic deposits. On the contrary, the slope of the Gulf of Naples in correspondence to the
Dohrn canyon, a kilometric feature crossing the Bay does not show magnetic anomalies,
confirming that this canyon deeply erodes sedimentary units supplied by the palaeo-Sarno
river mouth.

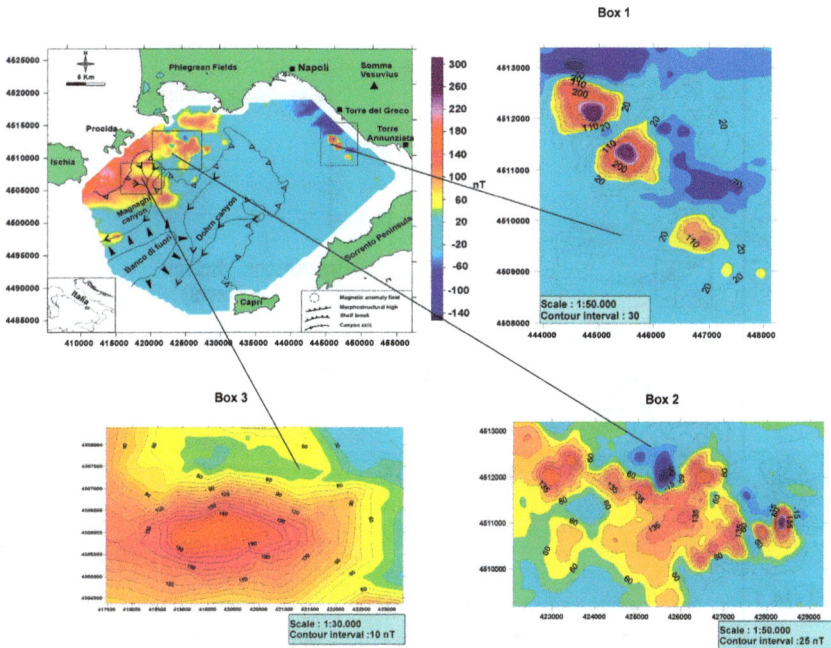

Fig. 27. High resolution magnetic anomaly map of the Naples Bay (modified after Aiello et
al., 2005). Box 1 represents a magnetic anomaly area located offshore the Somma-Vesuvius
volcanic complex, while Box 2 represents another magnetic anomaly area located offshore
the Phlegrean Fields volcanic complex. Box 3 represents a magnetic anomaly area located on
the continental slope of the gulf, at the Magnaghi canyon's head.

The interpreted map of the magnetic anomalies in the Gulf of Pozzuoli (fig. 23) has allowed to distinguish positive and negative magnetic anomaly areas. The inner continental shelf of the Gulf of Pozzuoli is regarded as negative magnetic anomalies and the correlation with the volcanic structures evidenced by the Sparker data is not clear. This is probably due to the necessity to record a densely-spaced magnetic survey, in order to identify on marine magnetics the volcanic dykes shown by seismic profiles. The volcaniclastic unit identified on seismic profiles does not seem to produce significant magnetic signatures, probably due to its composition (tuffs rather than lavas).

In order to describe some morphological features of the study area, elevation versus average slope plots have been used, allowing to highlight where steep and flat areas occur (Moore & Mark, 1993). Such a plots have been derived by DEMs and slope maps through an opportunely built routine. Their use helps to identify morphological domains through the individuation of elevation/slope pairs, attributed to specific domains. The calculation has been carried out using a depth window of 1 m and then evaluating average slope value of all DEM cells, that fallen inside each window (fig. 28). A median filter (25 points window) was applied to smooth the progress of plot examination. The Ischia volcanic emerged and submerged volcanic edifice includes several morphological ranges, each one characterized by a well-defined elevation interval vs. average slope (fig. 28).

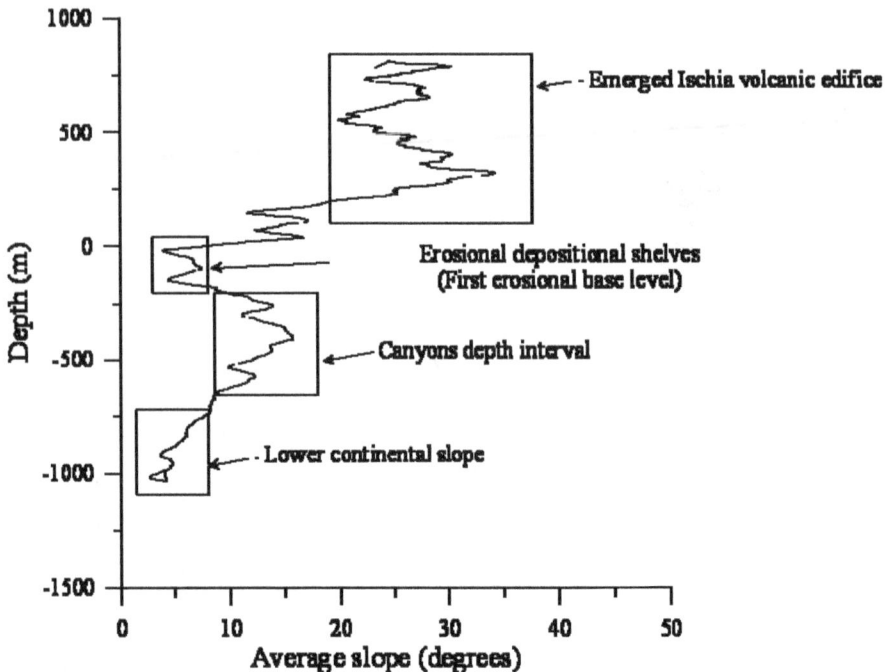

Fig. 28: Sketch diagram showing elevation intervals versus average slopes offshore the Ischia island.

The following morphological ranges have been identified (fig. 28):

A) The Ischia outcropping "central" edifice, at depth>0, characterized by a slope range of about 20°-40° on average;

A2) An intermediate stage, that acts as an according layer towards the continental shelf;

B) The continental shelf, between the coastline and the – 140-150 m (200 m in some cases) isobath (slopes 3°/8°, on average);

C) The upper continental slope, located between the platform edge and the 650 m isobath (slopes 8°/20°, on average);

D) The lower continental slope, deeper than 650 meters in depth (average slopes 0°/8°).

These domains include several morphological elements, each representing a tectonic and/or sedimentary process or a volcanic event. On the shelf terraces of abrasion and/or deposition, relic morphologies of volcanic edifices, canyons and gullies can be recognized. The depositional shelf break is partially eroded at the head of some canyons. Contrary to what recorded in normal depth distribution, it has been outlined the increasing of dip angles in the lower portion of the A physiographic unit, probably due to basal faulting of the Mt. Epomeo resurgent block. Submarine canyons are present on A and C units, acting as a morphological link between ranges. Debris avalanches develop between these volcanic features both in the southern and in the northern sides; on the contrary lateral collapses that characterize this area seems to be originated within morphological protrusion.

7. References

Acocella, V.; Funiciello, R. & Lombardi S. (1997). Active tectonics and resurgence in Ischia island (southern Italy). *Il Quaternario (Italian Journal of Quaternary Sciences)*, Vol. 10, No.2, pp. 427-432.

Acocella, V. & Funiciello, R. (1999). The interaction between regional and local tectonics during resurgent doming: the case of the island of Ischia, Italy. *Journal of Volcanology and Geothermal Research*, Vol. 88, pp. 109-123.

Acocella, V.; Funiciello, R., Marotta, E., Orsi G. & De Vita, S. (2004). The role of extensional structures on experimental calderas and resurgence. *Journal of Volcanology and Geothermal Research*, Vol. 129, pp. 199-217.

Aiello, G.;, Marsella, E. & Sacchi, M. (2000). Quaternary structural evolution of Terracina and Gaeta basins. *Rendiconti Lincei Scienze Fisiche e Naturali*, Vol. 9, No.11, pp. 41-58.

Aiello, G,; Budillon, F., Cristofalo, G., D'Argenio, B., de Alteriis, G., De Lauro, M., Ferraro, L., Marsella, E., Pelosi, N., Sacchi, M. & Tonielli, R. (2001). Marine geology and morphobathymetry in the Bay of Naples, In: *Structures and Processes of the Mediterranean Ecosystems*, F.M. Faranda, L. Guglielmo & G. Spezie (Eds.), 1-8, Springer Verlag Italy.

Aiello, G.; Angelino, A., Marsella, E., Ruggieri, S. & Siniscalchi, A. (2004). Carta magnetica di alta risoluzione del Golfo di Napoli (Tirreno meridionale). *Bollettino della Società Geologica Italiana*, Vol. 123, pp. 333-342.

Aiello, G.; Angelino, A., D'Argenio, B., Marsella, E., Pelosi, N., Ruggieri, S. & Siniscalchi, A. (2005). Buried volcanic structures in the Gulf of Naples (Southern Tyrrhenian sea,

Italy) resulting from high resolution magnetic survey and seismic profiling. *Annals of Geophysics*, Vol.48, No6, pp. 883-897.

Aiello, G.; Conforti, A., D'Argenio, B. & Putignano, M.L. (2008). Explanatory Notes to the Geological Map n. 465 "Isola di Procida". Scale 1: 50.000. APAT, Department of Soil Defense, National Geological Survey of Italy, Editorial House Systemcart, Rome, Italy.

Aiello, G.; Marsella, E., Di Fiore, V. & D'Isanto, C. (2009a). Stratigraphic and structural styles of half-graben offshore basins in Southern Italy: multichannel seismic and Multibeam morpho-bathymetric evidences on the Salerno Valley (Southern Campania continental margin, Italy). *Quaderni di Geofisica*, Vol.77, pp. 1-33.

Aiello, G.; Budillon, F., Conforti, A., D'Argenio, B., Putignano, M.L. & Toccaceli, R,M. (2009b). Explanatory Notes to the Geological Map n. 464 "Isola d'Ischia". Scale 1: 25.000. APAT, Department of Soil Defense, National Geological Survey of Italy, Editorial House Systemcart, Rome, Italy.

Aiello, G.; Marsella, E. & Passaro S. (2009c). Submarine instability processes on the continental slopes off the Campania Region (Southern Tyrrhenian sea, Italy): the case history of Ischia island (Naples Bay). *Bollettino di Geofisica Teorica Applicata*, Vol.50, No2, pp. 193-207.

Aiello, G.; Marsella E & Ruggieri S. (2010). Three-dimensional magneto-seismic reconstruction of the "Torre del Greco" submerged volcanic structure (Naples Bay, Southern Tyrrhenian sea, Italy): Implications for Vesuvius' marine geophysics and volcanology. *Near Surface Geophysics*, Vol.8, No1, pp. 17-31.

Aiello, G.; Cicchella, A.G., Di Fiore, V. & Marsella F. (2011). New seismo-stratigraphic data of the Volturno Basin (northern Campania, Tyrrhenian margin, southern Italy): implications for tectono-stratigraphy of the Campania and Latium sedimentary basins. *Annals of Geophysics*, Vol. 54, No3, pp. 265-283.

Argnani, A.; & Trincardi, F. (1990). Paola slope basin: evidence of regional contraction on the Eastern Tyrrhenian margin. *Memorie della Società Geologica Italiana*, Vol.44, pp. 93-105.

Arnò, V.; Principe, C., Rosi, M., Santacroce, R. & Sheridan M.F. (1987). Eruptive history. In: *Somma-Vesuvius*, R. Santacroce (Ed.), Quaderni De La Ricerca Scientifica, CNR, Italy.

Balducci, A.; Vaselli, M. & Verdini G. (1985). Exploration well in the Ottaviano permit, Italy. *European Geothermal Update, Proceedings 3rd International Seminar on the Results of the EC Geothermal Energy Research*, Reidel, Dordrecht.

Barra, D.; Cinque, A., Italiano, A. & Scorziello R. (1992) Il Pleistocene superiore marino di Ischia: Paleoecologia e rapporti con l'evoluzione tettonica recente. *Studi Geologici Camerti*, Suppl. 1, pp. 231-243.

Bernabini, M., Latmiral, G., Mirabile, L. & Segre, A.G. (1973). Alcune prospezioni sismiche per riflessione nei Golfi di Napoli e Pozzuoli. *Rapp. Comm. Int. Mer. Medit.*, Vol.21, pp. 929-934.

Bigi, G.; Cosentino, D., Parotto, M., Sartori, R. & Scandone, P. (1992). Structural Model of Italy. *Monografie Progetto Finalizzato Geodinamica*, CNR, Roma, Italy.

Boccaletti, M., Ciaranfi, N., Cosentino, D., Deiana, G., Gelati, R., Lentini, F., Massari, F., Moratti, G., Pescatore, T.S., Ricci Lucchi, F. & Tortorici, L. (1990). Palinspastic restoration and paleogeographic reconstruction of the peri-Tyrrhenian area during the Neogene. *Palaeogeography, Palaeoclimatology, Palaeoecology*, Vol.77, No1, pp. 41-42.

Brocchini, D.; Principe, C., Castradori, D., Laurenzi, M.A. & Gorla L. (2001). Quaternary evolution of the southern sector of the Campanian Plain and early Somma-Vesuvius activity: insights from the Trecase 1 well. *Mineralogy and Petrology*, Vol. 73, pp. 67-91.

Bruno, P.P.G. & Rapolla, A. (1999). Study of the sub-surface structure of Somma-Vesuvius (Italy) by seismic reflection data. *Journal Volcanology and Geothermal Research*, Vol. 92, No3-4, pp. 373-387.

Bruno, P.P.G.; de Alteriis, G. & Florio, G. (2002). The western undersea section of the Ischia volcanic complex (Italy, Tyrrhenian sea) inferred from marine geophysical data. *Geophysical Research Letters*, Vol.29, No9, pp. 1-4.

Calcaterra, D., De Riso, R., Evangelista, A., Nicotera, M.V., Santo, A. & Scotto Di Santolo, A. (2003). Slope instabilities in the pyroclastic deposits of the Phlegrean district and the carbonate Apennine (Campania, Italy). *International Workshop on Occurrence and Mechanisms of Flows in Natural Slopes and Earthfills IW-Flows 2003*, Sorrento, May, 14-16, 2003.

Capuano, P.; Gasparini, P., Zollo, A., Virieux, J., Casale M. & Yeroyanni M. (2003). The Internal Structure of Mt. Vesuvius. *Liguori Editore*, Napoli, ISBN88-207-3503-2, pp. 1-591.

Cassano, E. & LaTorre, P. (1987). Geophysics. In: *Somma-Vesuvius*, R. Santacroce (Ed.), Quaderni De La Ricerca Scientifica, CNR, Italy.

Castellano, M., Buonocunto, C., Capello, M. & La Rocca, M. (2002). Seismic surveillance of active volcanoes: the Osservatorio Vesuviano seismic network (OVSN-Southern Italy). *Seismology Research Letters*, Vol.73, pp. 177-184.

Chiesa, S., Cornette, Y., Forcella, F., Gillot, P.Y., Pasquarè, G. & Vezzoli, L. (1985). Carta Geologica dell'Isola d'Ischia. *Monografie Progetto Finalizzato Geodinamica*, CNR, Roma, Italy.

Chiesa, S.; Civetta, L., De Lucia, M., Orsi, G. & Poli, S. (1987). Volcanological evolution of the island of Ischia, In: *The volcanoclastic rocks of Campania (Southern Italy)*, P. De Girolamo (Ed.), Rendiconti Acc. Sc. Fis. e Mat. in Napoli Special Issue, pp. 69-83.

Chiocci, F.L. & de Alteriis, G. (2006). The Ischia Debris Avalanche: first clear submarine evidence in the Mediterranean of a volcanic island prehistorical collapse. *Terra Nova*, Vol.18, No3, pp. 202-209.

Christie-Blick, N. (1991). Onlap, offlap and the origin of unconformity-bounded depositional sequences. *Marine Geology*, Vol.97, pp.35-56.

Civetta, L., Gallo, G. & Orsi, G. (1991). Sr and Nd isotope and trace element constraints on the chemical evolution of the magmatic system of Ischia (Italy) in the last 55 ky. *Journal of Volcanology and Geothermal Research*, Vol.46, pp.213-230.

Colantoni, P., Del Monte, M., Fabbri, A., Gallignani, P., Selli, R. & Tomadin L. (1972). Ricerche geologiche nel Golfo di Pozzuoli. *Quaderni De La Ricerca Scientifica*, CNR, Vol.83, pp. 26-71.

Cole, R.H. (1965). Underwater Explosions. *Dover Publications*, New York.

Corradi, N.; Ferrari, M.; Giordano, F., Giordano, R., Ivaldi, R. & Sbrana, A. (2009) SAM source and D-Seismic system:The use in Marine Geological Mapping C.A.R.G and P.n.r.a projects. *27th IAS Meeting of Sedimentologists*, Alghero (Italy), pp. 85-90.

Correggiari, A., Roveri, M. & Trincardi, F. (1992). Regressioni forzate, regressioni deposizionali e fenomeni di instabilità in unità progradazionali tardo-quaternarie. *Giornale di Geologia*, Vol.54, pp.19-36.

D'Argenio, B., Pescatore, T.S. & Scandone P. (1973). Schema geologico-strutturale dell'Appennino meridionale (Campania e Lucania). *Quaderni dell'Accademia Nazionale dei Lincei*, Problemi Attuali di Scienza e Cultura, Vol.183, pp. 49-72.

D'Argenio, B.; Aiello, G., de Alteriis. G., Milia, A., Sacchi, M. et al. (2004). Digital Elevation Model of the Naples Bay and adjacent areas, Eastern Tyrrhenian sea, In: *Mapping Geology in Italy*, E. Pasquarè & G. Venturini (Eds.), APAT, National Geological Survey of Italy, Spec. Vol. SELCA, Florence, 22-28.

de Alteriis, G., Tonielli, R., Passaro, S. & De Lauro, M. (2005). Isole Flegree (Ischia e Procida). Batimetria dei fondali marini della Campania. Scala 1:30.000. *Liguori Editore*, Napoli.

de Alteriis, G., Insinga, D.D., Morabito, S., Morra, V., Chiocci, F.L., Terrasi, F., Lubritto, C., Di Benedetto, C. & Pazzanese, M. (2010) Age of submarine debris avalanches and tephrostratigraphy offshore Ischia island, Tyrrhenian sea. *Marine Geology*, V. 278, pp. 1-18.

De Astis, G., Pappalardo, L. & Piochi, M. (2004). Procida volcanic history: new insights into the evolution of the Phlegrean Volcanic District (Campania, Italy). *Bulletin of Volcanology*, Vol.66, pp. 622-641.

De Vita, S.; Esposito, B. & Mirabile, L. (1979). Criteri di Progetto di Sparker a cortina per sismica ad alta risoluzione. *Atti del convegno Scientifico Nazionale Progetto Finalizzato Oceanografia e Fondi Marini*.

De Vita, P., Agrello, D. & Ambrosino, F. (2006). Landslide susceptibility assessment in ash-fall pyroclastic deposits surrounding Mount Somma-Vesuvius. Application of geophysical surveys for soil thickness mapping. *Journal of Applied Geophysics*, Vol.59, pp. 126-139.

De Vita, P., Celico, P., Di Clemente, E. & Rolandi, M. (2007). Engineering geological models of the initial landslides occurred on 30 April 2006 at the Mount of Vezzi (Ischia island). *Italian Journal of Engineering Geology and Environment*.

Di Fiore, V., Aiello, G. & D'Argenio, B. (2011). Gravity instabilities in the Dohrn canyon (Bay of Naples, Southern Tyrrhenian sea): potential wave and run-up (tsunami) reconstruction from a fossil submarine landslide. *Geologica Carpathica*, Vol.62, No1, pp.55-63.

Di Girolamo, P. & Stanzione, D. (1973). Lineamenti geologici e petrologici dell'isola di Procida. *Rendiconti Soc. Italiana Mineralogia Petrologia*, Vol.29, pp. 81-125.

Di Girolamo, P., Ghiara, M.R., Lirer, L., Munno, R., Rolandi, G. & Stanzione, D. (1984). Vulcanologia e petrologia dei Campi Flegrei. *Bollettino della Società Geologica Italiana*, V.103, pp. 349-413.

Di Maio, R., Piegari, E., Scotellaro, C. & Soldovieri M.G. (2007). Tomografie di resistività per la definizione dello spessore e del contenuto d'acqua delle coperture piroclastiche a M.te di Vezzi (Isola d'Ischia). *Italian Journal of Engineering Geology and Environment*.

Di Nocera, S., Matano, F., Rolandi, G. & Rolandi R. (2007). Contributo sugli aspetti geologici e vulcanologici di Monte di Vezzi (Isola d'Ischia) per lo studio degli eventi franosi dell'Aprile 2006. *Italian Journal of Engineering Geology and Environment*

Dvorak, J.J. & Mastrolorenzo, G. (1991). The mechanisms of recent vertical crustal movements in Campi Flegrei caldera, southern Italy. *Geol. Soc. Am. Special Paper,* Vol. 263.

Edgerton, H.E. & Hayward, G.G. (1964). The boomer sonar source for seismic profiling. Journal of Geophysical Research, Vol. 68, pp. 3033-3042.

Esposti Ongaro, T., Neri, A., Todesco, M. & Macedonio, G. (2002). Pyroclastic flow hazard at Vesuvius from numerical modelling II. Analysis of local flow variables. *Bulletin of Volcanology,* Vol.64, pp. 178-191.

Fabbri, A., Argnani, A., Bortoluzzi, G., Correggiari, A., Gamberi, F., Ligi, M., Marani, M., Penitenti, D., Roveri, M. & Trincardi, F. (2002). Carta geologica dei mari italiani alla scala 1:250.000. Guida al rilevamento. *Presidenza del Consiglio dei Ministri, Dipartimento per i Servizi Tecnici Nazionali, Servizio Geologico,* Quaderni serie III, Vol.8, pp. 1-93.

Finetti, I. & Morelli, C. (1973). Esplorazione sismica per riflessione nei Golfi di Napoli e Pozzuoli. *Bollettino di Geofisica Teorica Applicata,* Vol.16, pp. 175-222.

Forcella, F., Gnaccolini M. & Vezzoli, L. (1981). Stratigrafia e sedimentologia dei depositi piroclastici del settore sud-orientale dell'Isola d'Ischia. *Rivista Italiana Paleontologia Stratigrafia,* 87, pp. 329-366.

Fusi, N., Mirabile, L., Camerlenghi, A. & Ranieri, G. (1991). Marine geophysical survey of the Gulf of Naples (Italy): relationship between submarine volcanic activity and sedimentation. *Memorie della Società Geologica Italiana,* 47, pp. 95-114.

Gabbianelli, G., Tramontana, M., Colantoni, P. & Fanucci F. (1996). Lineamenti morfostrutturali e sismostratigrafici del Golfo di Patti (Margine nord-siciliano). In: *Caratterizzazione ambientale marina del sistema Eolie e dei bacini limitrofi di Cefalù e Gioia,* F.M. Faranda & P. Povero (Eds.), Data Report, pp.443-454.

Galdi, A., Giordano, F., Sposito, A. & Vultaggio, M. (1988) Misure geomagnetiche nel Golfo di Pozzuoli: Metodologia e risultati. *Atti del 7° Convegno GNGTS-CNR,* Vol.3, pp. 1647-1658.

Galloway, W.E., Dingus, W.F. & Paige R.E. (1991). Seismic and depositional facies of Paleocene-Eocene Wilcox Group submarine canyon fills, Northwest Gulf Coast, USA, In: *Seismic Facies and Sedimentary Processes of Submarine Fans and Turbidite Systems,* P. Weimer & M.H. Link (Eds.), Springer-Verlag, pp. 247-271.

Gillot, P.Y., Chiesa, S., Pasquarè, G. & Vezzoli, L. (1982). 33.000 yr. K/Ar dating of the volcano-tectonic horst of the isle of Ischia, Gulf of Naples. *Nature,* Vol.229, pp.242-245.

Guadagno, F.M. & Mele, R. (1995) La fragile isola d'Ischia. *Geologia Applicata e Idrogeologia,* Vol.30, No1, pp.177-187.

Helland Hansen, W. & Gjelberg, J.G. (1994). Conceptual basis and variability in sequence stratigraphy: a different perspective. *Sedimentary Geology,* Vol.92, pp.31-52.

Hernandez Molina, F.J., Somoza, L., Rey, J. & Pomar, L. (1994). Late Pleistocene-Holocene sediments on the Spanish continental shelves: model for high resolution sequence stratigraphy. *Marine Geology,* Vol.120, pp. 120-174.

Knott, S.T. & Hersey, J.B. (1956). Interpretation of high resolution echo-soundings techniques and their use in bathymetry, marine geophysics and biology. *Deep Sea Research*, Vol.4, pp. 36-44.

Insinga, D., Molisso F., Lubritto, C., Sacchi, M., Passariello, I. & Morra, V. (2008). The proximal marine record of Somma-Vesuvius volcanic activity in Naples and Salerno bays, Eastern Tyrrhenian sea, during the last 3 kyrs. *Journal of Volcanology and Geothermal Research*, Vol. 177, pp. 170-186.

Isaia, R., Marianelli, P. & Sbrana, A. (2009). Caldera unrest prior to intense volcanism in Campi Flegrei (Italy) at 4.0 ka B.P.: Implications for caldera dynamics and future eruptive scenarios. *Geophysical Research Letters*, Vol. 36, doi: 10.1029/2009GL040513.

Latmiral, L., Segre, A.G., Bernabini, M. & Mirabile, L. (1971). Prospezioni sismiche per riflessione nei Golfi di Napoli e Pozzuoli ed alcuni risultati geologici. *Bollettino della Società Geologica Italiana*, Vol.90, pp.163-172.

Luongo, G., Cubellis, E. & Obrizzo, F. (1987). Ischia: storia di un'isola vulcanica. *Liguori Editore*, Napoli.

Luongo, G., Cubellis, E. & Obrizzo, F. (1997). Storia e strumenti per un Museo Vulcanologico, In: *Mons Vesuvius*, G. Luongo (Ed.), Fiorentino Editore, Napoli, pp. 383-408.

Marani, M., Taviani, M., Trincardi, F., Argnani, A., Borsetti, A.M. & Zitellini, N. (1986). Pleistocene progradation and postglacial events of the NE Tyrrhenian continental shelf between the Tiber river delta and Capo Circeo. *Memorie della Società Geologica Italiana*, Vol.36, pp. 67-89.

Martinson, D.G., Pisias, N.G., Hays, J.D., Imbrie, J., Moore, T.C. & Shackleton, N.J. (1987). Age dating and orbital theory of the Ice Ages: development of a high resolution 0 to 300.000 year chronostratigraphy. *Quaternary Research*, Vol.27, No1, pp. 1-29.

Mastrolorenzo, G., Palladino, D., Vecchio, G. & Taddeucci, J. (2002). The 472 AD Pollena eruption at Somma-Vesuvius, Italy and its environmental impact at the end of Roman Empire. *Journal of Volcanology and Geothermal Research*, Vol.113, pp. 19-36.

Mele, R. & Del Prete, S. (1998). Fenomeni di instabilità dei versanti in Tufo Verde del Monte Epomeo (Isola d'Ischia, Campania). *Bollettino della Società Geologica Italiana*, Vol.117, No1, pp. 93-112.

Milia, A. (1998) Stratigrafia, strutture deformative e considerazioni sull'origine delle unità deposizionali oloceniche del Golfo di Pozzuoli (Napoli). *Bollettino della Società Geologica Italiana*, vol. 117, pp. 777-787.

Milia, A. & Torrente, M.M. (2000). Fold uplift and syn-kinematic stratal architectures in a region of active transtensional tectonics and volcanism, Eastern Tyrrhenian sea. *Geological Society of America Bulletin*, Vol.112, pp.1531-1542.

Milia, A. & Torrente, M.M. (2003). Late Quaternary volcanism and transtensional tectonics in the Bay of Naples, Campanian continental margin, Italy. *Mineralogy and Petrology*, vol.79, pp. 49-65.

Milia, A., Mirabile, L., Torrente, M.M. & Dvorak J.J. (1998). Volcanism offshore of Vesuvius volcano (Italy): Implications for hazard evaluation. Bulletin of Volcanology, 59, 404-413. *Journal of the Geological Society of London*, Vol.160, pp.309-317.

Milia, A.; Molisso, F., Raspini, A., Sacchi, M. & Torrente, M.M. (2008). Syneruptive features and sedimentary processes associated with pyroclastic currents entering the sea:

the AD 79 eruption of Vesuvius, Bay of Naples, Italy. *Journal of the Geological Society,* V.165, No4, pp. 839-848.

Mirabile, L., (1969). Prime esperienze di stratigrafia sottomarina eseguite presso l'Istituto Universitario Navale. *Annali IUN,* Vol. 38.

Mirabile, L., Nicolich, R., Piermattei, R. & Ranieri, G. (1989). Identificazione delle strutture tettonico-vulcaniche dell'area flegrea: sismica multicanale del Golfo di Pozzuoli. *Atti del 7° Convegno GNGTS,* Vol.2, Roma, Italy, pp. 829-838.

Mirabile, L., Fevola, F., Galeotti, F., Ranieri, G.& Tangaro, G. (1991). Sismica monocanale ad alta risoluzione con sorgente multi spot di tipo sparker: applicazione ai dati di tecniche di deconvoluzione. *Atti del 10° Convegno GNGTS-CNR,* Roma, Italy.

Molisso, F., Insinga, D., Marzaioli, F., Sacchi, M. & Lubritto, C. (2010). Radiocarbon dating versus volcanic event stratigraphy: age modelling of Quaternary marine sequences in the coastal region of the Eastern Tyrrhenian sea. *Nuclear Instruments and Methods in Physics Research B,* Vol. 268, pp.1236-1240.

Mongardi, S., Correggiari, A. & Trincardi, F. (1995). Regional drape deposits in a Quaternary turbidite succession. Inferences from high resolution study of the Late Quaternary drape of the sea floor of the Paola basin (Tyrrhenian sea). *16th IAS European Sedimentological Meeting,* Aix-le-bains, France, p. 106.

Moore, J.G. & Mark, R.K. (1992). Morphology of the island of Hawaii. *GSA Today,* Vol. 2, pp. 257-262.

Orsi, G., Gallo, G. & Zanchi, A. (1991). Simple-shearing block resurgence in caldera depressions. A model from Pantelleria and Ischia. *Journal of Volcanology and Geothermal Research,* Vol.71, p.249-257.

Orsi, G., Piochi, M., Campajola, L., D'Onofrio, A., Gialanella, L. & Terrasi, F. (1996). [14]C geochronological constraints for the volcanic history of the island of Ischia (Italy) over the last 5000 years. *Journal of Volcanology and Geothermal Research,* Vol.71, p.249-257.

Paoletti, V.; Fedi, M. Florio, G., Supper, R. & Rapolla, A. (2004). The new integrated aeromagnetic map of the Phlegrean Fields Fields volcano and surrounding areas. *Annals of Geophysics,* Vol. 47, No 5, pp. 1569-1580.

Passaro, S. (2005). Integrazione di dati magnetici e batimetrici in aree vulcaniche e non vulcaniche: esempi dall'isola d'Ischia e dal Banco di Gorringe (Oceano Atlantico). *PhD Thesis,* Università di Napoli "Federico II".

Pescatore, T.S. & Rolandi, G. (1981). Osservazioni preliminari sulla stratigrafia dei depositi vulcanoclastici del settore sud-occidentale dei Campi Flegrei. *Bollettino della Società Geologica Italiana,* Vol. 100, pp. 233-254.

Poli, S., Chiesa, S., Gillot, P.Y., Gregnanin A. & Guichard, F. (1987). Chemistry versus time in the volcanic complex of Ischia (Gulf of Naples, Italy). *Contributions to Mineralogy and Petrology,* Vol.95, No3, pp.322-335.

Poli, S., Chiesa, S., Gillot, P.Y., Guichard, F. & Vezzoli, L. (1989). Time dimension in the geochemical approach and hazard estimation of a volcanic area: the isle of Ischia case (Italy). *Journal of Volcanology and Geothermal Research,* Vol.36, pp. 327-335.

Posamentier, H.W. & Allen, G.P. (1993). Variability in the sequence stratigraphic model: effects of local basin factors. *Sedimentary Geology,* Vol.86, pp.91-109.

Posamentier, H.W. & Vail, P.R. (1988). Eustatic control on clastic deposition II – sequence and system tracts models, In: *Sea level changes: an integrated approach*, C.K. Wilgus, B.S. Hastings et al. (Eds.), SEPM Special Publication., Vol.42, pp.125-154.

Posamentier, H.W., Erskine R.D. & Mitchum, R.M. (1991). Models for Submarine Fan Deposition within a Sequence Stratigraphic Framework, In: *Seismic Facies and Sedimentary Processes of Submarine Fans and Turbidite Systems*, P. Weimer & M.H. Link (Eds.), New York, Springer Verlag, pp. 127-136.

Posamentier, H.W., James, D.P., Allen, J.P. & Tesson, M. (1992). Forced regressions in a sequence stratigraphic framework: concepts, examples and exploration significance. *AAPG Bulletin*, Vol.76, pp. 1687-1709.

Ranieri, G. & Mirabile, L. (1991). Ricerca ed applicazione di metodi geofisici al rilievo sperimentale della struttura medio-profonda dell'area flegrea con uso di sorgenti sismiche water-gun. *Annali Istituto Universitario Navale di Napoli*, Vol.63

Rittmann, A. (1930). Geologie der Insel Ischia. *Ergn vur Vulk*, Berlin.

Rittmann, A. (1948) Origine e differenziazione del magma ischitano. *Schweiz Miner Petrogr Mitt.*,Vol. 28, pp. 643-698.

Rosi, M. & Santacroce, R. (1983) The A.D. 472 "Pollena" eruption: volcanological and petrological data for this poorly known plinian type event at Vesuvius. *Journal of Volcanology and Geothermal Research*, Vol.17, pp. 249-271.

Rosi, M. & Sbrana, A. (1987). Phlegrean Fields. *Quaderni De La Ricerca Scientifica*, CNR, Vol.114, No9, 175 pp.

Rosi, M., Sbrana A. & Vezzoli, L. (1988a). Tephrostratigraphy of Ischia, Procida and Campi Flegrei volcanic products. *Memorie della Società Geologica Italiana*, 41, pp. 1015-1027.

Rosi, M., Sbrana, A. & Vezzoli, L. (1988b) Stratigraphy of Procida and Vivara islands. *Bollettino GNV*, Vol.4, pp. 500-525.

Sacchi, M., Insinga, D., Milia, A., Molisso, F., Raspini, A., Torrente, M.M. & Conforti, A. (2005) Stratigraphic signature of the Vesuvius 79AD event off the Sarno prodelta system, Naples Bay. *Marine Geology*, Vol.222-223, pp. 443-469.

Saccorotti, G., Ventura, G. & Vilardo, G. (2002). Seismic swarms related to diffusive processes: the case of Somma-Vesuvius volcano, Italy. *Geophysics*, Vol.67, pp.199-203.

Santacroce, R. (1987). Somma-Vesuvius. *CNR, Quaderni De La Ricerca Scientifica*, Vol.114, pp.1-175.

Scarpa, R., Tronca, F., Bianco, F.. & Del Pezzo, E. (2002). High resolution velocity structure beneath Mount Vesuvius from seismic array data. *Geophysical Research Letters*, Vol.29, pp.204-219.

Scarpati, C., Cole, P. & Perrotta, A. (1993). The Neapolitain Yellow Tuff – A large volume multiphase eruption from Campi Flegrei, Southern Italy. *Bulletin of Volcanology*, Vol.55, pp.343-356.

Secomandi, M., Paoletti, V., Aiello, G., Fedi, M., Marsella, E, Ruggieri, S., D'Argenio, B., Rapolla, A. (2003). Analysis of the magnetic anomaly field of the volcanic district of the Naples Bay. *Marine Geophysical Researches*, 24, 207-221.

Sheridan, M.F., Barberi, F., Rosi, M. & Santacroce, R. (1981). A model for Plinian eruption of Vesuvius. *Nature*, Vol.289, pp. 282-285.

Sheridan, M.F. (1982) Application of computer assisted mapping to volcanic hazard evaluation of surge eruptions: Vulcano, Lipari and Vesuvius. *Journal of Volcanology and Geothermal Research*, Vol.17, pp. 187-202.

Sigurdsson, H., Cashdollar, S. & Sparks, S.R.J. (1982). The eruption of Vesuvius in AD79: reconstruction from historical and volcanological evidence. *American Journal of Archaeology*, Vol.86, pp.39-51.

Tesson, M., Allen G.P. & Ravenne, C. (1993). Late Pleistocene shelf perched lowstand wedges on the Rhone continental shelf. In: *Sequence Stratigraphy and Facies Associations*, H.W. Posamentier, C.P. Summerhayes, B.U. Haq & G.P. Allen (Eds.), IAS Special Publication, No18.

Todesco, M., Neri, A., Esposti Ongaro, T., Papale, P. , Macedonio, G. & Santacroce, R. (2002) Pyroclastic flow hazard at Vesuvius from numerical modelling I. Large scale dynamics. *Bulletin of Volcanology*, Vol.64, pp. 155-177.

Tramontana, M., Colantoni, P. & Fanucci, F. (1995). Risultati preliminari delle indagini morfologico-sedimentologiche condotte nell'ambito del progetto EOCUMM94, In: *Caratterizzazione ambientale marina del sistema Eolie e dei bacini limitrofi di Cefalù e di Gioia*, F.M. Faranda (Ed.), Data Report, 1995, pp. 331-338.

Trincardi, F. & Field, M.E. (1991). Geometry, lateral variation and preservation of downlapping regressive shelf deposits: Eastern Tyrrhenian sea margin, Italy. *Journal of Sedimentary Petrology*, Vol.61, pp.775-790.

Trincardi, F. & Normark, W.R, (1988). Sediment waves on the Tiber prodelta slope: interaction of deltaic sedimentation and currents along shelf. *Geomarine Letters*, Vol.8, pp. 149-157.

Trincardi, F., Correggiari, A. & Roveri, M. (1994). Late Quaternary transgressive erosion and deposition in a modern continental shelf: the Adriatic semienclosed basin. *Geomarine Letters*, Vol.14, pp.41-51.

Trincardi, F., Cattaneo, A. & Correggiari, A. (2003). Growth of the modern Po delta and prodelta system. *COMDELTA Conference*, Aix En Provence, France, p. 141.

Vail, P.R., Mitchum, R.M. & Thompson, S. (1977) Seismic stratigraphy and global changes of sea level, Part 3, relative changes in sea level from coastal onlap, In: *Seismic Stratigraphy – Applications to Hydrocarbon Exploration*, C.E. Payton (Ed.), AAPG Mem. 26, pp.63-81.

Vail, P.R.., Hardenbol J. & Todd, R.G. (1984) Jurassic unconformities, chronostratigraphy and sea level changes from seismic stratigraphy and biostratigraphy, In: *Interregional unconformities and hydrocarbon accumulation*, AAPG Mem. 36, 129-144.

Vezzoli, L. (1988). Island of Ischia. *Quaderni De La Ricerca Scientifica*, CNR., Roma.

Zollo, A., Gasparini, P., Biella, G., De Franco, R., Buonocore, B., Mirabile, L., De Natale, G., Milano, G., Pingue, F., Vilardo, G., Bruno, P.P.G., De Matteis, R., Le Meur, H., Iannaccone, G., Deschamps, A., Virieux, J., Nardi, A., Frepoli, A., Hunstad, I., Guerra, I. (1996) 2D seismic tomography of Somma-Vesuvius: description of the experiment and preliminary results. *Annals of Geophysics*, Vol.39, pp. 471-486.

Zollo, A., Gasparini, P., Virieux, J., Biella, G., Boschi, E., Capuano, P., De Franco, R., Dell'Aversana, P., De Matteis, R., De Natale, G., Iannaccone, G., Guerra, H., Le Meur H. & Mirabile, L. (1998). An image of Mt. Vesuvius obtained by 2D seismic tomography. *Journal of Volcanology and Geothermal Research*, Vol.82, pp.161-173.

Zollo, A., Gasparini, P., Virieux, J., Biella, G., Boschi, E., Capuano, P., De Franco, R., Dell'Aversana, P., De Matteis, R., De Natale, G., Iannaccone, G., Guerra, H., Le Meur H. & Mirabile, L. (2003). An image of Mt. Vesuvius obtained by 2D seismic tomography, In: *The Internal Structure of Mt. Vesuvius*, P. Capuano, P. Gasparini, A. Zollo, J. Virieux, R. Casale & M. Yeroyanni (Eds.), Liguori Editore, Napoli, pp, 75-104.

2

Medium to Shallow Depth Stratigraphic Assessment Based on the Application of Geophysical Techniques

Roberto Balia

University of Cagliari, Dipartimento di Ingegneria del Territorio,
Italy

1. Introduction

In strict terms, the word "stratigraphy" refers to the study and description of a natural succession of more-or-less parallel layers, or strata, of sedimentary rocks. However, in the fields of environmental engineering and engineering geology, the term "stratigraphy" assumes also a general and broader meaning, since it very often refers to a generic underground sequence of not always sedimentary and not only natural materials.

That said, the importance of an adequate knowledge of the site stratigraphy in engineering and environmental problems is well known. Geotechnical studies for building foundation design, waste landfill design or pre-reclamation assessment, aquifers monitoring and evaluation, and sea water intrusion control, are among the most common activities in which at least some aspects, namely thickness, composition and hydrogeology of the unconsolidated cover, depth to bedrock and conditions of the latter, must be clarified at the best. As far as the investigation depth is concerned, it could range from few meters – few tens of meters in geotechnical and waste landfill studies, to few hundreds of meters in regional hydrogeological studies and in the assessment of the fresh-water/sea-water relationships along the coastal belts.

In all the above situations, classical geological and hydrogeological surveys, integrated with direct investigations such as shallow excavations, and adequately deep and properly distributed pits and bore holes, can provide the required information.

However, this strategy can imply both technical and economical concerns, mainly regarding the distribution and quantity of direct surveys.

Actually in several, simple cases (e.g.: very small extension of the study area; limited lateral variations, that is 1D conditions, where only qualitative information is required), surface geological data along with a very small amount of direct investigations can be more than enough.

Conversely, when the stratigraphic assessment is the premise of a more complex and relevant work covering relatively large areas characterized by complex geological conditions, the following questions arise: first, what degree of accuracy is needed in the assessment of the underground stratigraphy? Second, as a consequence of the answer to the first question and also based on the depth to the target, what type of direct investigation is more appropriate and, for instance in the case of bore holes, how are they to be distributed? Third, are the technical requirements consistent with a reasonable budget? In this context, a valuable aid may be provided by the geophysical survey techniques.

As known, these techniques provide indirect information about geological, hydrogeological, geotechnical and environmental conditions, through the study of some physical characteristics of the subsurface. For instance, if you measure a high speed of propagation of elastic waves, it is most likely associated with consolidated rocks, while low speed values should correspond to loose materials; similarly, a relatively low electrical resistivity can be associated with the presence of aquifers, while very high values should correspond to hard, dry rocks. So, the gravity method is based on the density, the magnetic method on the magnetic susceptibility, the seismic methods on the acoustic properties, namely the density and the velocity of elastic waves, the electrical methods mainly on the electrical resistivity and so on. Both the theory and the practice of geophysical methods are widely treated in many text books of applied geophysics (e.g. Dobrin, 1976; Reynolds, 1997; Sharma, 1997; Telford et al., 1990).

In this chapter, on the basis of several case studies, we shall try and illustrate in what way geophysical techniques can contribute to the stratigraphic assessment of a site providing high-level information and contributing at least to rationally planning, if not completing avoiding, the drilling campaign. In all cases, the primary method of investigation has been that of reflection seismology employed at different scales. However, this method was prevalently preceded by a gravity survey, which is essential for the proper design of the acquisition parameters, and accompanied by other geophysical data and direct surveys, such as drillings and exploratory excavations. As known, the reflection seismic method owes its great development to the fact that it has been linked, historically, to the search for oil and gas. However, in the past three decades the data acquisition and processing techniques of this method have been progressively adapted to shallow targets. In the early eighties of the past century, the term "shallow reflection" was associated to targets at depths in the order of some hundreds of meters, and applications for depths of few tens of meters, or less, were conducted only at the experimental level.

Nowadays, the use of this method with targets at depths of few tens of meters and even of few meters, has become a technical reality. In the examples illustrated in the following sections, the maximum depth to the targets ranges from a few hundred meters to a few meters and therefore it can be said to be from a medium to a very shallow depth . For an adequate knowledge of principles, data acquisition and data processing for the reflection seismic method, refer to Dobrin (1976) and Yilmaz (1987).

2. Stratigraphic assessment of a coastal plain affected by groundwater salination

The coastal plain covered in this section is a fluvial valley that also includes a river delta (Balia et al., 2003). The surface geology of the plain and its surroundings is characterized, from bottom to top, by a Paleozoic metamorphic complex outcropping on the edges of the plain, and Pleistocene and Holocene sediments and alluvium, up to a few hundred meters thick, overlying the Paleozoic bedrock. Granites (Upper Carboniferous- Permian) outcrop a few kilometers from the edges of the valley. Before the geophysical surveys, the thicknesses of recent alluvium, ancient alluvium, and metamorphic complex in the plain were only estimated on the basis of morphology and surface geology. As regards hydrogeology, the surface water bodies are the river, its channels at the mouth, which are no longer connected with the river itself but contain incoming seawater, and several seasonal streams flowing down from the surrounding hills. Apart from the water occurring in the fractured Paleozoic rocks, from which a few small ephemeral springs issue during the cooler months, groundwater is primarily in the alluvial deposits, and the dominant, qualitative theory was that two aquifers could be distinguished: a shallow phreatic aquifer extending down to a few tens of meters, and an undefined, deeper, confined aquifer, separated from the former by a clay layer from a few meters to several tens of meters thick. The lower boundary and deeper stratigraphy of the confined aquifer were poorly understood so far.

Due to the importance of understanding at the best the hydrogeological model of the plain, a relatively intensive application of geophysical techniques was used as a tool for elucidating a number of aspects of primary importance for the realistic modeling of salination and its evolutionary trend. Among these, the following were the most important: 1) conditions of shallow and deep salination; 2) structural model of the plain, including depth to Paleozoic basement; 3) stratigraphy of the Pleistocene-Holocene sedimentary cover; 4) relationships between the phreatic aquifer and the confined aquifer.

Therefore, the primary targets of the stratigraphic assessment by means of geophysical methods were the depth to the Paleozoic bedrock and the stratigraphy of the overlying Pleistocene-Holocene cover. For these purposes, the primary geophysical method was that of reflection seismology, although gravity and electrical methods were also employed. Thus, one seismic profile was positioned and designed, based on gravity data previously acquired and processed in the frame of the same project, and on preliminary tests. In detail, the acquisition geometry was designed for a target depth of 100-200 m. A 48-channel off-end spread of single 40 Hz geophones at 5 m spacing was used, with a minimum offset of 30 m and, consequently, a maximum offset of 265 m. The acquisition system was a 48-channel seismograph with a 60 Hz low-cut filter and a 600 Hz antialias (high-cut) filter. Record length was 500 ms (millisecond) and sampling interval 0.25 ms. Small dynamite charges (30-100 g) placed in 1.5-2 m boreholes at 5 m intervals were used as an energy source, giving a maximum nominal CMP (common midpoint) fold of 2,400%. In all, 172 shots were performed, obtaining a total seismic section length of 975 m.

The data quality was satisfactory and the dominant reflection frequency was in the order of 70-80 Hz. Processing included amplitude equalization, 40-120 Hz band-pass filtering, statics, CMP sorting, velocity analysis, NMO (normal moveout) correction, CMP stacking, and time-to-depth conversion. Further more or less sophisticated processing proved not strictly necessary and was not applied. Interval velocities were computed from stack velocities by means of the Dix equation and were used for time-to-depth conversion.

The depth section reported in figure 1 shows two main reflectors, both attributable to the Paleozoic basement. The upper one lies at a maximum depth of about 280 m (CMP trace #40) and emerges more or less regularly up to a depth of less than 100 m (CMP trace #250). The morphology of the lower one, which is still present in the northern side of the section, exhibits a high at CMP trace #275, at a depth of about 100 m.

Tectonic structures like faults and fracture zones are also present. The upper reflector is associated with the boundary between the Pleistocene-Holocene cover and the Paleozoic metamorphic rocks, while the lower reflector is associated with the transition from metamorphic rocks to granite. The velocity of Pleistocene-Holocene sediments and alluvium is in the order of 1,700-2,000 m/s and the average interval velocity between the two reflectors is 2,700 m/s. These values suggest that Pleistocene–Holocene sediments are fairly consolidated. Also, due to their relatively low velocity, Paleozoic metamorphic rocks should be relatively fractured and altered, at least in the upper part. The lack of a coherent signal in the lower part of the section, from CMP trace #270 to the northwest, may be attributed to relatively homogeneous granite.

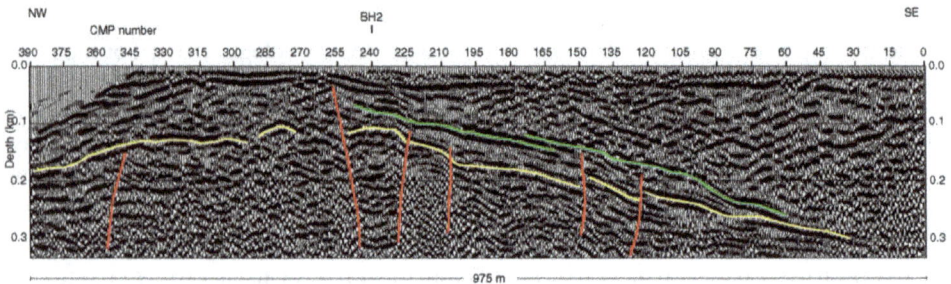

Fig. 1. Interpreted depth section of the P-wave reflection seismic profile SP2. CMP trace interval is 2.5 m . See text for description of reflectors. (After Balia et al., 2003)

Given the aim of the work, a detailed knowledge of the Pleistocene-Holocene cover was of primary interest. Thus, the data pertaining to the southernmost part of the seismic profile were processed separately, especially refining velocity analysis for shallower events. The corresponding time section is shown in figure 2. Sediments and alluvium overlying the bedrock are clearly stratified and show a low around CMP trace #100, with a maximum estimated depth of roughly 150 m. The latter structure may be associated with a paleovalley, probably related to the ancient course of the river. According to geological knowledge, reflector 1 in figure 2 (green color) corresponds to one boundary that separates Holocene materials with different characteristics (e.g. different density and velocity due to different compaction), and reflector 2 (yellow color) corresponds to the boundary between permeable Holocene alluvium and impermeable Pleistocene terraced alluvium. This suggests that mathematical modeling of the aquifers contained in the Holocene cover could be limited to a depth of 150-200 m below ground level. The total cost (planning, data acquisition, data processing and interpretation) of the seismic profile shown above is equivalent to that of two-three adequately deep boreholes. However these, even if distributed at the best, could not in any way guarantee the same complete information provided by the seismic profile.

Having solved the problem of the relationships between the cover and the Paleozoic basement, the relationships between the phreatic and the underlying confined was next. For this purpose, another reflection profile was carried out, but electrical resistivity and borehole data were also used for its hydrogeological interpretation.

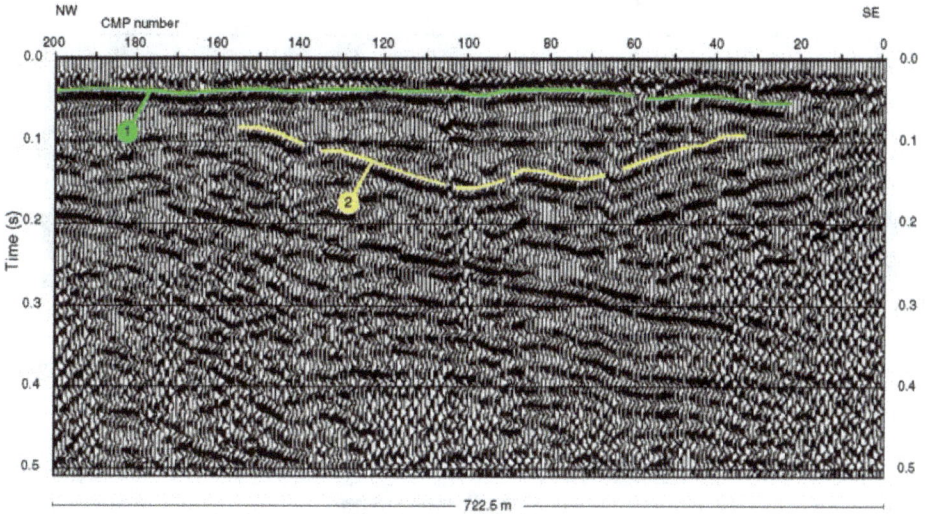

Fig. 2. Interpreted time section for the southernmost part of seismic profile SP2. See text for description of reflectors. (After Balia et al., 2003)

Concerning the seismic profile, it was designed for relatively shallow targets: data were acquired using a 24-channel off-end spread with a channel interval of 3 m and a 30 m in-line minimum offset. Single 50 Hz natural frequency geophones were used. The recording instrument was set with the following acquisition parameters: record length 200 ms, sample interval 0.25 ms, low-cut filter 70 Hz, high-cut filter 700 Hz. A shot-gun was used as energy source, with a shot interval of 3 m which, given the spread, gave a maximum CMP fold of 1,200%. The processing sequence included amplitude equalization, frequency filtering, statics, muting, CMP sorting, velocity analysis, NMO correction, CMP stacking, f-k migration, and time-to-depth conversion. The optimum stack velocity was about 1,700 m/s. The seismic section, shown in figure 3, exhibits two reflectors, not very easy to interpret. In order to perform an interpretation consistent with the real geological and hydrogeological conditions, it was decided to drill a calibration borehole (BH1 in figure 3), which was located in correspondence of the CMP position #182, that is less than a hundred meters away from the center of a vertical electrical sounding (VES9 in figure 3). The drilling capabilities allowed a depth of not more than 35.5 m, since drilling had to be stopped at the depth of 32.7 m from ground surface (about 28 m under sea level) because a hard layer of pebbles in a silty-sandy matrix containing high pressure saltwater was met. In spite of this, the obtained stratigraphy proved rather meaningful. It is shown in figure 4 compared with the corresponding resistivity and seismic columns. As can be seen, there is a close correlation among the following discontinuities:

-   transition from clay layer to layer made up of pebbles in a silty-sandy matrix and containing high pressure saltwater;
-   transition from 22 ohm-m to 3 ohm-m;
-   upper reflector (reflector 1 in Figure 3).

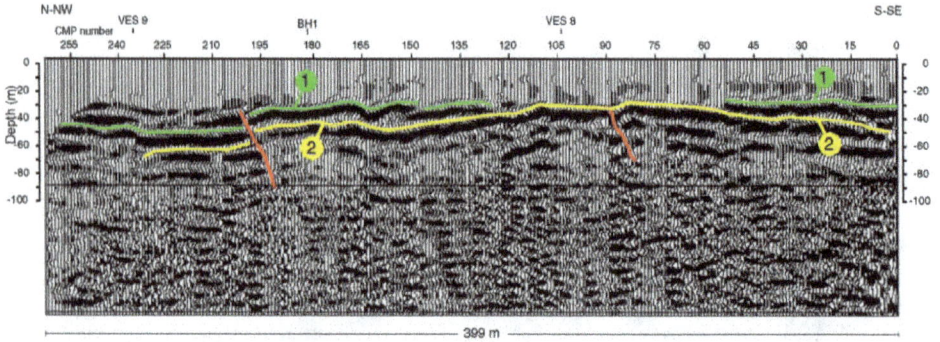

Fig. 3. Geophysical interpretation of the seismic section acquired for clarifying the relationships between the phreatic aquifer and the underlying confined aquifer. (After Balia et al., 2003)

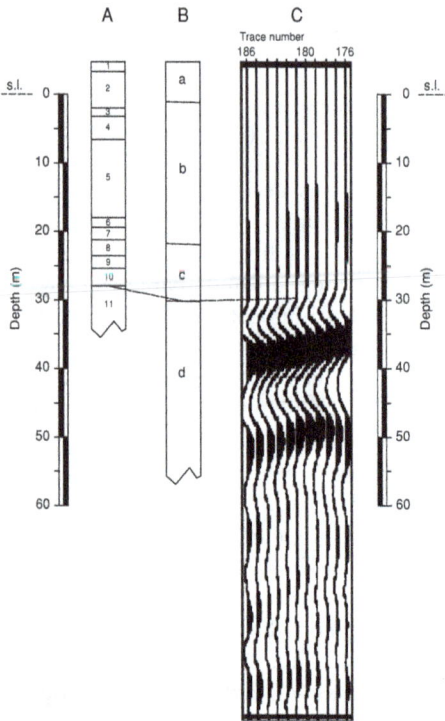

Fig. 4. Borehole stratigraphy (A) compared with the resistivity (B) and seismic (C) columns. Legend for the stratigraphic column (A): 1. clayey soil; 2. fine-coarse sand; 3. pebbles in a sandy matrix; 4. sand with microconglomerates and rare pebbles; 5. pebbles in a silty-sandy matrix; 6. coarse sand; 7. pebbles in a silty-sandy matrix; 8. silt and clay with medium-coarse sand; 9. pebbles; 10. thick clay with minor sand; 11. pebbles in a silty-sandy matrix. Legend for the resistivity column (B): a. 10-21 ohm-m; b. 5 ohm-m; c. 22 ohm-m; d. 3 ohm-m. (After Balia et al., 2003)

As regards the lower reflector (reflector 2 in Figure 3), it can only be said that it should represent the lower boundary of the layer of pebbles in silty-sandy matrix, that is of the shallowest unit of the confined aquifer.

In terms of the hydrogeological model and salination status, these results could be interpreted as follows. It is confirmed that there is a separation between the phreatic aquifer and the underlying confined aquifer and, also for this reason, they very probably have rather different histories. The former is actually affected by saltwater intrusion characterized by the present evolution depending on several factors, such as overexploitation, upstream dams, recent artificial channels that have been opened for fish-farming, and recurrent drought; while salination affecting the latter seems to be quite different and more likely to be related to vicissitudes that occurred in an ancient past when the seashore was situated several kilometers inland from its present position.

3. Stratigraphic assessment to evaluate water resources

In this second example, the problem is to ascertain the possibility of finding fresh groundwater under the Quaternary cover (Balia et al., 2008). The surface geology of the site (a coastal plain situated in a graben area) is characterized by the Paleozoic basement that outcrops on one edge of the plain and deepens very quickly towards the middle of the graben, and Pleistocene-Holocene sediments and alluvium. Apart from the surface and near-surface Pleistocene-Holocene sediments and alluvium, before the geophysical campaign the stratigraphy to volcano-metamorphic basement was substantially unknown.

The commonly accepted aquifer system model is as follows:

- a shallow, phreatic aquifer, hosted in recent alluvium, characterized by small depth to water and thickness varying in the range 10-30 m
- a deeper, multilayer aquifer, separated from the former by clay layers interbedded with gravels, with an overall maximum thickness of 20-25 m; this aquifer about 130 m thick, is partially and/or locally confined and hosted in alluvium characterized by a strongly variable permeability, so that it is rather irregular and discontinuous.

The whole hydrogeological conditions of the plain have been extensively studied, in the more or less recent past, even with the contribution of geophysical surveys (Balia et al., 2008, and references therein). In the following, the stratigraphic assessment below the Quaternary cover, that is below the already known aquifer system, is recalled. The need for this assessment was that the already known, near-surface aquifers were mostly exhausted or polluted by seawater, mainly due to overexploitation. Thus, two P-wave seismic reflection profiles were acquired and interpreted (Balia et al., 2008, and references therein). The depth-converted sections are shown in figure 5. Both sections exhibit five reflectors, numbered 1 to 5. Reflector 1 is associated to transition from near-surface Pliocene-Quaternary sediments and alluvium to Miocene sediments; this transition is situated at a depth of the order of 130-150 m, in good agreement with electrical and electromagnetic data, as well as with drillings.

Reflectors 2-4 are interpreted as transitions between different Miocene lithologies. Reflector 5 is not interpreted; however, given the depth, it is very likely to be associated to the transition between the Miocene sediments and the volcano-metamorphic basement. On the basis of the most widely accepted geological and structural scheme of the region, the latter should be made up of Oligocene andesites in its upper part, and then by the Paleozoic, metamorphic rocks and granite. As can be noted in both seismic sections, while reflectors 1 and 2 are almost flat and continuous, reflectors 3 to 5 are increasingly undulated and discontinuous, with evidence of faulting in the deepest layers, and this could mean first that the volcano-metamorphic basement is significantly fractured, thus being suitable for hosting aquifers and, second, that some tectonic event occurred during the Miocene, most probably just before or at the same time as the deposition of the second layer, that is the one bounded by reflectors 1 and 2.

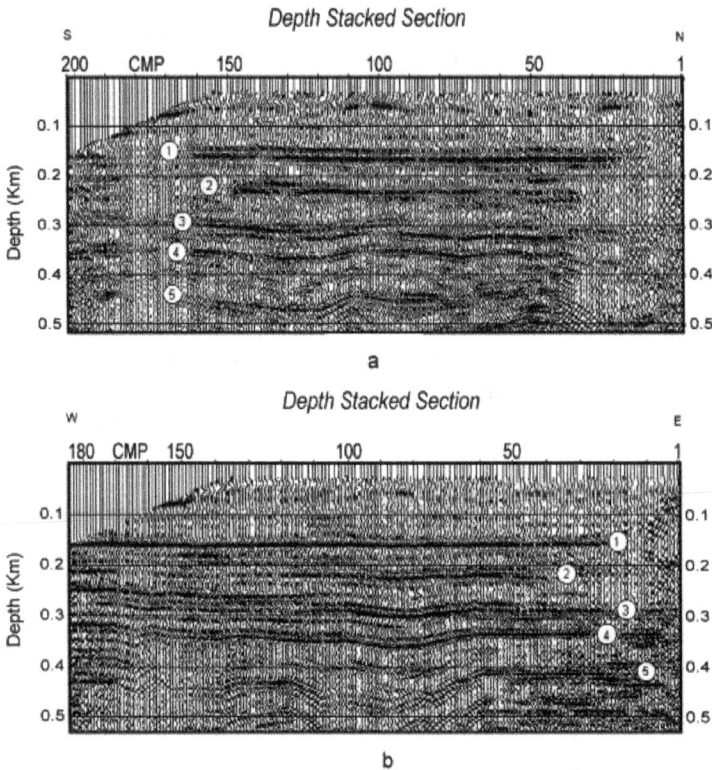

Fig. 5. S–N (a) and W–E (b) seismic depth-sections in the sample area of the plain. CMP trace interval is 2.5 m. The meaning of reflectors 1-5 is explained in the text. (After Balia et al., 2008)

In hydrogeological terms, the stratigraphic conditions described above indicate that, due to conspicuous thickness of impermeable Miocene sedimentary products, at least in the studied portion of the plain, the probability of finding freshwater at a depth of less than 350-400 m is very low, since it could be hosted only in the fractured, volcano-metamorphic basement. Actually this is not a propitious response, but will at least prevent wasting money on inadequate drillings.

Again it was necessary to check the condition of the two aquifers hosted in the Quaternary alluvium, and with regard to geophysical techniques, this was done by means of the electrical resistivity method, namely using the vertical electrical sounding (VES) technique.

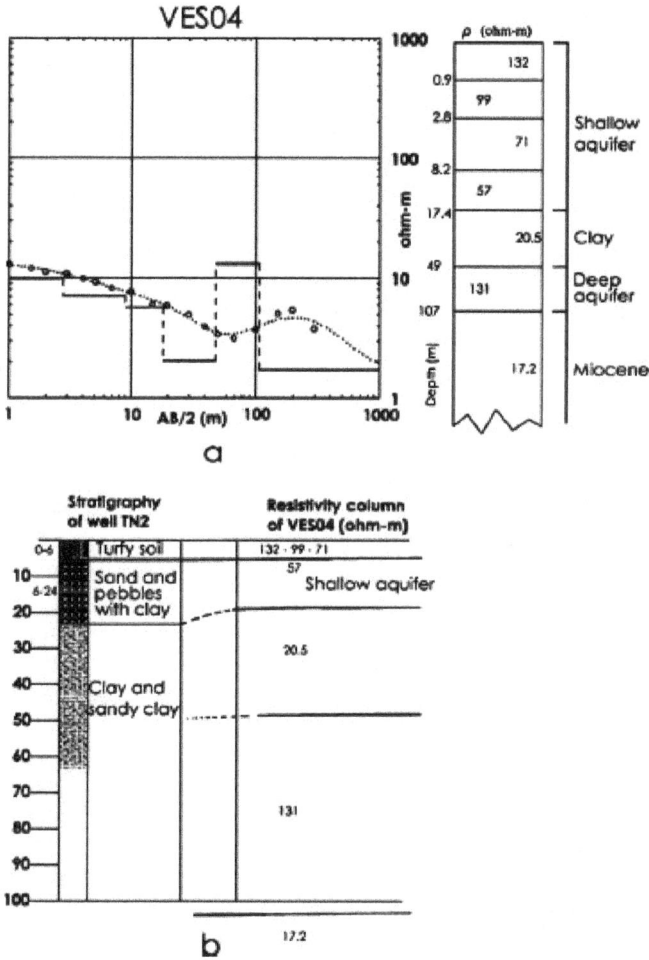

Fig. 6. Electrical resistivity curve (a–left) and interpreted resistivity column (a–right) compared with the stratigraphy from a borehole (b) (After Balia et al., 2008)

Figure 6 shows the apparent resistivity curve of the vertical electrical sounding VES04, acquired in the survey area, its interpretation in terms of true resistivity and thicknesses, and the comparison between the hydrogeological interpretation of geophysical results (top-right in figure 6), and the stratigraphic column of a well drilled in the vicinity of the VES centre). The correlation is good and, while the stratigraphy of the drilling, executed without core recovery, seems rather qualitative, the resistivity column shows several differentiations, that include the bottom of the clay layer (that is the top of the confined aquifer) and the transition to conductive Miocene materials. Drilling was stopped at a depth of 68 m, with a water flow rate of 20 liters/s, and the hydrostatic level rose close to the ground surface (Balia et al., 2008); these conditions confirmed that the deep aquifer is a confined one.

4. Assessment of a mine tailings basin by means of shallow reflection seismology and gravity

Old waste landfills represent a serious environmental problem not only for their polluting potential but also because very often they interfere with the expansion of urban areas. For these reasons, the need for site reclamation interventions is more and more felt. A site reclamation intervention should be designed and estimated carefully both in technical and economic terms, since the simple assumption of incorrect parameters is one of the major causes of inefficacious work and cost escalation. Therefore, prior to reclamation works, an accurate knowledge of the landfill is necessary, while in many cases the general information about old landfills is very poor, and even their horizontal extent and depth are inadequately known. Surface geophysical methods are non-invasive and can play an important role in delineating the waste geometry since they can provide highly detailed, widely extended and low-cost information.

In this section the case of a mine tailings pond is shown. The pond received the post-flotation wastes produced by several mines in the last decades of the past century. It extends for about 0.4 km², with a mean elevation at the top of 145 m above the sea level. The mine tailings, mainly made up of a calcareous matrix, have the consistence of a dense, soft, fine-grained, apparently homogeneous soil. Several dangerous substances are present in the tailings. The main threat to the environment is the possible interaction between the surface and ground waters, and the polluting liquids originating from the tailings, characterized by high concentrations of heavy metals. Moreover, the oxidation of sulfides to sulfuric acid leads to an acidic condition and speeds up metal dissolution processes. Apart from a thin cover of Quaternary alluvium, the geological environment of the small valley hosting the landfill is made up of Palaeozoic rocks, namely more or less fractured limestone and dolomite. Originally, the small valley was crossed by a stream flowing from the surrounding hills. Some boreholes were drilled in the basin, but their interval was relatively large, so that they did not allow an accurate assessment of the waste body geometry. The aim of the experiments at the tailings basin was to verify the effectiveness of some geophysical techniques in order to acquire information on the thickness of the landfill and the location of possible faults and fracture zones affecting the hosting Palaeozoic rocks, since they could represent a possible way for diffusion of the polluting substances. On the whole one gravity profile, one P-wave seismic reflection profile and one resistivity/IP profile were acquired. Shallow reflection data were acquired with the following apparatuses, parameters and geometry: acquisition system: 48-channel seismograph; geophones: single, vertical, undamped, 40 Hz natural frequency; energy source: 8 kg sledge-hammer with vertical stacking (1–3 shots/record); spread type: off-end, minimum offset 10 m, shot interval 1 m, channel interval 1 m; maximum CMP fold: 2400%; record length: 0.250 s, sampling interval: 0.00025 s; analogue filters: low-cut filter off, antialias filter 1,000 Hz. Data processing was performed with the following steps: field files editing and early mute application; sorting into CMP gathers; 60–300 Hz band-pass filtering; velocity analysis; NMO correction; CMP stack; noise attenuation (two-trace horizontal mixing). The overall quality of the seismic data was good and the velocity analysis, carried out by picking the hyperbolic patterns on the CMP gathers, revealed a rather low P-wave velocity field (240–260 m/s); the seismic section obtained through the processing sequence listed above is shown in figures 7 and 8.

Apart from the very shallow reverberations, that most probably depend on the water table, the seismic section is dominated by a very clear reflector whose depth is in the order of 10 m with respect to the ground level on the SE of the section, and then deepens to depths exceeding 15 m.

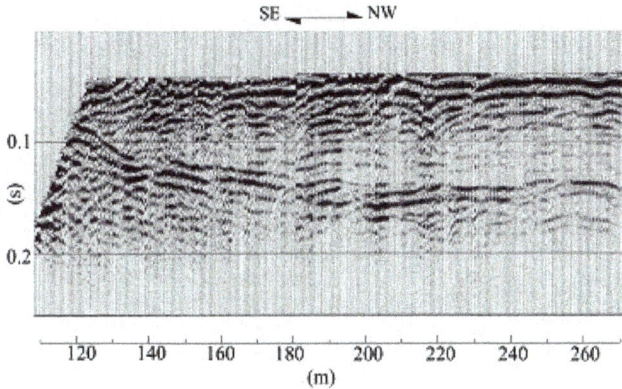

Fig. 7. Seismic time section at the mine tailings basin.

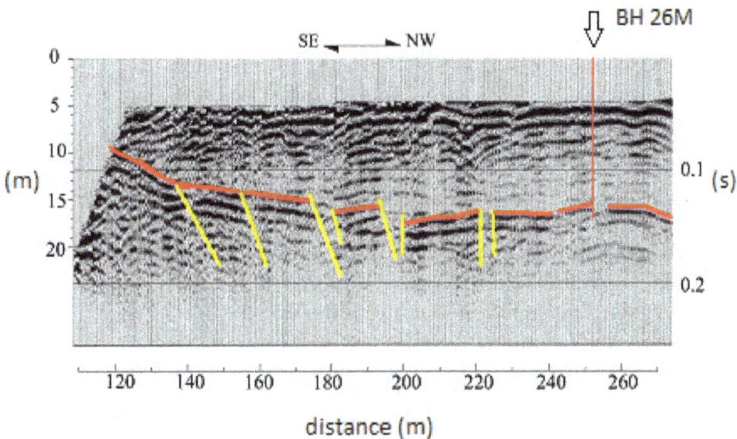

Fig. 8. The same section as in Figure 7 after interpretation. The main reflector (red) and several fractures (yellow) are enhanced. The position of borehole BH 26M is indicated by the arrow. (After Balia & Littarru, 2010)

This reflector is associated with the bottom of the basin, that is with the ancient ground surface of the valley, made up of Paleozoic shales, locally named Cabitza shales. The reliability of the seismic section in terms of depth to the bottom of the basin is testified by comparison with one borehole (BH 26M) at the progressive distance of 252 m along the line. The stratigraphy of this borehole is in figure 9 and shows the top of the Paleozoic shales at a depth of 17 m with respect to the present ground surface, in perfect agreement with the depth deduced from the seismic section.

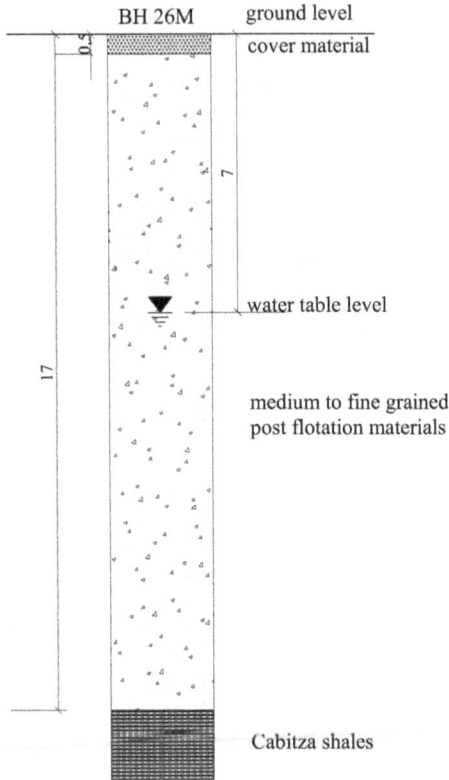

Fig. 9. Stratigraphic column of borehole BH 26M. (After Balia & Littarru, 2010)

Actually, it would not be possible to deduce the true morphology at the bottom of the basin only based on the few available boreholes. In addition, the seismic section also shows both the position and the distribution of the fractures affecting the basement, which is very useful information for the reasons mentioned above.

With regard to other geophysical methods used in this case, while the induced polarization and resistivity measurements have provided poor results, probably because of the strong electrical heterogeneity of the materials dumped in the basin, the gravity survey proved useful both designing the data acquisition parameters and to verify the results of the seismic survey. Figure 10 shows the result of the gravity anomaly modelling carried out assuming as a constraint the depth to the landfill bottom obtained by the seismic reflection survey, and allowing the density contrast between the filling material and the host rocks to vary. For the landfill, a 2.5D mass distribution model was assumed and the best fitting between the experimental and the computed gravity anomaly was obtained for a density contrast of -360 kg/m^3, meaning that since the host rocks have a density of about 2,500 kg/m^3, the filling material should have a density of about 2,140 kg/m^3. This is very reasonable and suggests that the good reflection coefficient corresponding to the reflector, and thus the good acoustic impedance contrast, is mainly attributable to the strongly different elastic characteristics between the two media and not to density contrast.

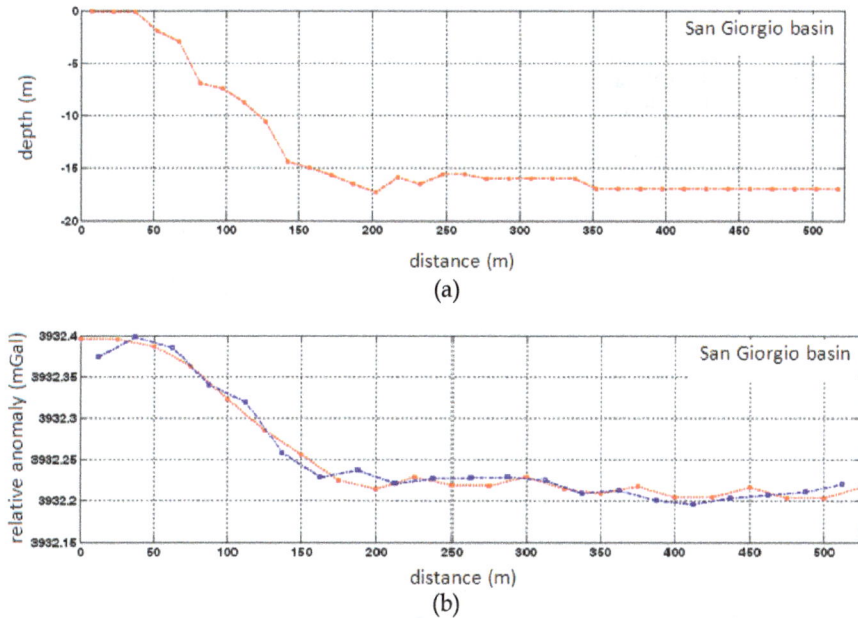

Fig. 10. Gravity model of the tailings pond: (a) depth to the bottom of the landfill as deduced from the seismic section and the outcrops of the basement; (b) measured (blue line) and computed (red line) gravity anomaly. The best fitting was obtained adopting a density contrast of -360 kg/m^3 for the filling material with respect to the hosting rocks. (After Balia & Littarru, 2010)

5. A much more complicated situation

Undoubtedly, the assessment of the mine tailings pond discussed in the previous paragraph was a relatively easy problem, mainly due to flat topography, target at depths just slightly exceeding 10 m from the ground surface, and good contrast of physical characteristics between the filling and the basement. The present case concerns an old municipal solid waste landfill where mixed wastes were dumped for about 20 years, from the end of the 1960s to the end of the 1980s (Balia & Littarru, 2010). The geological environment of the site is made up of Quaternary sediments with a local thickness in the order of 5 m, overlying Oligocene-Miocene volcano-sedimentary rocks. In the area, shallow aquifers are generally scarce and of poor quality: the Quaternary sediments host a phreatic aquifer, while some brackish water lenses can be found in the Oligocene-Miocene rocks. Very likely, the waste disposal excavation had a depth that was equal to the Quaternary sediment thickness, which means that wastes were partially placed in the phreatic aquifer.

One gravity profile, several vertical electrical soundings (VES), two seismic refraction and one reflection profiles were acquired on the landfill. However, since the results of both electrical and seismic refraction measurements were of poor quality, only the gravity and P-wave seismic reflection surveys, located as in figure 11, will be presented. Data were acquired along a roughly west–east oriented profile. Gravity and topographic measurements were taken at 24 station points placed at 10 m intervals, for a total profile length of 230 m.

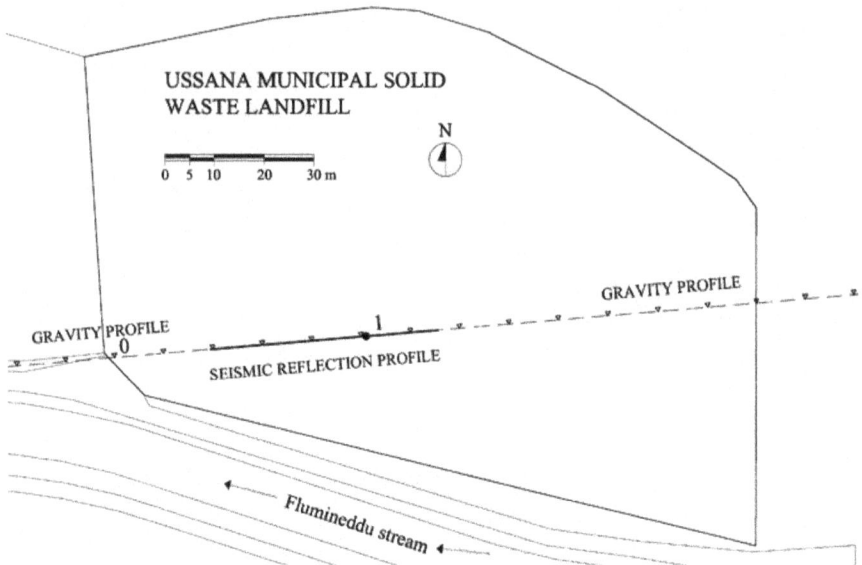

Fig. 11. Position map of the geophysical surveys at the solid waste municipal landfill. Gravity station points are indicated by small triangles; the seismic profile position is enhanced with a bold line; the location of the reference point and of the test pit are indicated with 0 and 1 respectively. (After Balia & Littarru, 2010)

The westernmost gravity station was placed 50 m west of a reference point close to the presumed western border of the landfill, and the easternmost station was 180 m east of the reference point. Figure 12 shows the relative gravity anomaly obtained after reductions, calculated assuming a mean density of 2,200 kg m³, considered representative of the rocks of the study area. A negative gravitational effect with a maximum amplitude of −0.32 mGal is rather clear, as well as its correlation with the presumed boundaries of the landfill.

Seismic reflection data were acquired with the following apparatuses, parameters and geometry: acquisition system: 48-channel seismograph; geophones: single, vertical, undamped, 50 Hz natural frequency; energy source: 5 kg sledge-hammer with vertical stacking (5–10 shots/record); spread type: off-end, minimum offset 0.8 m, shot interval 0.4 m, channel interval 0.4 m; maximum CMP fold: 2,400%; record length 0.128 s, sampling interval 0.000 25 s; analogue filters: low-cut filter off; antialias filter 1,000 Hz. On the whole 94 field records were acquired, corresponding to 234 CMP positions at 0.2 m intervals.

Data were characterized by a poor signal-to-noise ratio. Noise was mainly coherent, but abrupt quality changes from one record to the next also indicated the presence of a major, apparently random noise component; since external sources of noise were totally absent (no wind, no road traffic, no factories), this was attributed to strong variations in both subsoil and surface conditions from one shot point to the next. Velocity analysis turned out to be the most demanding operation and eventually the velocity function was derived by means of a CVS (constant velocity stack) analysis, with velocity increasing in the range 400–490 m/s at a 10 m/s step. The full processing sequence included the following steps: field files editing; low-cut 120 Hz filtering; sorting into CMP gathers; velocity analysis; dynamic corrections; CMP stack; two-trace horizontal stack (trace mixing). Eventually, the stack section obtained

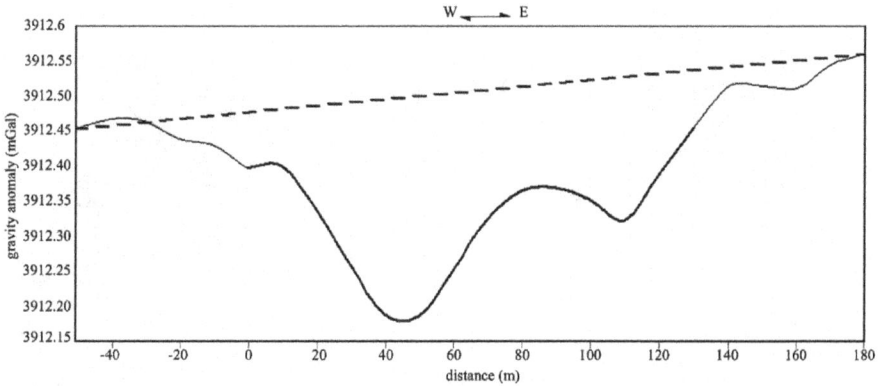

Fig. 12. The gravity anomaly generated by the waste landfill. The dashed line represents the long-wavelength field; the central part of the anomaly, enhanced with a bold line, corresponds to the presumed position of the landfill. (After Balia & Littarru, 2010)

by applying a dynamic (NMO) correction based on a velocity of 460 m/s was assumed as reliable. This section, shown in figure 13, exhibits a reflector made up of a normal-polarity, minimum-phase response generated by a positive reflection coefficient. The reflector is relatively undulated and exhibits some discontinuities; time resolution is not always good, so that time picking can be easily affected by errors of some milliseconds. The two-way reflection time along the entire section ranges between 0.014 s and 0.025 s, with a weighted mean value of 0.0178 s. An abrupt deepening of the reflector is clearly evident between the progressive distances of 37 and 40 m. Since no intermediate reflections were resolved, the mean depth of the reflector was estimated using the mean reflection time and the stack velocity (460 m/s), obtaining a value of 4.1 m. In order to verify the nature of the reflector, one test pit was excavated on the seismic profile, at the progressive distance of 52 m. This direct investigation showed a clayey capping layer with a mean thickness of 0.7 m overlying a layer of loose, dry waste about 1.45 m thick. Below the latter, wet to saturated waste, more compacted than the overlying one, mixed with sand and gravel, with a mean thickness of 0.6 m and lying on the natural bottom was found. Consequently, the local thickness of the landfill, from the ground surface to the bottom, was found to be in the order of 2.75 m, as against the 4.1 m estimated by means of the seismic data alone, adopting the average reflection time and the stack velocity for time-to-depth conversion.

However, the real reflection time corresponding to the pit position was 0.014 s. Using this time, and once more the stack velocity of 460 m/s, a depth of about 3.2 m was found. This depth is very close, though not equal, to the depth of 2.75 m estimated from the pit. Apart from the experimental errors that obviously occur when taking measurements on a real and complex model, this small — though significant — discrepancy could also be explained by the fact that the stack velocity is generally inadequate for time-to-depth conversion except for the one-layer case, which was not the case at hand, as shown by the excavation.

To refine the model of the landfill taking into account the results of the pit, first the P-wave velocity of the first layer in the vicinity of the pit itself was measured and proved to be of 430 m/s. Then, the following equation system was solved

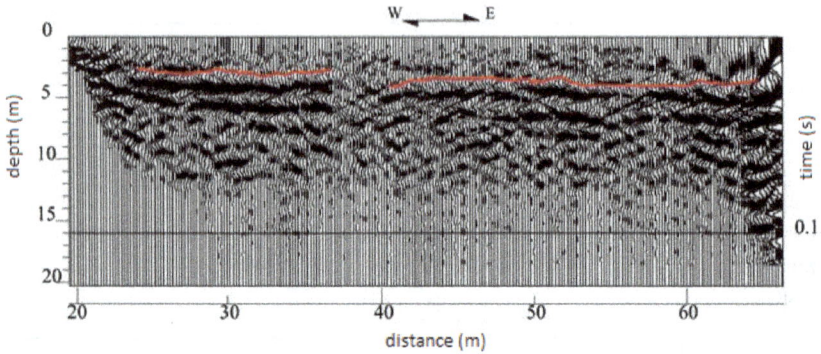

Fig. 13. Stack section (P-wave) at the municipal solid waste landfill. CMP trace interval is 0.2 m, and the reflector is enhanced by a red line. The section refers to the interval between the progressive distances of 20 m and 66 m of figure 12. (After Balia & Littarru, 2010)

$$V_{rms} = ((VI1^2 * t1 + VI2^2 * t2 + VI3^2 * t3)/(t1 + t2 + t3))^{1/2} \qquad (1)$$

$$tr = t1 + t2 + t3 \qquad (2)$$

$$t1 = h1/VI1 \qquad (3)$$

$$t2 = h2/VI2 \qquad (4)$$

$$t3 = h3/VI3 \qquad (5)$$

Equation (1), written for the three-layer case, gives the root mean square velocity (V_{rms}) as a function of the interval velocities (VIn) and the respective one-way transit times along vertical paths (tn) for each of the three layers (e.g. Sheriff, 1984). Equation (2) simply says that the total one-way reflection time (tr) is the sum of the one-way transit times pertaining to the three layers. Equations (3)–(5) give the one-way transit times as functions of the layer thicknesses (hn) and the respective interval velocities. In this case, the following quantities were known and assumed as constraints:

- the root mean square velocity, approximated with the stack velocity: $V_{rms} = 460$ m/s;
- the velocity of the first layer, measured at the surface in the vicinity of the pit: $VI1 = 430$ m/s;
- the layer thicknesses, measured inside the pit: $h1 = 0.7$ m; $h2 = 1.45$ m; $h3 = 0.6$ m;
- the total two-way reflection time, picked in the seismic section across the pit: $tr = 0.014$ s
- the two-way transit time for the first layer was computed: $t1 = h1/VI1 = 0.00326$ s.

Therefore, the only unknowns were the interval velocities of the second and third layers, for which the computation provided the following estimates: $VI2 = 288$ m/s and $VI3 = 1{,}809$ m/s. These velocity estimates are absolutely coherent with the respective layers: loose, dry waste the first, and wet to saturated waste, compacted and mixed with sand and gravel the second. The P-wave velocity in the bedrock (Vb) was deduced from seismic refraction data with estimated values in the range 2,500–3,600 m/s. Therefore, the seismic model for the waste landfill is that shown in figure 14: of course, it is not valid for the entire landfill, but only for the surroundings of the pit.

In this case, the gravity survey provided a good result essentially for picking out the boundaries of the landfill, while the shape of the anomaly could not be assumed as

0.70 m	Clayey cap	VI1 = 430 m/s
1.45 m	Dry waste	VI2 = 288 m/s
0.60	m Wet-saturated waste	VI3 = 1809 m/s
	Natural bedrock	Vb = 2500 – 3600 m/s

Fig. 14. The model of the landfill corresponding to the underground condition in the surroundings of the test pit, with layer thickness (left) and P waves velocity (right). (After Balia & Littarru, 2010)

representative of the morphology of the bottom. As a matter of fact, a comparison between figures 12 and 13 shows that between the progressive distances of 45 m and 66 m the absolute value of the gravity anomaly decreases eastwards, while in the same interval and direction the reflection time, and thus the depth to the bottom, increases. This means that on the eastern side of the above interval wastes become thicker but also much heavier than on the western side, thus suggesting that an accurate gravity modelling for the entire landfill is not allowed, or at least is very complicated. Concerning the seismic reflection experiment, it has been verified that the attainable resolution was not sufficient for picking out the layering inside the landfill, but only the bottom could be found. In this condition, time-to-depth conversion based only on seismic data provided results affected by an overestimate error in the order of 15–20%, as shown by the test pit. The test pit also explained why both VES techniques and refraction seismology failed. The VES technique because of the very high variability, both laterally and vertically, of the physical characteristics of the shallow layers, which caused very noisy and incoherent data. Refraction seismology clearly failed because of the velocity inversion between the first and the second layer. However, the seismic refraction data provided the P-wave velocity in the bedrock. This is a typical example of a situation in which a reliable site assessment cannot be carried out by means of geophysical techniques alone.

6. Conclusions

With most of the case histories referred to in the previous paragraphs of this chapter it has been demonstrated that it is possible to synthesize a piece of relevant stratigraphic information in a wide range of situations with the fundamental support of geophysical surveys.

In the first case this information concerned depth to Paleozoic basement, thickness and layering of Pleistocene-Holocene sedimentary cover, and relationships between the two shallow aquifers. Information at medium depth (100-300 m) was obtained by means of seismic reflection surveying, while shallow information was deduced from a combination of electrical and seismic reflection methods, completed by one borehole.

In the second case, where the aim was to predict the actual possibility of finding fresh water below the already known aquifer system, the stratigraphy of the study area was assessed down to depths in the order of 400-500 m by means of reflection seismology, and the shallow conditions were studied mainly by means of electrical resistivity measurements.

The third case, the one concerning the mine tailings basin, is a typical example of how, in favorable conditions, a single method can provide the desired information.

Conversely the last example, the municipal solid waste landfill, shows very clearly that when the underground condition is very poor, chaotic, with random variations of physical characteristics, geophysical techniques should be used with caution and the data they provide must be analyzed carefully before drawing any conclusions.

In this chapter, the most relevant geophysical method has undoubtedly been that of reflection seismology. However, beyond the results, the general significance of all the above examples lies essentially in the success of the combination of different geophysical techniques, which proved effective and useful in reducing unknowns and ambiguity. Gravity surveying, used as a starting point in all the geophysical works referred to in this chapter, provides information, at least qualitative, on bedrock morphology and structures, and allows an accurate design of reflection surveys, especially as regards spread geometry, acquisition parameters, profile position and orientation. Electrical and electromagnetic methods proved highly useful for the stratigraphic assessment at depths of less than 100-150 m.

Finally, a very important aspect concerns the planning of exploratory drillings. Clearly, when good geophysical results are available, one can perform rational borehole positioning, make preliminary estimates of the depths to be reached, and ultimately save time and costs.

7. Acknowledgements

Thanks are due to all co-authors of the papers that constitute the basis for this chapter, and to all members of the geophysical field crew at the Dipartimento di Ingegneria del Territorio, University of Cagliari, Italy.

8. References

Balia, R., Ardau, F. Barrocu, G. Gavaudò, E. & Ranieri, G. (2008) Assessment of the Capoterra coastal plain (southern Sardinia, Italy), by means of hydrogeological and geophysical studies. *Hydrogeology Journal*, Vol.17, pp. 981-997, DOI 10.1007/s10040-008-0405-z.

Balia, R., Gavaudò, E., Ardau, F. & Ghiglieri, G. (2003) Geophysical approach to the environmental study of a coastal plain. *Geophysics*, Vol.68, pp. 1446–1559, ISSN 0016-8033

Balia, R. & Littarru, B. (2010) Geophysical experiments for the pre-reclamation assessment of industrial and municipal waste landfills. *J. Geophys. Eng.*, Vol.7, pp. 64-74, DOI 10.1088/1742-2132/7/1/006.

Dobrin, M.B. (1976) *Introduction to geophysical prospecting*. McGraw-Hill, ISBN 0-07-017195-5, New York, USA

Reynolds, J.M. (1997) *An introduction to applied and environmental geophysics*. John Wiley & Sons, ISBN0-471-95555-8, Chichester, UK

Sharma, P.V. (1997) *Environmental and engineering geophysics*. Cambridge University Press, ISBN 0 521 57240 1, Cambridge, UK

Sheriff, R.E. (1984) *Encyclopedic Dictionary of Exploration Geophysics*. Society of Exploration Geophysicists, ISBN 0-931830-31-3, Tulsa, OK, USA

Telford, W.M., Geldart, L.P. & Sheriff, R.E. (1990) *Applied geophysics*, Cambridge University Press, ISBN 0-521-33938-3, Cambridge, UK

Yilmaz, O. (1987) *Seismic data processing*. Society of Exploration Geophysicists, ISBN 0931830400, Tulsa, OK, USA

Orbital Control on Carbonate-Lignite Cycles in the Ptolemais Basin, Northern Greece – An Integrated Stratigraphic Approach

M.E. Weber[1], N. Tougiannidis[1], W. Ricken[1],
C. Rolf[2], I. Oikonomopoulos[3] and P. Antoniadis[3]
[1]University of Cologne
[2]Leibniz Institute for Applied Geosciences
[3]National Technical University of Athens
[1,2]Germany
[3]Greece

1. Introduction

Establishing the time frame is crucial for most geoscientific investigation. Without proper time control, past geologic processes cannot be inferred appropriately, nor can their dynamics be understood adequately. Rhythmic changes in sedimentary cycles hold the key to establishing precise and high-resolution chronologies. The concept goes back to theoretical considerations first published by Milankovitch (1941). His calculations showed that changes in earth's orbital geometry lead to changes in the seasonal and latitudinal distribution of incoming solar radiation (insolation). Three main periods are responsible for these insolation changes, eccentricity (the shape of the orbit around the sun; with periods of 413 kyr, 123 kyr, and 95 kyr), obliquity (the tilt of the axis; changing at a period of 41 kyr), and precession (the wobbling spin of the axis with periods of 19 kyr and 23 kyr). He argued that these changes caused the waning and waxing of polar ice sheets. More than three decades later, Hays et al. (1976) and Imbrie et al. (1984) provided proof the cyclic changes of the earth energy budget were large enough to be preserved in marine sediment. Theoretical calculation of Berger (1976) and Berger and Loutre (1991) supported the Milankovitch theory and provided templates for orbital variability for the last couple of million yeas. Henceforth, cyclic changes in sediment strata were used to develop detailed orbital chronologies by assigning sedimentary cycles to orbital cycles.

Magnetic polarity stratigraphy is a necessary and independent tool to retrieve chronometric information for time series covering millions of years. It is based on polarity changes of the Earth's magnetic field measured over the oceans and correlated to dated magnetic polarity reversals found on land (Heirtzler et al., 1968). The presently accepted geomagnetic polarity timescale (GPTS) is the one from Cande and Kent (1995). Langereis & Hilgen (1991) and Hilgen et al. (1995) provided the first astronomical age scale for the Pliocene Capo Rosello sections in Sicily based on the GPTS. Their work was further substantiated by Krijgsman et al. (1995). For the Late Neogene, the GPTS is rather precise thanks to accurate radiometric dating methods such as $^{40}Ar/^{39}Ar$ (Kuiper et al., 2004).

A first, low-resolution, orbital-based chronology for the continental Ptolemais Basin (Fig. 1) was introduced by Steenbrink et al. (1999) and van Vugt et al. (2001; 1998). They ascribed cyclic changes of carbonates and lignites to orbital variability: maxima in lake carbonate correlate to maxima in insolation (and minima in precession). Accordingly, lignite maxima correlate to minima in insolation (i.e., beige layers in Capo Rosello). Also, they carried out extensive radiometric dating using $^{40}Ar/^{39}Ar$ (e. g., Kuiper, 2003) and delivered a high-quality composite stratigraphy for the Ptolemais Basin (Steenbrink et al., 2006). Also, Weber et al. (2010) published an orbital-based chronology for the two Upper Miocene lignite quarries Vegora and Lava. Therefore, there is robust and reliable stratigraphic information available we could base our investigations on.

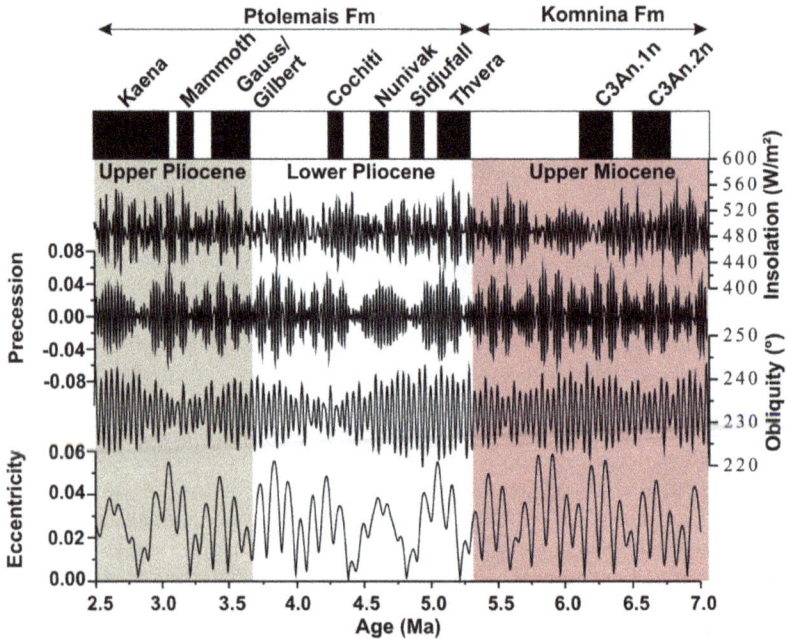

Fig. 1. Stratigraphic positions for the Ptolemais and Komnina Formations, northern Greece (e. g., Steenbrink, 2001), and magnetic chrons (black and white pattern; ages according to Cande & Kent, 1995) for the Upper Miocene to the Upper Pliocene. Curves display orbital variations of eccentricity, obliquity, precession, and the incoming solar radiation (insolation; all parameters calculated according to Laskar et al., 2004).

So far, most stratigraphic information comes from rather short outcrops in quarries and interpretations mainly rely on composite records. We investigated, for the first time, a long and continuous section (drilling KAP-107), covering the entire Lower Pliocene in a single borehole. Our goal therefore was to collect multiple high-resolution and continuous paleoclimate proxy data to establish both the orbital chronology and reconstruct paleoclimate variability during the Lower Pliocene, a time of extensive lignite formation in the Ptolemais Basin. The coal formation is sandwiched between two important events: the Upper Miocene Messinian Salinity Crisis, a time of severe constriction in the Mediterranean realm with multiple evaporation events from 5.96 – 5.33 Ma (Krijgsman et al., 1999), and the Upper Pliocene onset of northern hemisphere glaciation.

2. The Ptolemais basin

The Ptolemais Basin (Fig. 2) is a SSE – NNW elongated intramontane Basin. Together with two flanking mountain ranges, the Askion to the west and the Vermion to the east, it belongs to the Pelagonian Zone (Bornovas & Rondogianni-Tsiambaou, 1983). The Ptolemais Basin formed during the Late Neogene in northwestern Greece (Pavlides & Mountrakis, 1987) and contains lacustrine deposits of Upper Miocene to Quaternary age (up to 800 m thick; Anastopoulos & Koukouzas, 1972) with an extended Lower Pliocene alternation of lignites, clays, and marls. The depositional history reflects interaction of orbital forcing and tectonic movement (Steenbrink et al., 2006; Tougiannidis, 2009). Through continued Pleistocene extension, the Basin is further subdivided into the basins of Florina-Vevi, Amynteon-Vegora, Ptolemais, and Kozani-Servia (Antoniadis et al., 1994), where a number open pit mines and active coal mining fields are located (Tougiannidis, 2009). The deposits are highly fragmented due to fault tectonics.

Borehole KAP-107 was drilled for exploration purposes in 2006 by the Public Power Cooperation (PPC) in the Amynteon lignite field, 12 km northwest of the city of Ptolemaida (at 40°37'3" N and 21°37'20" E). It is 233 m long and has variable borehole diameters of 18 – 8 cm (top to bottom). We investigated the lowermost part between 72 m and 230 m in detail.

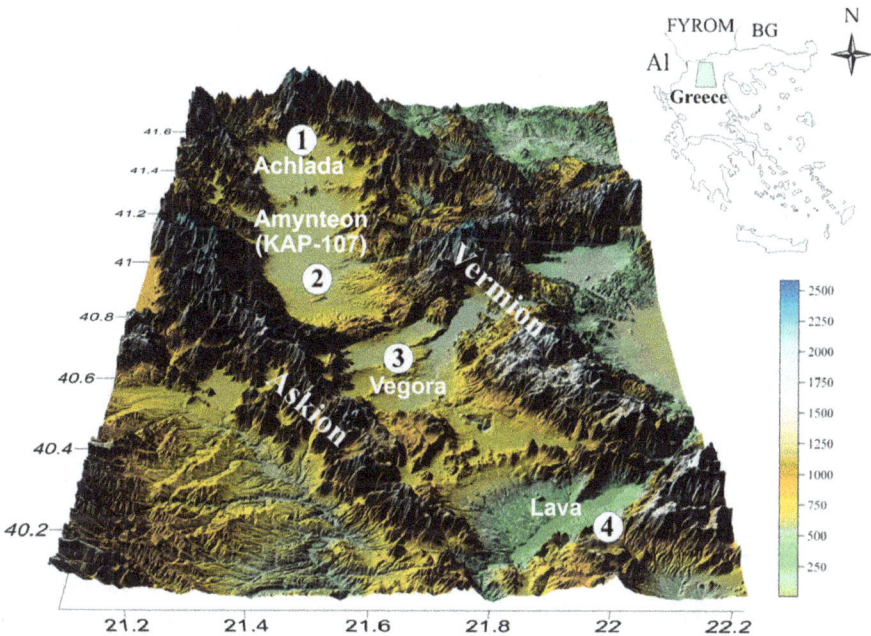

Fig. 2. 3-D location map of the Ptolemais Basin in northern Greece. The NNW – SSE elongated basin is flanked by the mountain ranges Askion in the west and Vermion in the east. Numbers indicate the four research sections referred to in this study: 1 – Achlada, 2 – Amynteon with borehole KAP-107, 3 – Vegora, and 4 – Lava. Data from sections Vegora and Lava are published in Weber et al. (2010). For digital elevation data see http://srtm.csi.cgiar.org.

3. Analytical methods

The methods of this study rely primarily on high-resolution non-destructive color measurements. As a fundamental sediment property color is often used for lithologic differentiation and to determine sedimentological structures, facies etc. We measured a total of approximately 16,500 samples non-destructively for color variability using a Minolta Chromatometer CM − 2002. This hand-held system is easy to use in the field. There is hardly any maintenance and measurements can be conducted quickly and cost effective. Measurements were made on clean and fresh (scraped with a spatula) surfaces at 1-cm resolution, using the CIELAB color model "L*-a*-b*". The system provides three color values for each measurement (details see Weber, 1998): the L* axis (the black-white color component), also known as lightness or grey value; the a*axis (the green/red component); and the b*axis (the yellow/blue component). Together, the three parameters describe coordinates in a spherical system (16 million possible variations). Measurements of C* describe the chroma (colorfulness). The difference between two successive color coordinates (ΔE^*ab) was calculated as $\Delta E^*ab = \sqrt{(\Delta L^*)^2 + (\Delta a^*)^2 + (\Delta b^*)^2}$. Since ΔE^*ab contains the variability of all three color components, we refer to it as the whole color difference.

Paleomagnetic measurements were made on discrete samples in the lab. We determined natural remanent magnetization (NRM) and alternating-field demagnetization (AF) to generate the characteristic remanent magnetization (ChRM) (Tougiannidis, 2009). A total of 200 completely oriented diamagnetic sample cubes of 12 cm^3 were retrieved at 80-cm intervals on average. NRM was measured in x and y direction using a cryogenic spin magnetometer (2G Enterprises). We applied AF before inclination and declination was measured and used orthogonal projections (Zijderveld, 1967) to correct the values for noise before using them for magnetostratigraphy.

We used the Analyseries software (Paillard, 1996) to perform astronomical tuning experiments and to construct age-depth models. Orbital parameters for the time frame 6 – 3 Ma were calculated in 1-kyr increments (see Fig. 1) for eccentricity, obliquity, precession, using the solutions provided by Laskar et al. (2004). Orbital insolation was calculated for the month of June at 40°N to reflect the approximate energy budget of the site.

We used ESALAB (Weber et al., 2010) to study of the resulting time series for frequency pattern and to compare them to orbital time series. ESALAB conducts both bulk and evolutionary spectral analysis (ESA). The program relies on the Lomb (1976) and Scargle (1982, 1989) algorithms, and provides an estimate of the spectrum by fitting harmonic sine and cosine components to the data set. This has two decisive advantages: the input data can be unequally spaced and the resulting spectra are rather robust and of high resolution. While performing the Fourier transformation, both window length and step size are freely adjustable. Also, the window type is selectable among Hanning, Haming, Blackman, sin^2, and boxcar. The output consists of graphic files for bulk and ESA spectra and tabulated data. The sample increment of 1 cm puts the Nyquist frequency (the highest frequency detectable) at 0.5 cm^{-1}, i.e., the spatial resolution for bulk and ESA is 2 cm.

4. Ground-truth stratigraphy

Ground-truth stratigraphy for borehole KAP-107 relies on paleomagnetic data (see Table 1). The site shows a flickering pattern of revers and normal polarity (Fig. 3). Inclination and

Orbital Control on Carbonate-Lignite Cycles in the Ptolemais Basin, Northern Greece –
An Integrated Stratigraphic Approach

65

declination values vary -4 – -67° and 359 – 6° for revers polarity, respectively, and 1 – 84° and 358° for normal polarity. Pronounced reversals occur at 141.1 m, 147.55 m, 165.14 m, 167.55 m, 172.88 m, 178.92 m, 191.24 m, 203.35 m, 215.1 m, and 220.93 m. Although the sample resolution of 80 cm (see above) is relatively low, the combination of inclination and declination changes provides a trustworthy reversal pattern that can be applied rather confidently. With the exception of 165.14 – 167.55 m, all remaining sections of normal polarity can confidently be assigned to chrons Gauss/Gilbert, Cochiti, Nunivak, Sidjufall, and Thvera (Fig. 3).

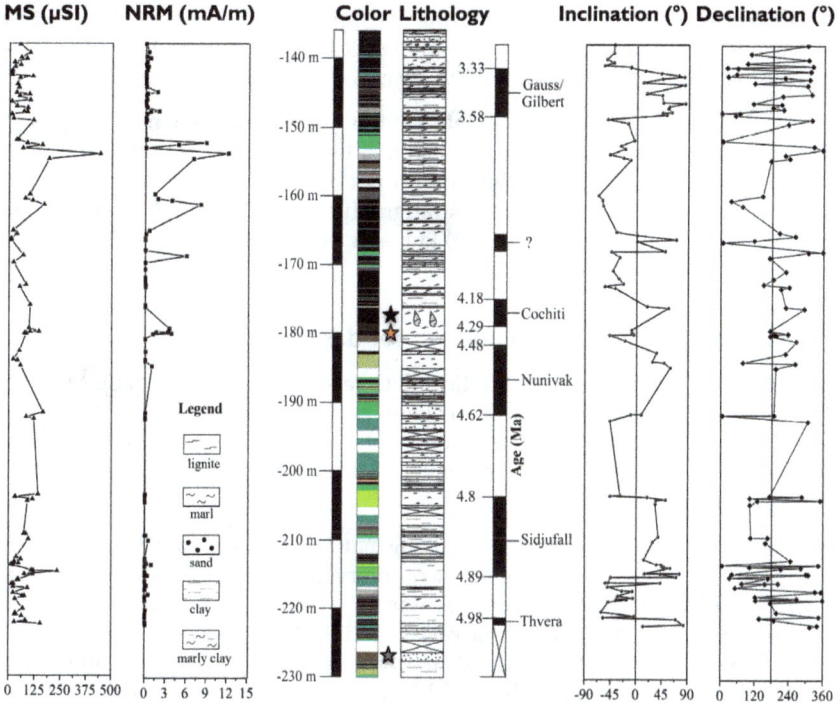

Fig. 3. Paleomagnetic ground-truth stratigraphy for borehole KAP-107. From left to right are magnetic susceptibility (MS), natural remanent magnetization (NRM), visual color, lithology, magnetic chron assignation, inclination, and declination. Age assignations are according to the age scale of Krijgsman et al. (1999). Black star represents the Neritina marker bed. This shell horizon belongs to the Theodoxus Member and has an age of roughly 4.4 Ma (orange star) according to $^{40}Ar/^{39}Ar$ dating of Steenbrink (2001). Grey star shows a glauconite horizon that has an approximate age of 5 Ma. Details see text.

Our age assignation is corroborated by additional sedimentologic and biostrati–graphic evidence. Unpublished records of the Federal Institute for Geosciences and Natural Resources (BGR, Hannover, Germany) from 1960 – 1970 mention a glauconite horizon just above the onset of sedimentation in the entire Ptolemais Basin, after the termination of the Messinian Salinity Crisis. We detected this layer at borehole KAP-107 in 228 m (Fig. 3), where the magnetic reversal pattern indicates the top of the Thvera chron (dated to

4.98 Ma). Also, we found the Neritina marker (*Theodoxus Macedonicus* WENZ, 1943) bed in roughly 174 – 178 m. This shell horizon belongs to the Theodoxus member and has an age of roughly 4.4 Ma (Steenbrink, 2001), which is in good agreement to 4.48 – 4.29 Ma deduced from our magnetic dating for this depth section.

Borehole KAP-107 was drilled to the depth where lignite deposition commenced in the Ptolemais Basin. According to our magnetic dating, the lowermost part of the core (underneath 222 m) belongs to the Thvera subchron and is hence somewhat older than 4.98 Ma. These stratigraphic results are inline with Steenbrink et al. (2006), who dated the onset of extensive lignite sedimentation approximately to the base of the Thvera subchron at 5.23 Ma. Hence there appears to be a lag time of approximately 100 kyr to the end of the Messinian Salinity Crisis that dates to 5.33 Ma (Krijgsman et al., 1999), and is marked by rapid re-flooding of the entire Mediterranean Basin (Garcia-Castellanos et al., 2009).

In summary, borehole KAP-107 represents continuous sedimentation during the entire Lower Pliocene. It covers a duration of 2 myr from 5.1 – 3.1 Ma. The resulting low-resolution age model comprises nine confident age control points (Table 1) from reversal dating and two additional evidences that support the stratigraphic assignation. Sedimentation rates are 14 – 2 cm/kyr, with an average rate of 6 cm/kyr, and a decreasing tendency towards the top of the borehole.

5. Cyclic variability of high-resolution color data

We measured a total of five sediment-optical data sets (L*, a*, b*, C, ΔE*ab). Most of them characterize facies changes and show striking cyclic variability (Figs. 4 and 5). The advantage of using a number of different sediment proxy records is that various aspects of changing sediment composition and supply and the relation to climate forcing can be addressed simultaneously.

Specifically sediment lightness shows a striking pattern of highs and lows with brighter (marl-rich) and darker (clay-rich or lignite-rich) intervals, and the coal seams as the darkest parts. Therefore, L* is generally a good indicator for either calcium carbonate (high values) contents (see also Weber, 1998) or organic carbon and/or lignite (low values). The similarity to the whole color difference ΔE*ab reveals that, for the most part, changes in L* dominate color variability. Initial spectral analysis of L* in the depth domain revealed sedimentary cycles ranging from 0.4 to 2.4 m, with a dominance centered around 1 – 1.4 m, and a tendency to become shorter with shallower core depths.

Color values a*, b*, and C* show lower-amplitude cycles and a clear increase in values to the top (Fig. 4). Specifically the sections shallower than approximately 92 m indicate an elevated plateau. Higher b* values reflect elevated contents of yellowish iron oxides (e. g., above 95 m core depth), most likely caused by the presence of hematite and goethite (see Weber et al., 2010). Low b* values refer to bluish components, abundant in bitumen or coal (mostly the sections underneath 135 m), or provided by sulfides.

The red-green variation (color value a*) provides an indication for the redox conditions, although the signal might be influenced by diagenetic overprint. Oxic conditions are indicated for core sections shallower than 95 m; the depths underneath show low-amplitude variations between slightly reduced and slightly oxic. The co-variability of color component C* throughout the record reveals that a* values primarily determine the chroma (Fig. 4).

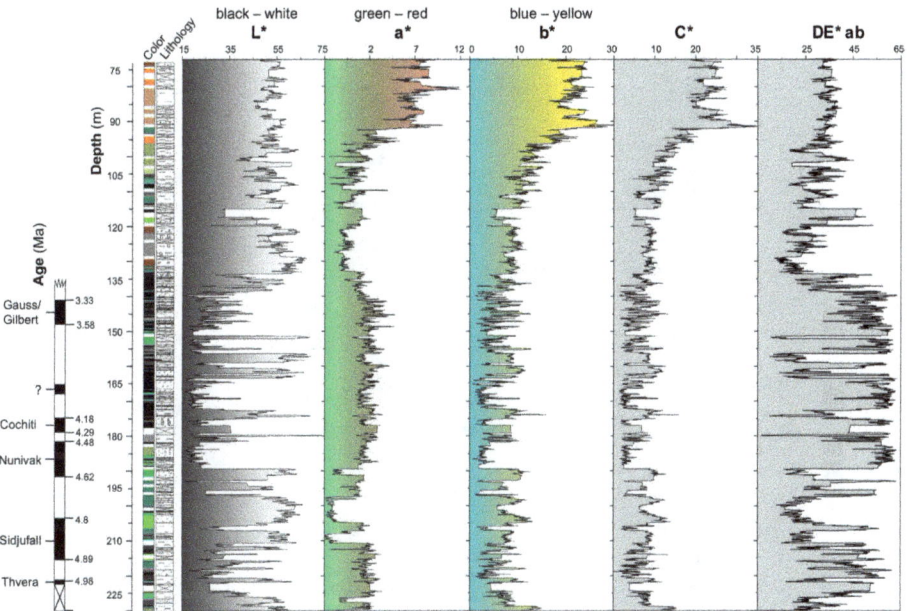

Fig. 4. Non-destructive color data from borehole KAP-107. From left to right are paleomagnetic reversals, visual color, lithology, sediment lightness (L*), red-green component (a*), yellow-blue component (b*), chroma (C*), and whole color difference (ΔE*ab). Note that, only for graphical considerations, all high-resolution data have been smoothed using an 11-point Gaussian filter. Note further that L* provides an estimate for either organic carbon or carbonate; a* indicates the redox state, and b* yields information about the iron oxide content.

6. Astronomical tuning

The process by which past variations in earth's orbit are correlated to the cyclic variability of sediment parameters is called astronomical or orbital tuning. In the Ptolemais Basin, we tuned the cyclic lignite marl alternations to orbital time series, using the algorithms of Laskar et al. (2004), and relying on the magnetic polarity time scale of Krijgsman et al. (1999) as ground-truth stratigraphy (Fig. 5).

Accordingly, we first generated a low-resolution age model for borehole KAP-107 by using the nine magnetostratigraphic fix points of Table 1 (see blue dots in Fig. 5) and converting core depths into ages linearly between tie points. As a tuning target, we used the insolation curve for the month of June at 40°N according to Laskar et al. (2004). Since magnetostratigraphy was only available for the lignite-bearing sections (135 – 223 m), we conducted the experiments only for the time interval 5.3 – 3.1 Ma.

Given the magnetostratigraphic boundaries and sedimentary cycles of 0.4 to 2.4 m with a dominance of 1 – 1.4 m (see above), calculated sedimentation rates vary from 2 to 12 cm/kyr with an average of 6 cm/kyr. This translates into a cycle length of roughly 1.2 m per precession (insolation) cycle, and 5 – 6 m per eccentricity cycle. Exactly these two frequencies are dominant in the L* and ΔE*ab records as lower and higher amplitude cycles, respectively (Fig. 5).

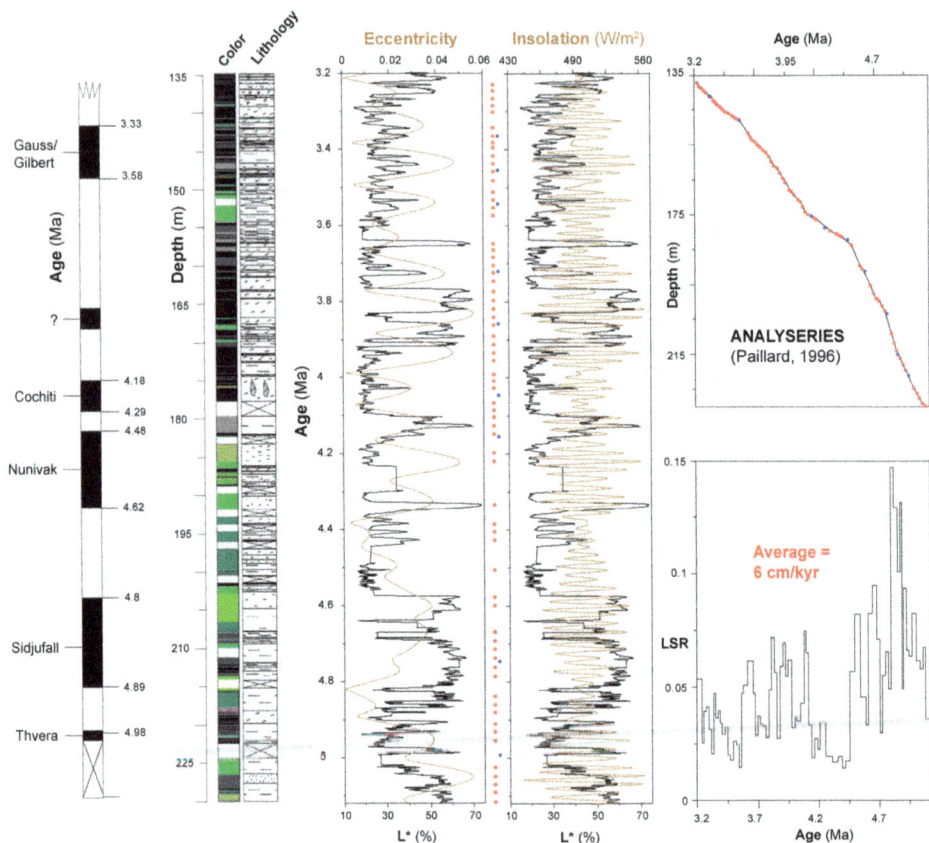

Fig. 5. Orbital tuning of borehole KAP-107. Sediment lightness (L*) of borehole KAP-107 tuned versus orbital insolation From left to right are magnetic reversal pattern, visual color, lithology, sediment lightness L* versus orbital eccentricity, and L* versus orbital insolation at 40°N for the month of June (data from Laskar et al., 2004). Histograms to the right give age-depth relationship (top) and linear sedimentation rates (LSR) between tuned tie points (bottom). Note that carbonate maxima were tuned to insolation maxima (red dots) to concur with the procedure introduced by Steenbrink et al. (2006; 2000). Blue dots give ages of ground-truth stratigraphy.

Steenbrink et al. (2000) pointed out that dark-colored marls correspond to relatively dry periods, whereas light-colored marls represent more humid periods (lake-level highstands). Since humid climate in the Mediterranean occurred during insolation maxima (e. g., Emeis et al., 2000), we used the insolation curve for the month of June at 40°N according to Laskar et al. (2004) and tuned individual carbonate maxima between the magnetostratigraphic tie points to insolation maxima. Thereby, we increased the resolution of the age model to an additional 70 age control points without violating the ground-truth boundaries provided by Table 1. Weber et al. (2010) followed the strategy of Steenbrink et al. (2006; 2000) and tuned insolation minima to dark intervals. However, the KAP-107 record contains broad lows and

Depth (m)	Age (Ma)	Remark	Sed. rate (cm/kyr)	Depth (m)	Age (Ma)	Remark	Sed. rate (cm/kyr)
137.39	3.231	ITP	5.3	170.62	4.068	ITP	4.3
138.37	3.249	ITP	2.5	171.39	4.086	ITP	7.5
138.86	3.269	ITP	3.9	172.78	4.105	ITP	6.5
139.57	3.287	ITP	4.1	174.23	4.127	ITP	3.3
140.31	3.304	Ga/Gilb (t)	3.1	174.96	4.149	ITP	1.9
141.10	3.330	ITP	2.1	175.54	4.180	Cochiti (t)	2.7
141.42	3.345	ITP	4.7	176.05	4.198	ITP	2.2
142.44	3.367	ITP	3.6	176.54	4.221	ITP	3.4
143.04	3.384	ITP	4.0	178.92	4.290	Cochiti (o)	1.9
143.61	3.398	ITP	3.4	179.80	4.335	Neritina (t)	1.7
144.31	3.418	ITP	2.9	180.65	4.385	ITP	1.9
144.93	3.439	ITP	3.2	181.06	4.406	ITP	1.4
145.60	3.460	ITP	2.0	181.37	4.428	ITP	1.7
146.05	3.483	ITP	1.8	182.27	4.480	Nunivak (t)	5.7
146.60	3.514	ITP	2.6	183.76	4.506	Neritina (o)	8.2
147.12	3.534	ITP	2.5	189.55	4.577	ITP	4.5
147.64	3.555	ITP	1.4	190.56	4.599	ITP	3.3
147.91	3.574	ITP	4.7	191.26	4.620	Nunivak (o)	8.2
148.18	3.580	G/Gilb(o)	5.1	195.27	4.669	ITP	9.5
151.71	3.649	ITP	6.1	197.50	4.692	ITP	7.1
152.80	3.667	ITP	4.7	199.06	4.714	ITP	3.0
153.65	3.685	ITP	2.9	199.76	4.738	ITP	5.5
154.27	3.707	ITP	3.7	201.07	4.761	ITP	6.5
155.00	3.726	ITP	3.4	202.60	4.785	ITP	4.9
155.75	3.748	ITP	3.3	203.35	4.800	Sidjufall (t)	14.7
156.60	3.774	ITP	2.8	209.03	4.839	ITP	12.9
157.27	3.799	ITP	4.9	211.83	4.860	ITP	10.1
158.35	3.821	ITP	7.2	213.92	4.881	ITP	13.1
159.88	3.842	ITP	5.6	215.10	4.890	Sidjufall (o)	4.9
161.02	3.862	ITP	2.7	216.14	4.911	ITP	9.4
161.84	3.892	ITP	5.9	218.07	4.932	ITP	6.6
163.12	3.914	ITP	6.9	219.63	4.955	ITP	6.3
164.62	3.936	ITP	5.5	221.20	4.980	Thvera (t)	8.3
165.74	3.956	ITP	6.2	224.93	5.025	Glauconite	6.1
168.01	3.993	ITP	3.2	226.31	5.048	ITP	5.8
168.56	4.010	ITP	3.7	227.60	5.070	ITP	6.7
169.27	4.029	ITP	3.5	229.01	5.091	ITP	3.6
				229.94	5.117	ITP	3.6

Table 1. Age model for borehole KAP-107. Integer numbers refer to ground-truth stratigraphy from paleomagnetic reversal dating (age scale according to Cande & Kent, 1995; Krijgsman et al., 1999), from $^{40}Ar/^{39}Ar$ dating of the Neritina bed and a glauconite horizon (van Vugt et al., 1998). Remaining age control points result from tuning color compent L* to orbital insolation (data from Laskar et al., 2004). Sedimentation rates are given in cm/kyr. Ga/Gilb refers to magnetic chrons Gauss/Gilbert. ITP stands for insolation tuning point.

sharp highs in the L* record, so the tuning the other half of the insolation cycle, i. e., insolation maxima to bright intervals, seemed more appropriate and accurate. In any way, we obtained about the same resolution in the tuned age models as in previous studies and the age models are congruent. To establish the correlation between climate proxy variability and orbital parameter in the time domain was rather easy and straight forward; hence, the resulting stratigraphic model appears robust and reliable with tuned age control points at (ideally) every insolation maximum (every 20 kyr; see Table 1). The only uncertainties resulted from some missing core sections, where we had to interpolate the ages (see Fig. 5).

The resulting high-resolution age model comprises an additional 70 age control points and indicates a preservation period for KAP-107 of almost precisely 2 myr (5.12 – 3.09 Ma). Sedimentation rates were quite variable over this rather long period. Specifically the lower part of the Ptolemais Formation (for formation names see Fig. 7) shows elevated and variable values during the Kyrio member, followed by very low values during the Theodoxus member, intermediate values during the Notio member, and low to intermediate values during the Anargyri member (Fig. 5). Compared to Upper Miocene sections Lava and Vegora, borehole KAP-107 indicates that the Amynteon Basin received only 20 – 30 % of the sediment material during the Lower Pliocene.

7. Spectral analysis

Spectral analysis in the depth domain is ambiguous when changes in sedimentation rate occur, because the amplitude of the response will be reduced or even be absorbed. Borehole KAP-107 exhibits quite some variability in sedimentation rates; hence we transformed the depth series first into the time domain using the ground-truth age control points (see blue dots in Fig. 5). Only then, we conducted bulk and evolutionary spectral analyses to decide about the following tuning procedure.

In the spectra we mainly found eccentricity and precession cycles (Fig. 6). This is not surprising because precession contributes a substantial part to the insolation forcing in low-to-mid latitudes. Orbital eccentricity modulates the amplitude of the precession cycle (Imbrie et al., 1993) and is therefore also important for the Ptolemais Basin (Steenbrink et al., 2006), specifically in the lower sedimentation-rate sites (Tougiannidis, 2009). Steenbrink et al. (2006) also found a robust obliquity signal in some parts of section Lava. However, obliquity has virtually no impact for borehole KAP-107, which is also not surprising since this frequency is related to the changing tilt of the earth and therefore mainly observable at higher latitudes (e. g., Weber et al., 2001), where it is mostly associated with the presence of larger ice sheets (e. g., Ruddiman, 2004).

Given the average sedimentation rates of 6 cm/kyr and the sampling increment of 1 cm, the average sample resolution is roughly 170 years. This resolution is enough to also be able to detect suborbital (millennial-to-centennial-scale) signals. However, the precession signal is so dominant that higher frequencies are not apparent in the time series, although rhythmic bedding in the cm to dm band can be observed in the outcrop. Either this type of bedding is autocyclic, i. e., it results from basin-internal processes that are not linked to orbital or solar forcing, or, more likely, post-sedimentary compaction operates different on the various facies types, thereby altering the depth-age relationship that has originally been established. Alternatively, different facies types may have been deposited at different rates. In any case, the age control points are only precession-controlled and hence cannot resolve these high-frequencies variations.

Orbital Control on Carbonate-Lignite Cycles in the Ptolemais Basin, Northern Greece –
An Integrated Stratigraphic Approach

71

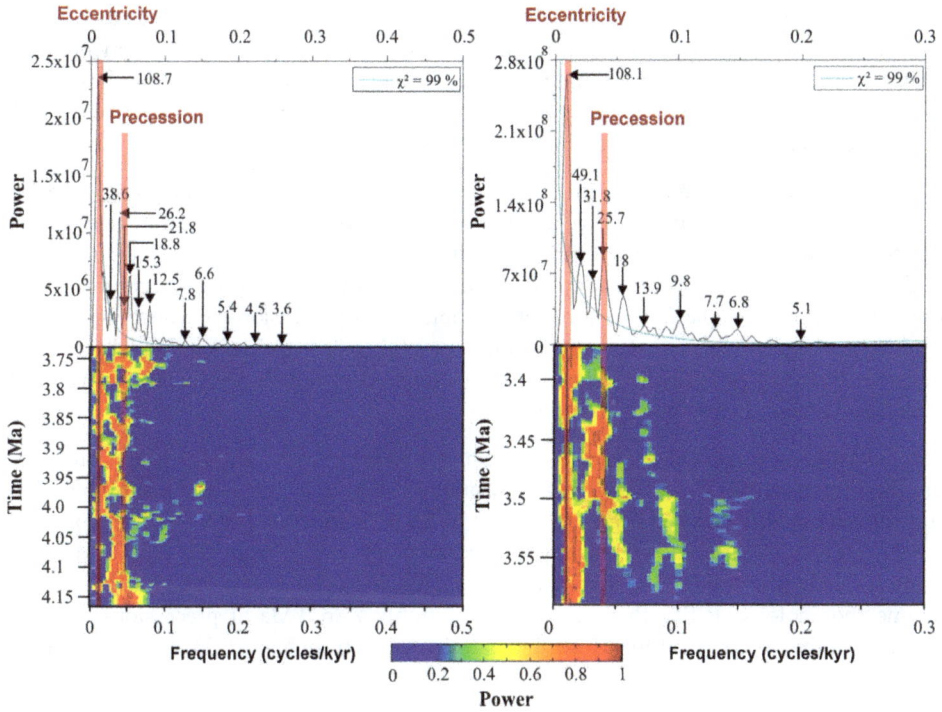

Fig. 6. Spectral analysis of color component L* for borehole KAP-107 for two time slices 4.2 –
3.7 Ma (left) and 3.6 – 3.35 Ma (right). Bulk spectra (top) were calculated using REDFIT
software (Schulz & Mudelsee, 2002). Green line shows the 99 % confidence interval.
Evolutionary spectral analyses (bottom) were calculated with ESALAB software (Weber et
al., 2010) with a window size of 100 kyr and a shift of 5 kyr from one analysis to the next.

8. Age model discussion and chronostatigraphic correlation

Rhythmic bedding of sedimentary sequences reflects mostly the response to cyclic changes
in earth's orbital geometry, namely eccentricity, obliquity, and precession. Changes in
orbital parameters lead to changes in insolation with respect to season and latitude. For a
given repetitive sedimentary succession we can use this relationship to establish a high-
resolution and precise chronology. This requires, however, other means of stratigraphic
control such as biostratigraphic markers or the orientation of magnetic grains, which are
locked immediately after the time of deposition. The magnetic reversal pattern, on the other
hand, can precisely be dated and yields the foundation of the geomagnetic polarity time
scale (GPTS, Cande & Kent, 1995). At least for the Late Neogene, the dating accuracy for
magnetic reversals is sufficient to establish orbital chronologies (e. g., Hilgen et al., 1995).

Obtaining independent stratigraphic information is crucial because there might also be
autocyclic processes (see above) that create rhythmic bedding. Here, single crystal $^{40}Ar/^{39}Ar$
dating using the laser fusion technique provided an essential step forward (e. g., Kuiper et
al., 2004).

Once a low-resolution age control is established, spectral analysis can reveal whether or not rhythmic bedding is related to orbital frequencies. Because this was the case, we were able to "tune" sedimentary cycles to orbital frequencies. This rather simple and straightforward approach yields a very powerful stratigraphic tool – the resulting time series provide age models of previously unmatched resolution and precision.

We also followed this strategy and dated borehole KAP-107 using magnetic polarity changes and additional stratigraphic evidence. The resulting low-resolution age model indicated spectral power on the eccentricity band and thereby showed that orbital forcing is the likely cause for rhythmic bedding. We used this information to tune the greyscale (L*) variability to orbital insolation. As a result, we obtained a precession-controlled age model over 2 myr of deposition in the Lower Pliocene, from 5.1 to 3.1 Ma. The age model provides, ideally, one control point every 1.2 m – enough for a very detailed reconstruction of the depositional history of any given environment.

In the next step, we applied the tuning method to a number of mining fields and outcrops from the Ptolemais Basin (see Fig. 7). Our chronology, again, is tied into the chronology of the composite record that has been established by Steenbrink et al. (1999) for the Ptolemais Basin. As a result, we obtained a complete sedimentary record from the Upper Miocene section Achlada (this study) and sections Lava and Vegora (Weber et al., 2010) to the Upper Pliocene (borehole KAP-107; this study), from roughly 7 to 3 Ma at precession-scale age control, with the exception of the Messinian Salinity Crisis from 5.9 to 5.33 Ma (Krijgsman et al., 1999).

Combined biomagnetostratigraphy and orbital tuning revealed that lacustrine sections Vegora and Lava from the central and southern Ptolemais Basin represent the period 6.85 – 6.57 Ma and 6.46 – 5.98 Ma at sedimentation rates of roughly 14 and 22 cm/kyr, respectively (Weber et al., 2010). Section Achlada from the northern edge of the Ptolemais, however, contains fluvial influence and shows less-convincing paleomagnetic evidence. According to Koukouzas et al. (2010; 2009) the Achlada deposits belong to the Komnina lignite sequence, which is why we tentatively correlated them to chron C3An.1n (Fig. 7).

Times of sediment deposition and lignite formation in the Ptolemais Basin were related to large-scale and global events. The Upper Miocene represented a global cooling (Billups, 2002; Billups et al., 2008) with lacustrine and fluvial input in the Ptolemais Basin (Steenbrink et al., 2006) and occasional lignite formation. The Messinian Salinity Crisis is characterized by at least partial desiccation in the Mediterranean realm (CIESM Workshop, 2007). Reworked alluvial to fluvial deposits dominated in the Ptolemais Basin. During the Lower Pliocene greenhouse warming accompanied the emerging Isthmus of Panama and enhanced oceanic overturning circulation (Brierley et al., 2009; Ravelo et al., 2004; Ravelo et al., 2006). Borehole KAP-107 covers this time of extensive lignite formation in the Ptolemais Basin.

During the Upper Pliocene continental ice sheets built up in the Northern Hemisphere between 3.6 and 2.4 Ma (Mudelsee & Raymo, 2005). A threshold towards full glacial to interglacial conditions occurred near 2.7 Ma (e. g., Ruggieri et al., 2009; Shackleton et al., 1984). This is broadly the time when lignite formation ceased in the Ptolemais Basin and coarser-grained, more oxic sediments were again deposited.

Orbital Control on Carbonate-Lignite Cycles in the Ptolemais Basin, Northern Greece –
An Integrated Stratigraphic Approach

73

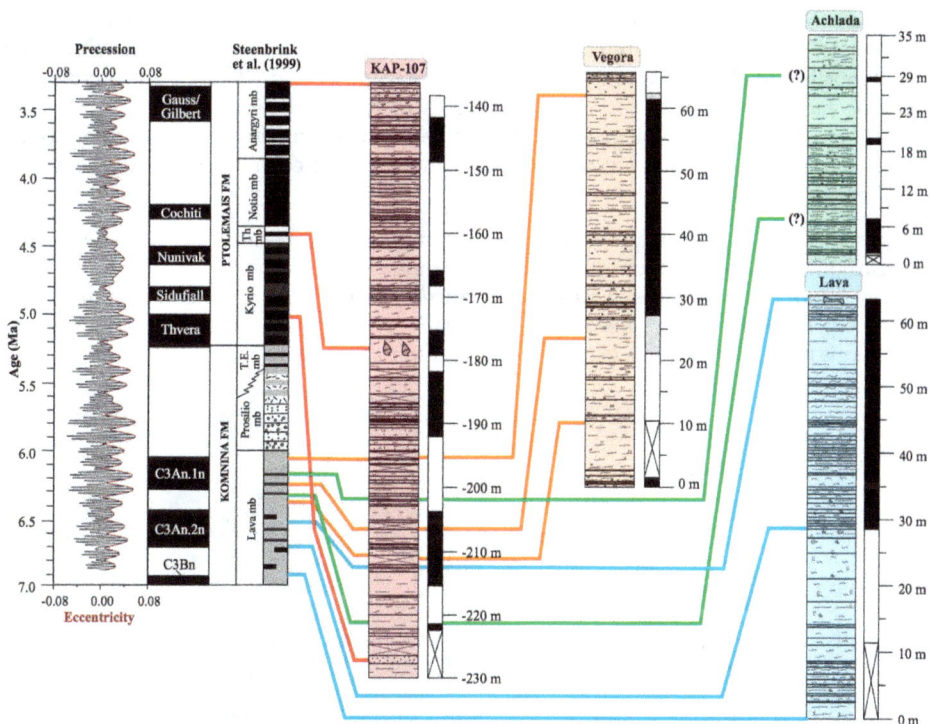

Fig. 7. Correlation of the four sites investigated in the Ptolemais Basin, spanning the time from roughly 7 Ma (chron C3Bn) to 3.3 Ma (Gauss/Gilbert boundary). From left to right are tuning targets (red – orbital eccentricity; black – orbital insolation at 40°N for the month of June (data from Laskar et al., 2004); stratigraphic classification according to Steenbrink et al. (1999), and Sites Amynteon (KAP-107), Vegora, Lava, and Achlada. For simplicity, only paleomagnetic reversal pattern is shown for each section. For detailed orbital tuning see Fig. 5 and Weber et al. (2010).

9. Photospectrometry as stratigraphic and paleoclimate tool

Initial attempts to identify rhythmic bedding and its relation to orbital forcing concentrated on counting individual beds manually and measuring their thicknesses (e. g., Hilgen et al., 1995). With the implementation of high-resolution rapid scanning techniques, photospectrometry has become very important (e. g., Weber, 1998). It provides a precise digital fingerprint of various geochemical and mineralogical properties that are related to climate and orbital variability. The data yield objective measurements and can be treated mathematically to analyze cyclic behavior and amplitudes of the climatic response. The use of color as high-resolution stratigraphic and paleoenvironmental tool has become increasingly important in recent years (e. g., Debret et al., in press).

The foundation for the KAP-107 tuning is sediment lightness (L*) or greyscale. Variations mimic rhythmic alternations of relatively bright marls and relatively dark clays or lignites. These changes indicate changes in humidity on precessional time scales. The Ptolemais

Basin is a continental setting in low- to mid-latitudes. Hence, changes in orbital insolation as the ultimate driver of humidity changes have a strong precessional component. Sediments were deposited in shallow water (Kaouras, 1989; van de Weerd, 1983) of an intramontane lacustrine basin. According to the pollen studies of Kloosterboer-van Hoeve et al. (2006) dark-colored marls of the Ptolemais Basin, which are mostly enriched in clay and/or organic carbon, correspond to relatively dry periods, whereas light-colored marls represent more humid periods.

Changes in humidity are expressed by color component L* because of the facies change from carbonate-rich (more humid) to clay- or lignite-rich (dryer). More arid conditions, on the other hand, exhibit increased redness (enhanced oxic conditions), whereas humid phases were likely more reduced. This relation is clearly shown in Fig. 8 (right), with separate populations for carbonate-rich, clay-rich, and lignite-rich sediment.

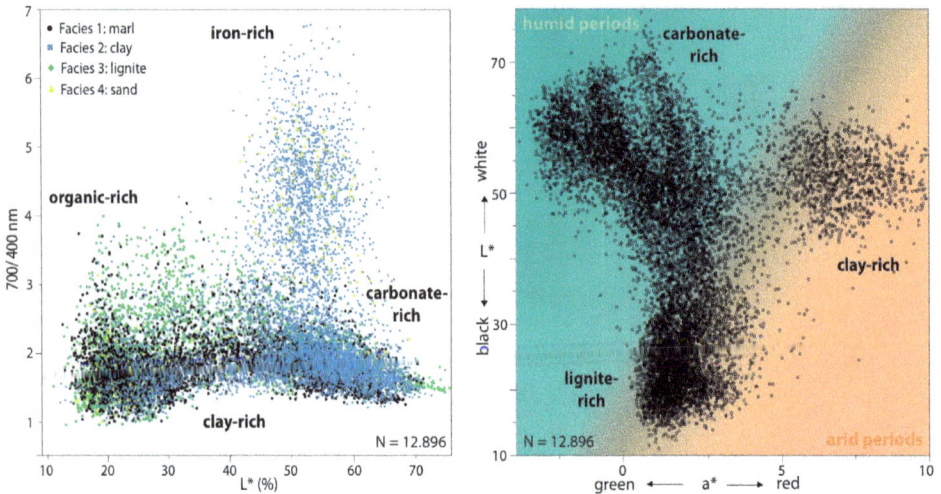

Fig. 8. Sediment color as high-resolution stratigraphic and paleoenvironmental tool. Left shows Q7/4 diagram (Debret et al., in press), i. e., the ratio of reflectance at 700 and 400 nm versus sediment lightness L*. Note separate populations for carbonate-rich, iron-rich, organic-rich, and clay-rich strata. Right shows L* versus a*. Note separate populations for clay-rich, carbonate-rich, and lignite-rich strata.

Debret et al. (in press) introduced the so-called Q7/4 diagram, displaying the ratio of the reflectance between 700 nm and 400 nm versus sediment lightness L* (Fig. 8, left). Here, lower ratios characterize clays and carbonates, whereas higher ratios indicate sediments rich in organic matter and iron. For the Ptolemais Basin this plot allows to distinguish separate populations for clay-rich, carbonate-rich, and lignite-rich strata with respect to the facies.

10. Summary and conclusions

This study presents a combined approach to establish geological time in sedimentary strata from magnetostratigrahy and orbital tuning. Polarity changes of the Earth's magnetic field provide the basis for the geomagnetic polarity timescale which is well dated for the Late Neogene using radiometric methods. Changes in earth's orbital geometry, on the other hand, cause changes in insolation, which can also be calculated rather precisely for the Late Neogene. Sedimentary successions are capable of archiving these cyclic changes as rhythmic bedding. We applied this concept of correlating sedimentary cycles to orbital cycles with the aid of magnetic polarity changes to Late Neogene strata from northern Greece.

Lower Pliocene borehole KAP-107 from the Ptolemais Basin shows alternations of marl-rich and clay- or lignite-rich strata. Using ground-truth stratigraphic markers and the magnetic reversal history, we established a low-resolution age model for the borehole spanning 5.1 – 3.1 Ma. Spectral analysis indicates that rhythmic bedding shows two distinct cycles, orbital eccentricity and precession. We used this information to tune the sedimentary record to orbital insolation. As a result, we obtained a high-resolution age model with, ideally, control points at every precession cycle (every 20 kyr). Through correlation of four research sections covering the Ptolemais Basin in a N – S transect, we obtained a complete sedimentary record from the Upper Miocene (7 Ma) to the Upper Pliocene (3 Ma), with the exception of the Messinian Salinity Crisis (5.96 – 5.33 Ma). Conducting the research with high-resolution photospetrometric scanning methods proved very helpful and allows for precise depth age assignations.

11. Acknowledgments

We thank the Public Power Corporation (Ptolemaida) for granting permission to work in their quarries. Also, we wish to thank Greek colleagues Mr. Panagiotis Nikolakakos (Director-general) and Dipl.-Geol. Antonis Nikou (Drilling Section), as well as Daniela Sprenk and Andreas Holzapfel (both University of Cologne) for technical assistance. Supplementary data to this paper are available at doi:10.1594/PANGAEA.770293 (thanks to Hannes Grobe and Rainer Sieger).

12. References

Anastopoulos, J. & Koukouzas, C. N. (1972). Economic geology of the southern part of the Ptolemais lignite basin (Macedonia,Greece). *Geol. Geophys. Res.*, Vol. 16, No. 1, pp. 1-189

Antoniadis, P. A., Blickwede, H. & Kaouras, G. (1994). Petrographical and Palynological analysis on a part (36,0 m – 51,0 m) of a Borehole of the Upper Miocene lignite deposit of Lava-Kozani, N.W. Greece. *Mineral Wealth*, Vol. 91, pp. 7-17

Berger, A. & Loutre, M. F. (1991). Insolation values for the climate of the last 10 million years. *Quaternary Science Reviews*, Vol. 10, No. 4, pp. 297-317

Berger, A. L. (1976). Obliquity and precession for the last 5,000,000 years. *Astron. Astrophys*, Vol. 51, pp. 127-135

Billups, K. (2002). Late Miocene through early Pliocene deep water circulation and climate change viewed from the sub-Antarctic South Atlantic. *Palaeogeography, Palaeoclimatology, Palaeoecology*, Vol. 185, No. 3-4, pp. 287-307

Billups, K., Kelly, C. & Pierce, E. (2008). The late Miocene to early Pliocene climate transition in the Southern Ocean. *Palaeogeography, Palaeoclimatology, Palaeoecology*, Vol. 267, No. 1-2, pp. 31-40

Bornovas, J. & Rondogianni-Tsiambaou, T. (1983). *Geological map of Greece (second edition)*, 1:500.000, Institute of Geology and Mineral Exploration, Athens

Brierley, C. M., Fedorov, A. V., Liu, Z., Herbert, T. D., Lawrence, K. T. & LaRiviere, J. P. (2009). Greatly Expanded Tropical Warm Pool and Weakened Hadley Circulation in the Early Pliocene. *Science*, Vol. 323, No. 5922, pp. 1714-1718

Cande, S. C. & Kent, D. V. (1995). Revised calibration of the geomagnetic polarity timescale for the Late Cretaceous and Cenozoic. *Journal of Geophysical Research*, Vol. 100, No. B4, pp. 6093-6095

CIESM Workshop (2007). The Messinian Salinity Crisis from mega-deposits to microbiology - A consensus report, in: Briand, F. (Ed.), CIESM Workshop Monographs, Monaco, 166 p.

Debret, M., Sebag, D., Desmet, M., Balsam, W., Copard, Y., Mourier, B., Susperrigui, A. S., Arnaud, F., Bentaleb, I., Chapron, E., Lallier-Vergès, E. & Winiarski, T. (2011). Spectrocolorimetric interpretatio of sedimentary dynamics: The new "Q7/4 diagram". *Earth-Science Reviews*, Vol. 109, pp.1-19

Emeis, K.-C., Sakamoto, T., Wehausen, R. & Brumsack, H.-J. (2000). The sapropel record of the eastern Mediteranean Sea - results of Ocean Drilling Program Leg 160. *Palaeogeography, Palaeoclimatology, Palaeoecology*, Vol. 158, pp. 371-395

Garcia-Castellanos, D., Estrada, F., Jiménez-Munt, I., Gorini, C., Fernàndez, M., Vergés, J. & De Vicente, R. (2009). Catastrophic flood of the Mediterranean after the Messinian salinity crisis. *Nature*, Vol. 462, No. 7274, pp. 778-781

Hays, J. D., Imbrie, J. & Shackleton, N. J. (1976). Variations in the Earth's Orbit: Pacemaker of the Ice Ages. *Science*, Vol. 194, No. 4270, pp. 1121-1132

Heirtzler, J. R., Dickson, G. O., Herron, E. M., Pitman, W. C., III & Le Pichon, X. (1968). Marine Magnetic Anomalies, Geomagnetic Field Reversals, and Motions of the Ocean Floor and Continents. *J. Geophys. Res.*, Vol. 73, No. 6, pp. 2119-2136

Hilgen, F. J., Krijgsman, W., Langereis, C. G., Lourens, L. J., Santarelli, A. & Zachariasse, W. J. (1995). Extending the astronomical (polarity) time scale into the Miocene. *Earth and Planet. Sci. Lett.*, Vol. 136, pp. 495-510

Imbrie, J., Berger., Boyle, E. A., Clemens, S. C., Duffy, A., Howard, W. R., Kukla, G., Kutzbach, J., Martinson, D. G., McIntyre, A., Mix, A. C., Molfino, B., Morley, J. J., Peterson, L. C., Pisias, N. G., Prell, W. L., Raymo, M. E., Shackleton, N. J. & Toggw. (1993). On the structure and origin of major glaciation cycles. *Paleoceanography*, No. 8(6), pp. 699-735

Imbrie, J., Hays, J.D., Martinson, D.G., McIntyre, A., Mix, A.C., Morley, J.J., Pisias, N.G., Prell, W.L., Shackleton, N.J. (1984). The orbital theory of Pleistocene climate: Support from a revised chronology of the marine $\delta^{18}O$ record, in: Berger, A.L., Imbrie, J., Hays, J., Kukla, G., Saltzman, B. (Eds.). D. Reidel, Dordrecht, pp. 269-305

Kaouras, G. (1989). *Kohlepetrographische, palynologische und sedimentologische Untersuchungen der pliozänen Braunkohle von Kariochori bei Ptolemais/NW-Griechenland*, PhD, 258p., Georg-August-Universität, Göttingen

Kloosterboer-van Hoeve, M. L., Steenbrink, J., Visscher, H. & Brinkhuis, H. (2006). Millennial-scale climatic cycles in the Early Pliocene pollen record of Ptolemais, northern Greece. *Palaeogeogr. Palaeoclimatol. Palaeoecol.*, Vol. 229, pp. 321-334

Koukouzas, N., Kalaitzidis, S. P. & Ward, C. R. (2010). Organic petrographical, mineralogical and geochemical features of the Achlada and Mavropigi lignite deposits, NW Macedonia, Greece. *International Journal of Coal Geology*, Vol. 83, No. 4, pp. 387-395

Koukouzas, N., Ward, C. R., Papanikolaou, D. & Li, Z. (2009). Quantitative Evaluation of Minerals in Lignites and Intraseam Sediments from the Achlada Basin, Northern Greece. *Energy & Fuels*, Vol. 23, No. 4, pp. 2169-2175

Krijgsman, W., Hilgen, F. J., Langereis, C. G., Santarelli, A. & Zachariasse, W. J. (1995). Late Miocene magnetostratigraphy, biostratigraphy and cyclostratigraphy in the Mediterranean. *Earth Planet, Sci. Lett.*, Vol. 136, pp. 475-494

Krijgsman, W. F., Hilgen, J., Raffi, I., Sierro, J. & Wilson, D. S. (1999). Chronology, causes and progression of the Messinian salinity crisis. *Nature*, Vol. 400, pp. 625-655

Kuiper, K. F. (2003). *Direct intercalibration of radio-isotopic and astronomical time in the Mediterranean Neogene*, PhD, 223p., Utrecht University, Utrecht

Kuiper, K. F., Hilgen, F. J., Steenbrink, J. & Wijbrans, J. R. (2004). 40^{Ar}/39^{Ar} ages of tephras intercalated in astronomically tuned Neogene sedimentary sequences in the eastern Mediterranean. *Earth and Planetary Science Letters*, Vol. 222, No. 2, pp. 583-597

Langereis, C. G. & Hilgen, F. J. (1991). The Rossello composite: a Mediterranean and global reference section for the Early to early Late Pliocene. *Earth Planet. Sci. Lett.*, Vol. 104, pp. 211-225

Laskar, J., Robutel, P., Joutel, F., Gastineau, M., Correia, A. C. M. & Levrard, B. (2004). A long-term numerical solution for the insolation quantities of the Earth. *Astronomy and Astrophysics*, Vol. 428, pp. 261-285

Lomb, N. R. (1976). Least-squares frequency analysis of unequally spaced data. *Astrophysics and Space Science*, Vol. 39, pp. 447-462

Milankovitch, M. (1941). *Kanon der Erdbestrahlung und seine Anwendung auf das Eiszeitenproblem* (edition), Belgrad

Mudelsee, M. & Raymo, M. E. (2005). Slow dynamics of the Northern Hemisphere glaciation. *Paleoceanography*, Vol. 20, pp. 1-14

Paillard, D. (1996). Macintosh Program Performs Time-Series Analysis, EOS.

Pavlides, S. B. & Mountrakis, D. M. (1987). Extensional tectonics of northwestern Macedonia, Greece, since the late Miocene. *J. Struct Geol.*, Vol. 9, No. 4, pp. 385-392

Ravelo, A. C., Andreasen, D. H., Lyle, M., Lyle, A. O. & Wara, M. W. (2004). Regional climate shifts caused by gradual global cooling in the Pliocene epoch. *Nature*, Vol. 429, pp. 263-267

Ravelo, A. C., Dekens, P. S. & McCarthy, M. (2006). Evidence for El Niño-like conditions during the Pliocene. *GSA Today*, Vol. 16, No. 3, pp. 4-11

Ruddiman, W. (2004). The Role of Greenhouse Gases in Orbital-Scale Climatic Changes. *Eos*, Vol. 85, No. 1, pp. 6-7

Ruggieri, E., Herbert, T., Lawrence, K. T. & Lawrence, C. E. (2009). Change point method for detecting regime shifts in paleoclimatic time series: Application to $\delta^{18}O$ time series of the Plio-Pleistocene. *Paleoceanography*, Vol. 24, No. 1, pp. 1-15

Scargle, J. D. (1982). Studies in astronomical time series analysis. II. Statistical aspects of spectral analysis of unevenly spaced data. *The Astrophysical Journal*, Vol. 263, pp. 835-853

Scargle, J. D. (1989). Studies in astronomical time series analysis. III. Fourier transforms, autocorrelation functions, and cross-correlation functions of unevenly spaced data. *The Astrophysical Journal*, Vol. 343, No. 133, pp. 874-887

Schulz, M. & Mudelsee, M. (2002). REDFIT: estimation red-noise spectra directly from unevenly spaced paleoclimatic time series. *Computer & Geosciences*, Vol. 28, pp. 421-426

Shackleton, N. J., Backmann, J., Zimmermann, H., Kent, D. V., Hall, M. A., Roberts, D. G., Schnitker, D., Baldauf, J. G., Desprairies, H., R., Huddlestun, P., Keene, J. B.,

Kaltenback, A. J., Krumsiek, K. A. O., Morton, A. C., Murray, J. W. & al, e. (1984). Oxygen isotope calibration of the onset of ice-rafted and history of glaciation in the North Atlantic region. *Nature*, No. 307, pp. 620-623

Steenbrink, J. (2001). *Orbital signatures in lacustrine sediments*, PhD Thesis, 167p., Utrecht University, Utrecht

Steenbrink, J., Hilgen, F. J., Krijgsman, W., Wijbrans, J. R. & Meulenkamp, J. E. (2006). Late Miocene to Early Pliocene depositional history of the intramontane Florina-Ptolemais-Servia Basin, NW Greece: Interplay between orbital forcing and tectonics. *Palaeoecology*, Vol. 238, pp. 151-178

Steenbrink, J., van Vugt, N., Hilgen, F. J., Wijbrans, J. R. & Meulenkamp, J. E. (1999). Sedimentary cycles and volcanic ash beds in the Lower Pliocene lacustrine succession of Ptolemais (NW Greece): discrepancy between 40Ar/39Ar and astronomical ages. *Palaeogeography, Palaeoclimatology, Palaeoecology*, Vol. 152, pp. 283-303

Steenbrink, J., van Vugt, N., Kloosterboer-van Hoeve, M. L. & Hilgen, F. J. (2000). Refinement of the Messinian APTS from sedimentary cycle patterns in the lacustrine Lava section (Servia Basin, NW-Greece). *Earth and Planetary Science Letters*, Vol. 181, pp. 161-173

Tougiannidis, N. (2009). *Karbonat- und Lignitzyklen im Ptolemais-Becken: Orbitale Steuerung und suborbitale Variabilität (Spätneogen, NW Griechenland). Sedimentologische Fallstudie unter Berücksichtigung gesteinsmagnetischer Eigenschaften*, PhD Thesis, 122p., Institute of Geology and Mineralogy, University of Cologne, Cologne

van de Weerd, A. (1983). Palynology of some upper Miocene and Pliocene Formations in Greece. *Geologisches Jahrbuch*, Vol. 48, pp. 3-63

van Vugt, N., Langereis, C. G. & Hilgen, F. J. (2001). Orbital forcing in Pliocene - Pleistocene Mediterranean lacustrine deposits: dominant expression of eccentricity versus precession. *Paleogeography, Paleoclimatology, Paleoecology*, Vol. 172, pp. 193-205

van Vugt, N., Steenbrink, J., Langereis, C. G., Hilgen, F. J. & Meulenkamp, J. E. (1998). Magnetostratigraphy-based astronomical tuning of the early Pliocene lacustrine sediments of Ptolemais (NW Greece) and bed-to-bed correlation with the marine record. *Earth and Planetary Science Letters*, Vol. 164, pp. 535-551

Weber, M. E. (1998). Estimation of biogenic carbonate and opal by continuous non-destructive measurements in deep-sea sediments: application to the eastern Equatorial Pacific. *Deep-Sea Research 1*, Vol. 45, pp. 1955-1975

Weber, M. E., Mayer, L. A., Hillaire-Marcel, C., Bilodeau, G., Rack, F., Hiscott, R. N. & Aksu, A. E. (2001). Derivation of $\delta^{18}O$ from sediment core log data: Implications for millennial-scale climate change in the Labrador Sea. *Paleoceanography*, Vol. 16, No. 0, pp. 1-12

Weber, M. E., Tougiannidis, N., Kleineder, M., Bertram, N., Ricken, W., Rolf, C., Reinsch, T. & Antoniadis, P. (2010). Lacustrine sediments document millennial-scale climate variability in northern Greece prior to the onset of the northern hemisphere glaciation. *Palaeogeography, Palaeoclimatology, Palaeoecology*, Vol. 291, No. 3-4, pp. 360-370

Zijderveld, J.D.A. (1967). A.C. demagnetization of rocks: analysis of results, in: Collinson, D.W.e.a. (Ed.), Methods in Palaeo-magnetism. Elsevier, Amsterdam, pp. 254-286

4

Ground Penetrating Radar: A Useful Tool for Shallow Subsurface Stratigraphy Characterization

Giovanni Leucci

Institute for Archaeological and Monumental Heritage,
National Council of Research – (CNR-IBAM),
Italy

1. Introduction

In a region such as the Salento, constituted by a flat surface with greatly expanded coverage of agricultural land, the studies performed on the stratigraphical evolution of different lithological units in the first meters of the soil has been based upon the analysis of cut – faces, quarries, cores and shallow trenches (Bossio et al., 1987; Bossio et al., 1992; Bossio et al., 1994; Bossio et al., 1998; Bossio et al., 1999; Ciaranfi et al., 1992; D'Alessandro, et al., 1994; Margiotta, 1999; Margiotta and Ricchetti, 2002; Palmentola, 1987). Data provided by such techniques are often one- or two-dimensional. They involve surveys that are time consuming and are patchy in terms of spatial coverage. Therefore the possibilities offered by the GPR to investigate the subsoil in a non-invasively way, and to obtain 3D maps of the subsurface itself, becomes of crucial importance for geologists. The successful obtained by GPR investigations on sedimentary rock stratigraphy is well documented in literature (Annan and Davis, 1989; van Overmeeren, 1998; Mills and Speece, 1997; Mokma et al., 1990; Nobes et al., 2001; Lapen et al., 1996; Baker, 1991; Beres, et al., 1995; Leucci et al., 2000; Carrozzo et al., 2000; Carrozzo et al., 2003). Since GPR holds enormous potential for such studies, it is appropriate to assess some key considerations for, i) field data acquisition, ii) raw data processing in order to enhance data display, iii) EM wave velocity measurements in order to characterize sediments response and to perform the time to depth conversion, iv) lithological interpretation of the GPR data set. This chapter attempts to give the steps required to acquired, process and interpret GPR data in a sedimentary rock environment.

GPR data were acquired in the Salento peninsula in two areas located near the city of Lecce, Italy (Fig. 1).

During last years lagoonal – continental and marine oligo-miocene deposits have been recognized in some areas of Salento leccese. Del Prete & Santagati (1972) described lagoonal – continental sediment underlying the well-known miocenic formation of Pietra leccese cropping out the " Vito Fazzi" hospital of Lecce. They referred this lagoonal deposit to Tortoniano (Miocene). Later, other important outcrops of these deposits were recognized near S. Maria al Bagno (Nardò, Lecce) by Bossio et al. (1992) and near Galatone (Lecce) by Colella (1994), respectively along the roadside of the Gallipoli – Lecce highway and along a

Fig. 1. Location of Lecce in south Italy (red point) and the surveyed areas (yellows points)

cut of Sud Est railway. Moreover Barbera et al. (1993) referred to late Oligocene a shallow marine calcarenite, rich in Scutelle, cropping out in a quarry near Galatone. Recently Bossio et al. (1999) recognized, not very far away from Lecce, two different informal units referred to the Oligo – Miocene transition, the Galatone Formation (lagoonal – continental deposits) and Lecce Formation (shallow marine deposits). Bossio et al. (2000) proposed to formalize the Galatone Formation.

Notwithstanding these researches, at the moment, the stratigraphical relationships between Galatone Formation and Lecce Formation in consequence of extended soil cover have not been defined.

GPR measurements has been carried out to define the geometrical relationship between these two units.

GPR measurements have been carrying out along a cut-face in order to assess the potential for imaging and characterising different lithological facies of this method and to choose the better antenna and set up. In this first phase some methodological aspects related to the data processing were considered. Particularly first one, although used in potential field and in seismic data processing, Discrete Wavelet Transform (DWT) based filtering procedures was used to GPR images for the particular problem of removing coherent noise (linearly and, mainly, horizontally correlated); second one some interesting approaches to increase resolution of radar signal were performed. Test indicate a 200 MHZ antenna to be a good compromise between resolution and depth penetration. For each litostratigraphic unit, in each of the two investigated areas, velocity analises using Common Depth Point (CDP) and Wide Angle Reflection and Refraction (WARR) techniques were also performed in order to characterize the lithological faces and to convert time in depth.

2. GPR background theory

GPR is an EM geophysical method for high-resolution detection, imaging and mapping of subsurface soils. In principle, and just to introduce the subject, the GPR can be viewed as composed by a central unity, a transmitting and a receiving antenna, and a computer. The central unity generates electromagnetic pulses that are radiated into the soil by the transmitting antenna. Rigorously, the pulses are radiated in all the directions, but most energy is radiated within a conic volume under the antenna, as shown in Fig. 2. When the electromagnetic waves meet any buried discontinuity (a buried object, or also the interface between two geological layers, a cavity, a zone with different humidity etc.), they are scattered in all the directions (the intensity of the scattered power is not spatially uniform, but depends on the scattering target) and so partly also toward the receiving antenna.

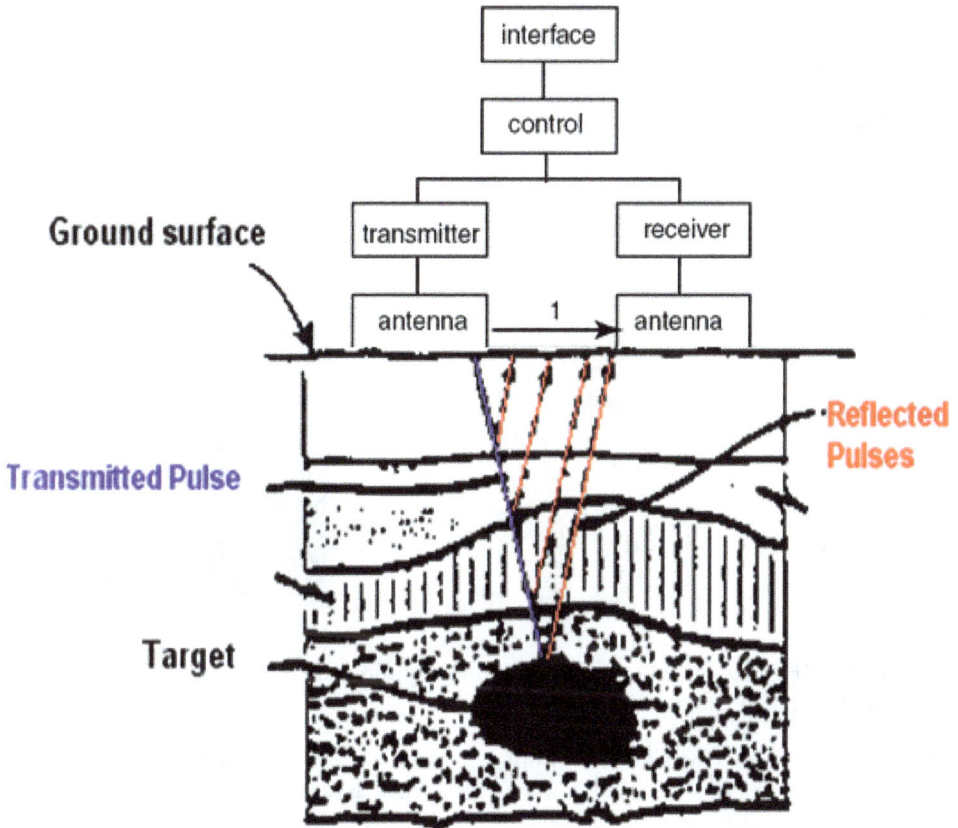

Fig. 2. A block diagram of a GPR system. The interface module enable the user to enter the system parameters, and displays and records the data. The control unit generates the timing signals so that all of the components operate in unison. This unit also does some preliminary data processing. The pulse travel paths in order of arrival are direct air wave, direct ground wave, and reflections.

Usually the transmitting and the receiving antenna are incorporated in a rigid structure and move together. In modern systems, the gathered signal is represented in real time on the screen of the computer, and is stored in the hard disk memory of the computer. It is implicit that the equipment of a GPR also includes suitable cables to connect the central unit, the antennas and the computer, and also a device to provide energy in the field (Fig. 3), usually, the antennas are also equipped with an odometer that allows to measure the covered distance. The energy is usually supplied by rechargeable batteries in the form of a zero frequency electrical voltage, and the central unit transforms this energy into a signal in the microwave frequency range.

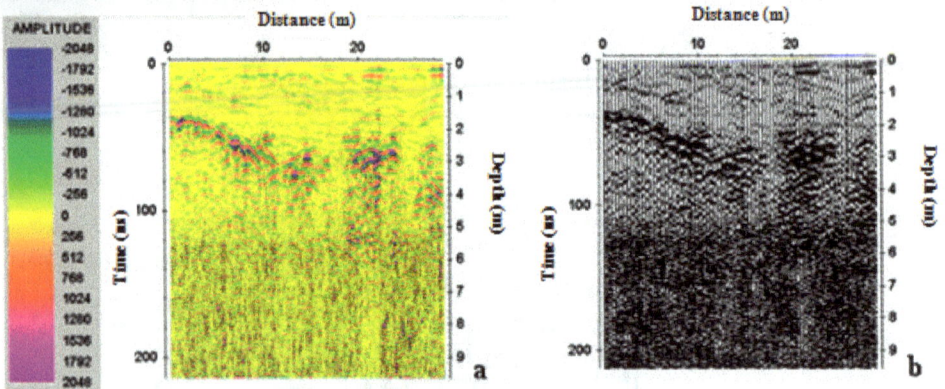

Fig. 3. A 200 MHz antenna with the Sir 2 system GPR unit and resulting radargrams: a) line scan visualization; b) wiggle trace visualization.

Most of the returned signals in radar profile are reflections from subsurface discontinuities, although other types of waves may also be present. Wave types such as a direct airwave, a critically refracted airwave and a direct ground wave generally appear as well, as predicted by the Ray Theory and simple geometrical relations. In a reflection profile, the principal reflections are generally more or less immediately identifiable, as can be seen, for example, in Fig. 3a and 3b. Note that most of the signals in the profiles are reflections except the two topmost, which are two direct waves from the transmitter to the receiver, one in the air and the other in the ground.

In certain common conditions during GPR investigations, in addition to reflections, the EM waves undergo diffractions from small inhomogeneities and objects. Diffractions that can be identified as hyperbolas in the time section occur in two cases: when the dominant wavelength, λ, in the radar pulse is larger than the dimensions of the diffractions source, and when waves are diffracted from sharp edges. The physical relation between the velocity, v, wavelength, λ , and frequency, f, of an EM wave is given by the equation (Conyers and Goodman, 1997):

$$v = \lambda f \tag{1}$$

According to eq. 1, if, for example, a GPR signal were transmitted at a center frequency of 100 MHz into geological environment with an average propagation velocity of 0.1 m/ns, the local dominant wavelength of the propagating signal would be approximately 1 m. Therefore, diffraction patterns would be obtained from objects or inhomogeneities that are smaller than 1 m.

The resolution of a GPR image is controlled by the sharpness of the focus of the system. The resolution is defined by the Rayleigh criterion (Reynolds, 1998) as the ability to distinguish between two close signals obtained during the GPR mapping, before their separate identity is lost and they appear to be one event. The range resolution, $d\lambda$, can be practically defined as the half-wavelength of the GPR signal in the geological medium (Conyers and Goodman, 1997). Processing methods such as deconvolution can enhance the range resolution below a quarter of the wavelength (Widess, 1973). For example, the calculated average basic vertical resolution for a 100 MHz center frequency mapping of a 0.1 m/ns environment is about 0.5 m (0.125 m). A reflecting horizon may vary laterally in dielectric constant, thus changing the reflection coefficient, or stop laterally, as a result of faulting or absence of deposition (e.g., channel sands). Horizontal (or spatial) resolution refers to the ability to detect the lateral changes in reflectors, such as those caused by faults or facies changes. In this case, the reflected energy that arrives at the receiver antenna does not come from a single point of incidence, but from a circular zone on the reflector. If t is the two-way time of a reflection, fc the frequency of a radar wave and v the velocity, the first Fresnel zone radius Fr from which most energy comes, is (Reynolds, 1998):

$$F_r \sim 0.5 \text{ v } (t/f_c)^{1/2} \tag{2}$$

The derivation of the Fresnel zone radius approximation for GPR is exactly analogous for seismic waves, although in reality, since GPR systems generally use directional dipole antennas, the EM sheaf of waves forms the shape of an elliptical cone (the long axis is perpendicular to the dipole). According to equation 2, if the area of a reflector is greater than an area bordered by circular zone with radius F_r, its shape will be accurately mapped on the time section. However, if the areal extent of the reflector is smaller, diffraction patterns from the edges may dominate its shape. From equation 2, it can be understood that that spatial resolution decreases as a function of depth (e.g., with the increase of the time).

To illustrate if average propagation velocity of 0.1 m/ns, the calculated spatial resolution of a reflector is about 4.5 m ($2F_r$) at the depth of 10 m (i.e. t = 200 ns) achieved in the 100 MHz GPR profiles. This means that the reflector must be larger than 4.5 m, in order to be best mapped. In practice, the spatial resolution is substantially better. Sheriff and Geldart (1995), discusses an effective Fresnel zone as equal to half the size of the first Fresnel zone.

Therefore, it can be shown that when such a reflector occupies only 25 percent of the Fresnel zone, its reflected amplitude decreases only by 40 percent. This result emphasizes the fact that even reflectors with lateral dimensions of 1 m (about $\frac{1}{4}$ F_r) can be clearly detectable at a depth of 10 m, in conditions of fair signal to noise ratio.

For a correct interpretation of the GPR signal, it is important to have some estimation of the electromagnetic characteristics of the background medium (which means, depending of the case, the soil, the masonry, the pillar, the wooden log and so on). A complete characterization would mean a measure of the dielectric permittivity and of the magnetic permeability, both meant as complex quantities to account for losses (in particular, in this way the dielectric permittivity accounts for the electric conductivity too) and variable vs. the frequency to account for the dispersion. These quantities, in general are also functions of the buried point (i.e. the medium is not homogeneous), and possibly are tensor quantities instead of scalar ones (i.e. the medium is anisotropic).

Under the hypothesis of homogeneous, isotropic, non-magnetic and low-loss medium, the propagation velocity of the electromagnetic waves c is related to the relative dielectric permittivity of the medium ε_s by means of the relationship $c = \frac{c_o}{\sqrt{\varepsilon_s}}$ (Reynolds, 1998), where c_o is the propagation velocity of the electromagnetic waves in free space, about equal to $3*10^8$ m/s. ε_s is a dimensionless real quantity, whereas the complex (absolute) dielectric permittivity is meant as $\varepsilon_{eq} = \varepsilon_0 \varepsilon_s - j\frac{\sigma}{\omega}$, where ε_0 is the dielectric permittivity of the free space (equal to 8.854×10^{-12} Farad/m), σ is the electric conductivity of the background medium and ω is the circular frequency.

For many materials, expected values of the relative dielectric permittivity are tabled. Of course, it is usually better to measure the propagation velocity from the data, because the actual current values depends on several environmental variables (the water content, the compactness of the soil, the presence of mineral salts, possibly the temperature). However, the tabled values can be helpful in order to check the likelihood of the retrieved value. In table 1 some values are provided.

3. The site geological setting

The Salento peninsula is the southernmost part of the Apulia region (southern Italy). Apulia is the emerged part of a plate stretching between the Ionian Sea and the Adriatic Sea which constitutes the foreland of both Apenninic and Dinaric orogens. It comprises a Variscan basement covered by a 3-5 Km thick Mesozoic carbonate sequence (the Calcari delle Murge unit), and overlain by thin deposits of Paleocene (Bossio et al., 1992,1998,1999; Margiotta, 1999; Margiotta and Ricchetti, 2002), Neocene (Bossio et al., 1992,1994,1998) and Quaternary age (Bossio et al., 1987; Palmentola, 1987; D'Alessandro et al., 1994). The mid-southern part of Salento peninsula is marked by a wide endorheic area, bordered both toward the East and the West by degradated fault scarps which are the flanks of two narrow ridges lengthened in NNW-SSE direction. Marls, calcareous marls and calcarenites belonging to several Pleistocene sedimentary cycles extensively crop out in the endorheic area. These deposits cover a stratigraphic sequence compound by calcareous units whose age is

Material	Dielectric constant	Conductivity (mS/m)	Velocity (m/ns)
Air	1	0	0.3
Distilled water	80	0.01	0.033
Fresh water	80	0.5	0.033
Sea water	80	30,000	0.01
Dry sand	3-5	0.01	0.15
Saturated sand	20-30	0.1-1.0	0.06
Limestone	4-8	0.5-2	0.12
Shale	5-15	1-100	0.09
Silt	5-30	1-100	0.07
Clay	4-40	2-1,000	0.06
Granite	4-6	0.01-1	0.13
Salt (dry)	5-6	0.01-1	0.13
Ice	3-4	0.01	0.16

Table 1. Some properties of geological materials (Reynolds, 1998)

comprised between the Upper Cretaceous and Upper Pliocene. Two elongated depressions characterized the margin of the endorheic area. They are shaped on Lowest Pleistocene deposits (Bossio et al., 1987), up to 70 m thick, made by calcareous, bioclastic sandstones, locally clinostratified; they shade into bluish clayey marls toward centre of the area. The bottom of depressions is covered by thick sandy colluvial deposits. The most part of endorheic area is constituted by a flat surface, gently sloping northeastward, reaching 120 m of altitude at its SE part. The surface is shaped on white quartz sands that can be most likely referred to the Middle Pleistocene. To the north-west, a low relict cliff joins this surface to a wide marine terrace placed between 40-80 m of altitude whose deposits, made by coarse calcareous sandstones, lie transgressively on Lower Pleistocene sandy and clayey deposits (Fig.4) (D'Alessandro et al., 1994). The Salento peninsula is marked by a wide, deep aquifer hosted into the Mesozoic limestones which rests on sea-water intruded from the nearby coastal area (Ghyben-Herzberg principle). However, a number of shallow water tables occur in the most recent deposits. In particular, in the endorheic area several water tables can be found within the Lower Pleistocene calcareous sandstone and in the Middle Pleistocene sands even if their characteristics are not well known. However, a significative drainage from shallow water tables to the deep acquifer is most likely to occur along sub-vertical planes of greater hydraulic conductivity (Leucci et al., 2004).

Particularly in the surveyed area the first report on the existence of Cenozoic deposits "oligoalini" was done by Del Prete and Santagati (1972). These authors described in detail below the sedimentary succession known Miocene formation of Lecce stone, which recognized the character oligoalino for the presence of carbon levels and the association with ostracods, on the basis of these data was attributed to the Tortonian succession. The authors believed the Stone of Lecce in continuity of sedimentation over the deposits of fresh water.

With regard to the geology of the studied area the Galatone Formation emerges on a narrow strip (Fig. 5) roughly oriented NW - SE in the southern part of Lecce.

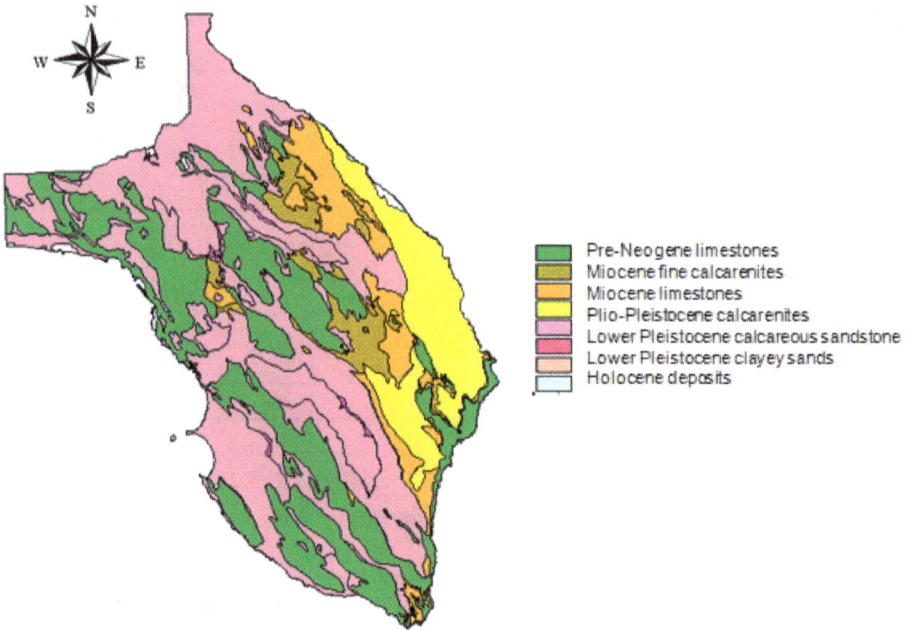

Fig. 4. The geological map of the Salento peninsula

Here is possible to observe an irregular alternation of layers composed of marl and calcareous marl and gray havana with obvious texture tending laminar micritic limestones and limestone thinly layered compact places also foundered.

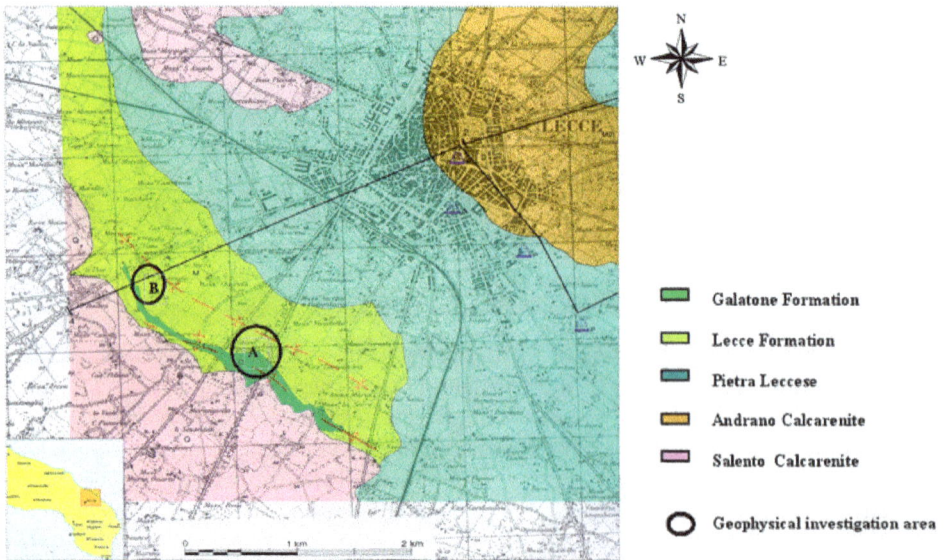

Fig. 5. Nord - West Lecce Provincia Geological Map (Bossio et al., 1999)

3. Field procedures

The field procedures, applied in the acquired data in the two selected areas located near Lecce (Fig. 5), have considerate some essential factors. These factors has summarized as follow.

3.1 The data acquisition mode

As affirmed by Davis and Annan (1989) and confirmed by Conyers and Goodman (1997), GPR data can be acquired either continuous or step mode. During acquisition of field data, the radar-transmission process is repeated many times per second as the antennas are pulled along the ground surface or moved in steps.

In continuous mode the antennae are dragged over the soil surface. In step mode the antennae are held a constant distance apart and moved progressively along the profile.

In rough terrain (where is present more vegetation) the continuous mode degrades data quality due to the changes in ground coupling as the height of the antennae above the ground will change through the survey (Woodward et al., 2003).

In the step mode the exact position of every traces taken on the profile is known. If the distance between the survey point is too great, steeply dipping and small features cannot be imaged (Woodward et al., 2003). The step size is a crucial choice because related to the traces number that would be collected. A low traces number making many structure unresolvable. Furthermore this process increase the acquisition time.

For the study presented in this paper GPR data were acquired in continuous mode.

3.2 Choice of frequency

Antenna frequency is a major factor in depth penetration. The higher the frequency of the antenna, the shallower into the ground it will penetrate (Davis and Annan, 1989). A higher frequency antenna will also see smaller targets. Antenna choice is one of the most important factors in survey design. The following table shows antenna frequency, approximate depth penetration and appropriate application.

Approximiate Depth Range	Primary Antenna Choice	Secondary Antenna Choice	Appropriate Application
0 - 0.5 m	1600 MHz	900 MHz	Structural Concrete, Roadways, Bridge Decks,
0 - 1 m	900 MHz	400 MHz	Concrete, Shallow Soils, Archaeology
0 - 9 m	400 MHz	200 MHz	Shallow Geology, Utilities, UST's, Archaeology
0 - 9 m	200 MHz	100 MHz	Geology, Environmental, Utility, Archaeology
0-30 m	100 MHz	Sub-Echo 40	Geologic Profiling
Greater than 30 m	MLF (80, 40, 32, 20, 16 MHz)		Geologic Profiling

Table 2. The approximate depth range with the antennae frequency choice (Leucci, 2008)

In order to choice the frequency antenna useful to the survey a GPR calibration profiles were performed using the 100, 200 and 500MHz center frequencies antennae. Calibration was carried out on the cut-face and GPR data were verified by direct comparison.

In Fig. 6 the photo (a) and the geological section of the cut – face is shown. The cut- face is located along the road Lecce – Gallipoli, southwest of Lecce (area B in Fig. 5). The cut has a variable depth (between about 4 and 1 m) compared to ground level. Lithological succession is characterized by an irregular alternation of compact micritic limestones, thinly laminated limestones. In particular in the stratigraphically lower portion of the section outcropping in the northern part of the embankment, marly layers are prevalent with thicknesses ranging from ten to forty centimeters thick limestone layers intercalated with the order of decimeters. In the southern part of the cut - face, the layers consist of limestones stratigraphically higher prevailing in places thinly laminated, with rare intercalations of marly levels. In the central part of the cut - face, which is the summit portion of the sequence, the layers are affected by many distortions that their identification is difficult when the limestone outcrop and nearly impossible at the marl. Still south along the cut - face, the layers reappear in their appearance than those with mostly typical of those calcareous marl.

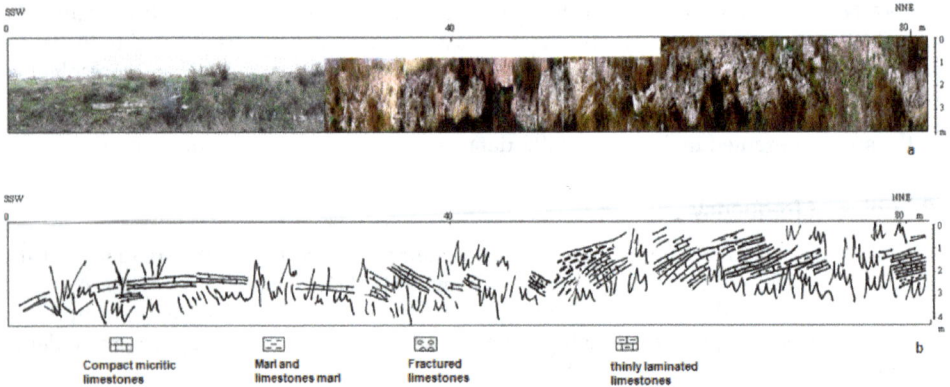

Fig. 6. The cut – face: a) photo; b) interpreted geological section

As mentioned previously, the layers are loosely bent: in particular it is recognizable the presence of an anticline with a large radius of curvature in the southern part of the cut - face, also at the bottom of the sequence, the layers plunge to the south - west.

From a preliminary analysis of GPR raw data seem that EM signal does not propagate beyond 100ns for the 100MHz and 200MHz antennas, and beyond 60ns for 500MHz antenna. To make more meaningful comparisons between the data for the three antennas on the two-way time window was set at 70ns (Fig. 7).

A comparison with the stratigraphical trends and therefore with the geological section (Fig. 6b) shown a good correlation between the raw radar sections and the stratification of the geological formations. GPR raw data show a good signal to noise ratio in areas where a thinly laminated limestones and micritic compact limestones are present. The signal to noise ratio is less good, as expected, in areas of intense deformation of the layers and in those areas where there are calcareous marl and marl.

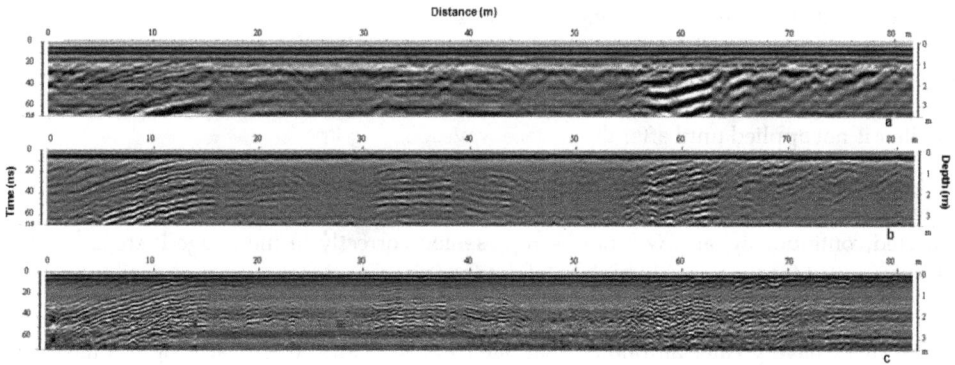

Fig. 7. The raw radar sections acquired on the cut – face: a) 100MHz; b) 200MHz; c) 500MHz antennae

The conclusion to make for this calibration survey is that for medium depth targets with relatively large sizes the choice of the antenna was not critical and either one between 200 and 500MHz would have done the job equally well. This means that for large targets buried 1 to 3 meters from the surface the 200MHz center frequency and the 500MHz center frequency are both adequate devices.

4. Processing methodology

One of the great advantages of the GPR method is the fact that the raw data is acquired in a manner that allows it to be easily viewed in real time using a computer screen. Often very little processing is required for an initial interpretation of the data, with most of the effort directed towards data visualization. On the other hand, depending on the application and target of interest, it may be necessary to perform sophisticated data processing, and many practitioners find that techniques common to seismic reflection such as migration can be applied. The outcome of processing is a cross-section of the subsurface EM properties, displayed in terms of the two-way travel time, i.e. the time taken for a wave to move from the transmitter to a reflector and return to the receiver. The amount of processing undertaken can range from basic, which allows rapid data output, to the more time consuming application of algorithms designed for use on seismic dataset (Ylmaz, 1987), which produce high quality output (Daniels et al. 1988; Conyers and Goodman, 1997). The processing sequence usually developed for GPR raw data is following done.

zero-time adjust (static shift) – During a GPR survey, the first waveform to arrive at the receiver is the air wave. There is a delay in the time of arrival of the first break of the air wave on the radar section due to the length of the cable connecting the antennae and the control unit. Therefore need to associate zero-time with zero-depth, so any time offset due to instrument recording must be removed before interpretation of the radar image.

Background removal filter (subtract average trace to remove banding) - Background noise is a repetitive signal created by slight ringing in the antennae, which produces a coherent banding effect, parallel to the surface wave, across the section (Conyers and Goodman, 1997). The filter is a simple arithmetic process that sums all the amplitudes of reflections that were recorded at the same time along a profile and divides by the number of traces summed the resulting composite digital wave, which is an average of all background noise, is then subtracted from the data set. Care must be taken in this process not to remove real linear events in the profile. The time window where the filter operates must be specified so that the filter is not applied until after the surface wave.

Horizontal (distance) stretch to get constant trace separation (horizontal normalization) – This correction need to remove the effects of non-constant motion along the profile. Data are collected continuously, and will not be represented correctly in the image if steps are not taken to correct for the variable horizontal data coverage.

Gain – Gain is used to compensate for amplitude variations in the GPR image; early signal arrival times have greater amplitude than later times because these early signals have not traveled as far. The loss of signal amplitude is related to geometric spreading as well as intrinsic attenuation. Various time-variable gain functions may be applied in an effort to equalize amplitudes of the recorded signals. The most commonly applied is an automatic gain control (AGC) that is a time – varying gain that runs a window of chosen length along each trace, point by point, finding the average amplitude over the length of the window about each point. A gain function is then applied such that the average at each point is made constant along the trace.

Topographic corrections – Surveyed elevation data are used to apply topography to the GPR survey profiles. Firstly trace windowing is applied to the data to remove all artefacts in the survey that arrived before the time zero arrivals. The actual elevation recorded along the GPR line are then entered into the data processing package and the time zero arrivals are hung from the topographic profile by applying a time shift to each individual trace.

Frequency filtering - Although GPR data are collected with source and receiver antennae of specified dominant frequency, the recorded signals include a band of frequencies around the dominant frequency component. Frequency filtering is a way of removing unwanted high and/or low frequencies in order to produce a more interpretable GPR image. High-pass filtering maintains the high frequencies in the signal but removes the low frequency components. Low-pass filtering does just the opposite, removing high frequencies and retaining the low frequency components. A combination of these two effects can be achieved with a band-pass filter, where the filter retains all frequencies in the pass band, but removes the high and low frequencies outside of the pass band.

Deconvolution - When the time-domain GPR pulse propagates in the subsurface, convolution is the physical process that describes how the propagating wavelet interacts with the earth filter (the reflection and transmission response of the subsurface). Deconvolution is an inverse filtering operation that attempts to remove the effects of the source wavelet in order to better interpret GPR profiles as images of the earth structure. Deconvolution operators can degrade GPR images when the source signature is not known. Deconvolution operators are designed under the assumption that the propagating source wavelet is minimum phase (i.e., most of its energy is associated with early times in the wavelet). This assumption is not necessarily valid for GPR signals. With GPR, the ground becomes part of the antennae, and the source pulse can vary from trace-to-trace and is not necessarily minimum phase. All filtering operations borrowed from seismic data processing must be applied with care as some of the underlying assumptions for elastic waves generated at the surface of the earth are not valid or are different for electromagnetic waves. For more see Ylmaz (1987)

Migration - Migration is a processing technique which attempts to correct for the fact that energy in the GPR profile image is not necessarily correctly associated with depths below the 2-D survey line.

As with deconvolution, migration can be seen as an inverse processing step which attempts to correct the geometry of the subsurface in the GPR image with respect to the survey geometry. For example, a subsurface scattering point would show up in a GPR image as a hyperbolic-shaped feature. Migration would associate all the energy in the wavelets making up the hyperbolic feature with the point of diffraction, and imaging of the actual earth structure (the heterogeneity represented by the point diffractor) would be imaged more clearly. Migration operators require a good estimate of subsurface EM wave velocity in order to apply the correct adjustments to the GPR image. For more see Ylmaz (1987).

F-K filter - Fourier transform techniques, or f-k filtering, i.e. by means of filters designed and applied in the frequency-wavenumber (or f-k) domain (Yilmaz, 1987). It is well known that a dipping line in the x-t domain maps to a line passing through the origin and with an orientation normal to the original line in the f-k amplitude spectrum. In other words, a line of constant apparent velocity corresponds to a line of constant slope in the f-k domain. In particular, horizontal lines map to the vertical direction, along the f-axis. Dipping events that overlap in the x-t domain can be separated in the f-k domain by their dips. This allows the elimination of certain types of unwanted energy from the data, representing linear coherent noise. Regardless of their location, lines with the same dip (parallel lines), map to the same radial line in the f-k amplitude spectrum, so that f-k filters could be effective for removing at the same time all undesired lines with the same slope, but impractical if one wants to remove only some of them instead of the whole family. Fan filters are generally used for dip filtering. In these cases the amplitude spectrum of the input is multiplied by a suitable function, the amplitude response of the filter consisting of ones in a fan-shaped zone and zeros elsewhere, to obtain the amplitude spectrum of the output, whereas the phase spectrum is left unchanged. Finally, the filtered signal is obtained by a two-dimensional inverse Fourier transform.

The Wavelet Transform – It is possible to decompose the radar signal into different scales where signal and certain noises may be effectively separated/isolated (multiresolution analysis) . Subsequent muting of the noise is easily achieved in the Wavelet Transform (WT) domain operating only on the scales where the offending noise appears.

5. Processing step on the GPR data acquired on cut-face

Data acquired on the cut- face were processed in order to enhance the signal to noise ratio. To optimize the selection of frequency filter a spectral analysis was performed on the radar sections and the average spectrum was calculated. Based on the results of this analysis was applied a bandpass filter. The coherent noise (horizontal band), present on the sections, probably due to the ringing of the antennae or reflections from obstacles visible on the surface located at a constant distance from the antenna itself, has been removed by a background removal filter.

Figure 8 shows the processed radar sections related to 100MHz (a), 200MHz (b) and 500MHz (c) center frequency antennae.

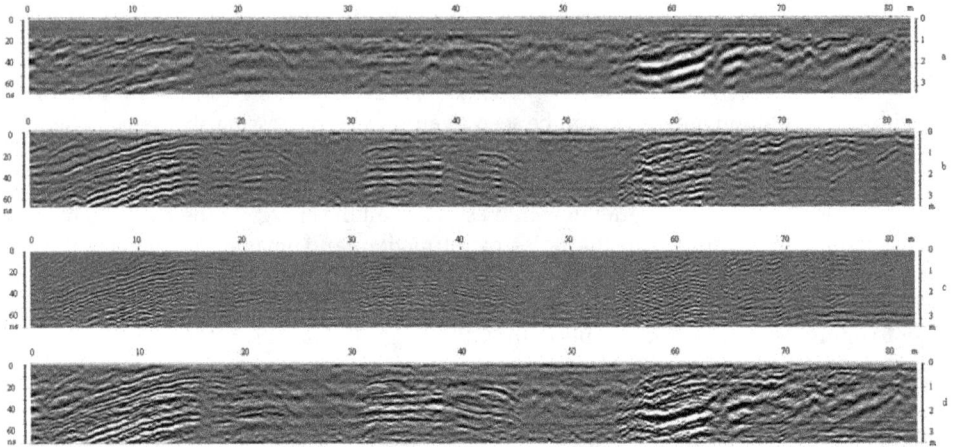

Fig. 8. The processed radar sections related to the profiles acquired on the cut – face: a) 100MHz; b) 200MHz; c) 500MHz antennae; d) sum of radar signals obtained with the 100, 200 and 500MHz antennae

The comparison between Fig. 6 and 8 shows:

a.   a general improvement of the signal to noise ratio over the entire section related to the 100MHz antenna. The horizontal noise was attenuated and, in the first 20ns, the EM reflections are more clear. These can be associated with the almost horizontal reflector visible between 15 and 35ns between 25-32m and 45-55m on the profile direction;

b.   in the 200MHz antenna section, the improvement of the signal to noise ratio is sensitive only in the outer zones, where the signal is still evident and corresponds to changes in stratigraphy. In the middle part of the processed radar section (between 45 and 55m) a very weak EM reflection is present at about 10ns;

c.   in the 500MHz antenna section, the signal to noise ratio is much improved in the area between 55 and 80m, where the layers trend is more evidenced; in the central part of the profile (30-45m), the signal present in the raw data between 30 and 40ns loses of evidence; the effect of the processing is positive in the two-way time window 6-20ns where is emphasize the superficial layers trend.

As known, increasing the frequency content of the spectrum is obtained by a narrowing of the package in the time domain with a consequent increase in temporal resolution. One technique to increase the length of the band in the case of radar signals was proposed by Malagodi et al. (1996). It consists of adding radar signals obtained with different antennae.

In order to reduce high frequency noise, resulting in continuous acquisition, the processed signals for the three antennas was added. The result is shown in Fig. 8d. Is possible to see:

a) a general improvement of the signal to noise ratio over the entire radar section; b) a sharper differentiation of the signal in areas with different EM, and therefore lithological, characteristics; c) a better correlation of the EM reflection events with the stratification.

In particular, the EM signal: i) in the area between 0 and 45m seems to allow the identification of the anticlinal evident in the geological section; ii) in the area between 45 and 55m shows weak EM reflections associated to very deformed limestone as seen in the more superficial geological section; iii) in the area between 55 and 80m shows a good correlation with the stratigraphical trend.

The slight differences in the geometry of the layers, especially between 0-40m present, are related to the non-perfect coincidence between the radar profile and the observed cut – face.

F-K and wavelet-based filtering techniques were applied to a part of radar section (Fig. 9a) refers to a test profile acquired with a 500 MHz. Some dipping reflections are barely

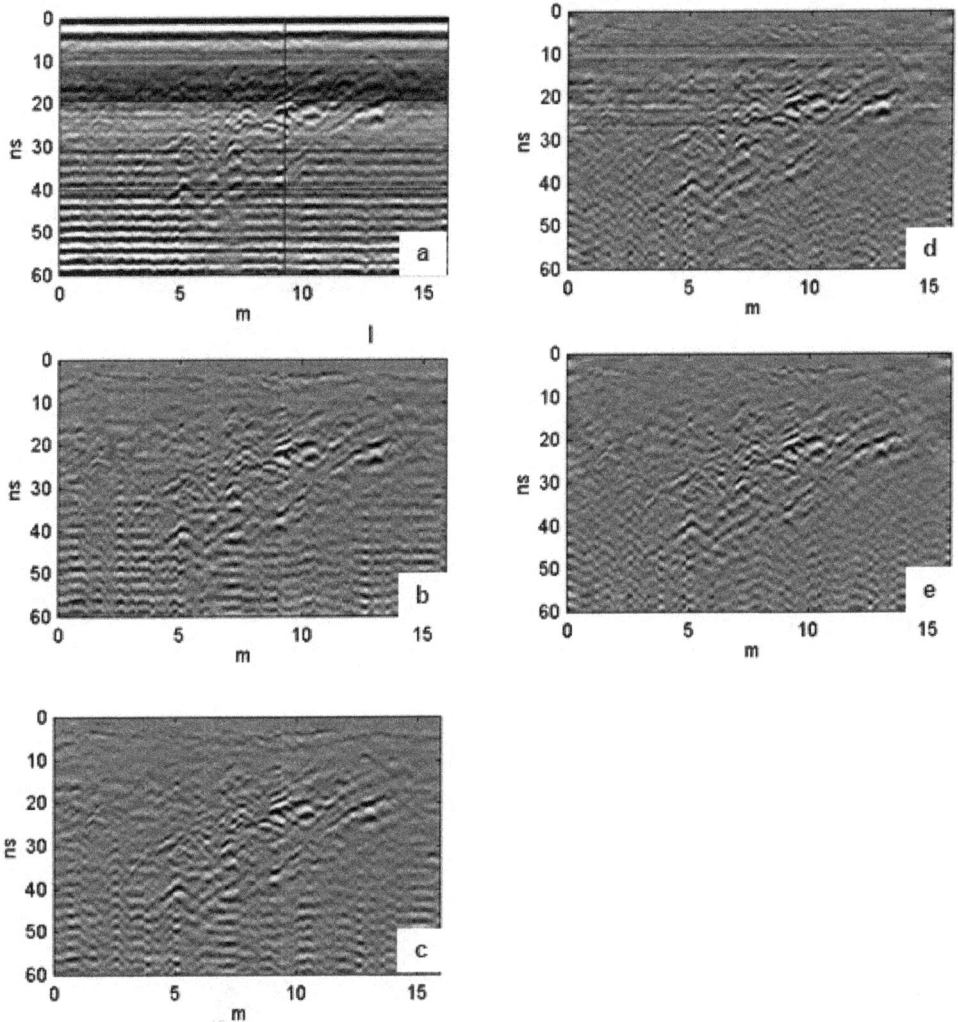

Fig. 9. Comparison of several type of filters applied on the radar section acquired on the cut – face: a) raw; b) background removal filter; c) F-K filter; d) 2D WT; e) 1D WT.

distinguishable through the variable-intensity system-ringing noise and the low-frequency background heavily contaminating the section.

It is clear that a background removal filter (Fig. 9b) is insufficient for removing completely the horizontal banding, due to its horizontal amplitude variation, whereas it could be too strong in the upper part, reducing the horizontal reflection continuity. Slightly better results, but quite similar to the background removal, gives the application of a F-K filter designed to remove the very steep dips and the horizontal noise (Fig. 9c). For comparison the results of two different filtering techniques using the 1D and 2D WT are shown in Figures 9d and 9e. Also in this case slightly better results gives the application of a 1D WT.

6. Velocity analysis

A detailed knowledge of the subsurface velocity field for the propagation of EM energy is critical to any through GPR processing. A number of method exist to estimate the EM wave velocity propagation in the subsoil (Fruhwirth et al., 1996; Conyers and Goodman 1997; Huisman *et al.* 2003). The common midpoint (CMP) and the wide angle reflection refraction (WARR) are the two method used in the field to estimate the EM wave velocity in the subsoil.

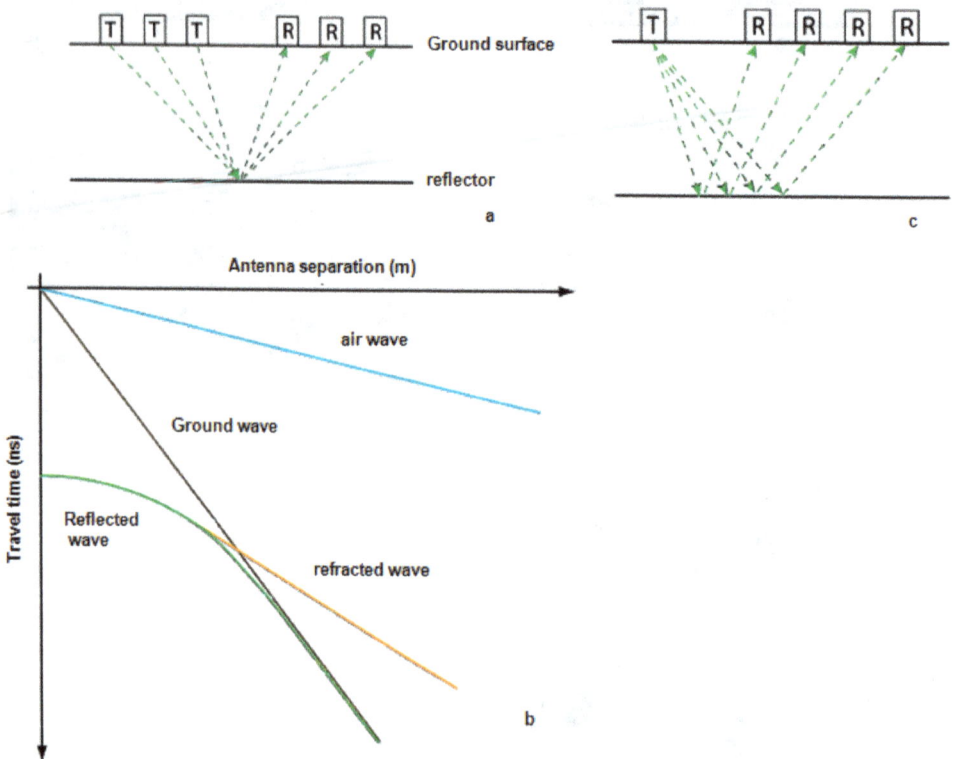

Fig. 10. The EM wave velocity measurements: a) CMP acquisition scheme; b) CMP radargram; c) WARR acquisition scheme

In a CMP measurement transmitter and receiver are moved away from each other in equidistant steps (Fig. 10a). At each position a trace is measured. This way, the reflected signal can be measured using a number of different angles. The resulting radargram displays the travel time as a function of the antenna separation (Fig. 10b).

Since air and ground wave travel directly between the transmitting and receiving antenna, there is a linear relationship between the travel time t of each wave and the antenna separation a with the constant of proportionality $1/v$:

$$t = a/v$$

with $v = c$ for the air wave and $v = c/(\varepsilon_r)^{1/2}$ for the ground wave. Due to their different velocities the slopes of both direct waves in the travel time diagram are different. Consequently, the propagation velocity v of the GPR wave through the soil can be determined directly from the radargram by estimating the slope of the ground wave. Since the ground wave travels near the soil-air interface it covers that soil section which is for example important for plant growth. The air wave travel time is usually applied during data processing as reference for calculating absolute travel times.

From a CMP measurement one can determine the reflector depth below the midpoint between the transmitting and the receiving antenna. From the reflection hyperbolas displayed in the travel time diagram relative permittivity and reflector depth can be determined independently. Plotting the measured data in a t^2-a^2-diagram, leads to a linear relationship between t and a:

$$t^2 = \frac{1}{v^2} a^2 + \frac{4h^2}{v^2}.$$

The propagation velocity of the electromagnetic wave can now be directly determined from the slope of line. The depth of the reflector can be directly inferred from the intersection of the line with the y-axis.

In contrast to a CMP measurement, in WARR measurement (Fig. 10c) only the transmitting or receiving antenna is moved along the measurement line while the other antenna stays stationary. In principle, a WARR measurement follows the same relationships concerning travel time as a CMP measurement. The difference is that the reflection point moves along the reflector. This is why a WARR measurement strictly is only applicable in the presence of horizontal or only slightly sloping reflectors and material properties are homogeneous.

EM-wave velocity can be more quickly and easily determined from the reflection profiles acquired in continuous mode, using the characteristic hyperbolic shape of reflection from a point source (Fruhwirth et al. 1996). This is a very common method of velocity estimation and it is based on the phenomenon that a small object reflects EM-waves in almost every direction.

The EM velocity analysis was performed on the cut-face in order to convert the two way travel time in depth. The EM velocity analysis have also contributed to characterize the lithology in the surveyed area. Two CDP, CDP1 and CDP2, located on the GPR profile acquired on the cut –face along the abscissa 35m and 49m respectively, were performed. The 500MHz center frequency antenna as transmitter and 200MHz center frequency antenna as receiver were used. This is possible because the frequencies bands of the two used antennae are partially overlapped (Conyers and Goodman, 1997).

To estimate the EM velocity the above described a^2-t^2 method was used. The EM velocity variations with depth obtained are typical of inhomogeneous media, and have obtained using the Dix method (Dix, 1950). The results are shown in Fig. 11. In Fig. 11 it possible to see the radar sections and the location of the CDP. Under the qualitative aspect, the velocity analysis confirms what has already emerged from the GPR survey: the area around 35m is characterized by the presence of different reflection hyperbolas confirming alternating layers highlighted by the survey also found in the processed radar section and in particular on the photo; the area around 49m is characterized by two hyperbolas of reflection that confirm an high homogeneity.

From the quantitative point of view, the velocity analysis shows a trend of EM velocities slightly decreasing with depth (from surface 10.5cm/ns to 9.0cm/ns at 40ns time depth). The Dix analysis seem to characterize a sequence of alternating limestone (v = 10.5cm/ns) and limestone - marl (v = 8.0cm/ns) layers in agreement with observations.

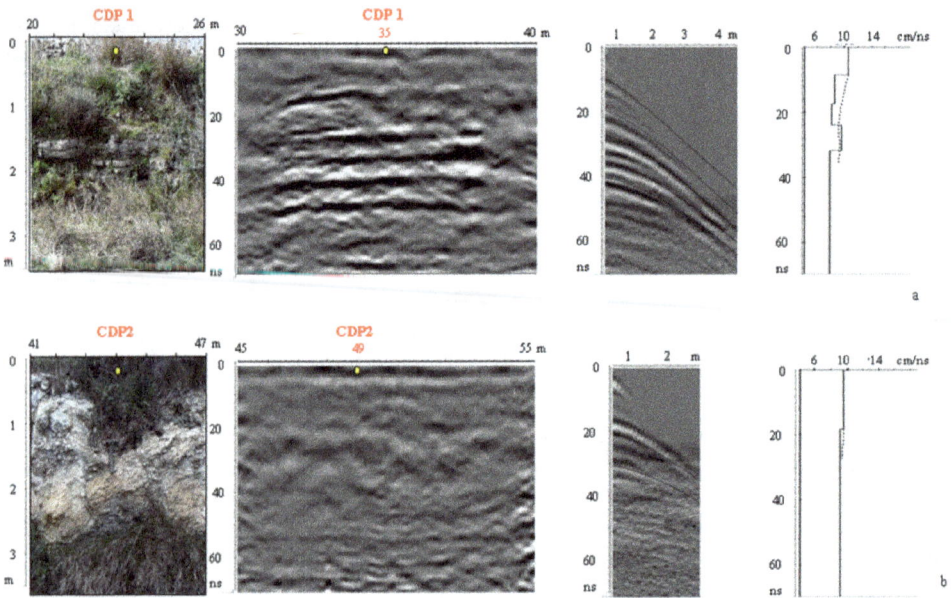

Fig. 11. The CMP results on a) Galatone Formation; b) Lecce Formation

The CDP2 is characterized by average EM velocity almost constant (10.0cm/ns). The velocity constant range, confirming the absence of layer surfaces that has emerged in the sections, even in this case the agreement with the observed is good.

7. The GPR survey on the Lecce and Galatone formation

GPR data were acquired using a Sir 2 with the 200MHz antenna. As shows in the data acquired on the cut – face the 200MHz antenna achieves a good compromise between resolution and penetration depth.GPR data were acquired in continuous mode. This technique allows large areas to obtain a good compromise between data quality and acquisition time. A critical parameter in this case may be the speed of drag the antenna: in fact the incremental sampling, of the tool, it provides data averaged over time, so in case of wavy structures, characterized by small wavelengths, a high-speed drive the antenna does not follow the undulations of the same detail (Conyers and Goodman, 1997).

The survey consists in 61 acquired radar profiles, most of them with a direction transverse to the line contact between the two geological formations (Fig. 12).

In the area A, measurements were made on profiles arranged in a grid of 11.2m x 25m (Fig. 12a).

The quality of raw data was good but it was not enough for an immediate interpretation due to the presence of an inevitable component of noise. The particular type of problem and the high resolution required, made data processing necessary.

Fig. 12. Location of the GPR profiles: a) A area; b) B area

Data processing was done using Reflex 6.0 software (Sandmeier, 2010).

A series of tests were carried out in frequency filtering on a sample of 15 radar sections, which was considered to better represent the entire set of data. The tests showed that the high-cut filters are effective in removing most of the noise, while low-cut filters also eliminate much of the useful signal. The low frequency noise, normally referred to the background related to the horizontal band on the radar sections that can cover the reflected events. It is due to the ringing of the antennae or reflections from obstacles visible on the surface located at a constant distance from the antenna (walls parallel to the profile or the operator pulls the antenna). To remove the background noise, the background removal filter was applied. The application of the migration has not been particularly incisive because on the field data were almost absent diffraction hyperbolas. Consequently, the migration was omitted from the processing of data. A block diagram of the processing performed to the GPR raw data sections and an example before and after processing is show in Figs 13 and 14.

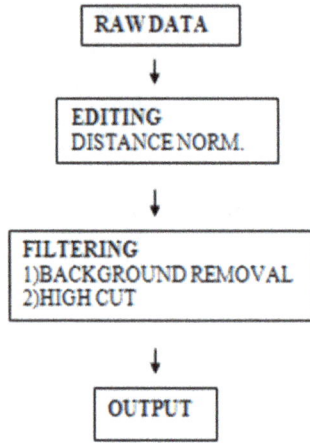

Fig. 13. Processing sequence for GPR data acquired in the areas A and B

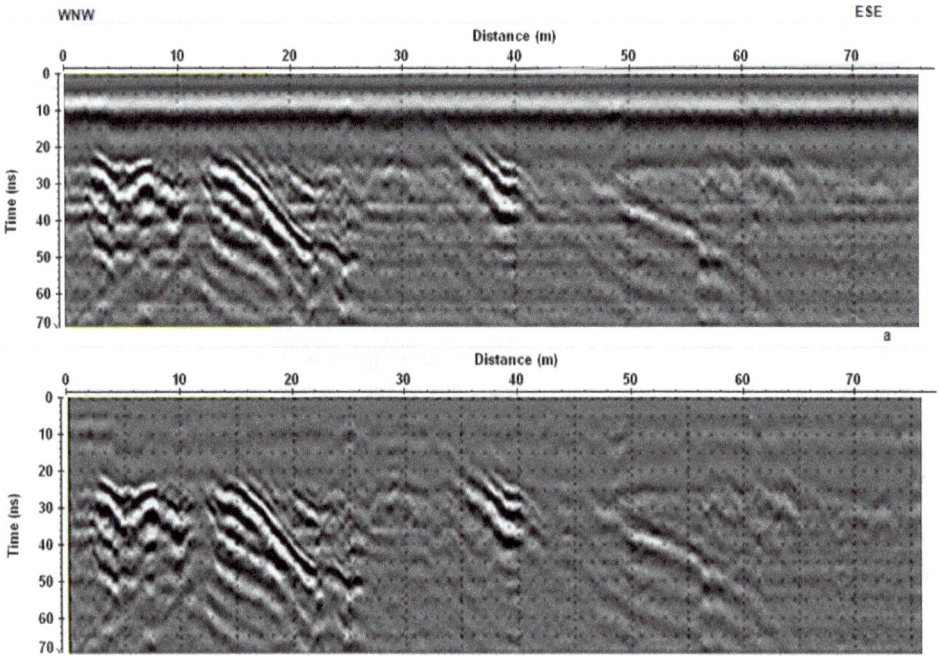

Fig. 14. Radar section related to the B3 profile acquired in area A: a) raw; b) processed

The high resolving power of the method GPR allowed to:

1.   in Galatone Formation
-     highlight the irregular alternation of layers with a thickness of about 15cm related to EM characteristics very different (corresponding to limestones and marly-clays);
-     identify two main systems of folds with a large radius of curvature and with a wavelength ranging from a few meters (5 or 6 m) to a few tens of meters (about 60m), the first system is characterized by folds with axes oriented in the approximately north direction and dipping a few degrees both to the north and south, the second folds around whose axis has direction NW - SE (Fig. 15);
-     recognize geometric discontinuities, probably related to faults, noted in the cut-face.

2.   in Lecce Formation
-     distinguish layers with thicknesses of about 50cm and highlight slight variations of electromagnetic properties of the medium, reflecting a general lithological homogeneity of the sediments;
-     show that the layers are everywhere weak sub-horizontal or diving to the north (as found in area B).

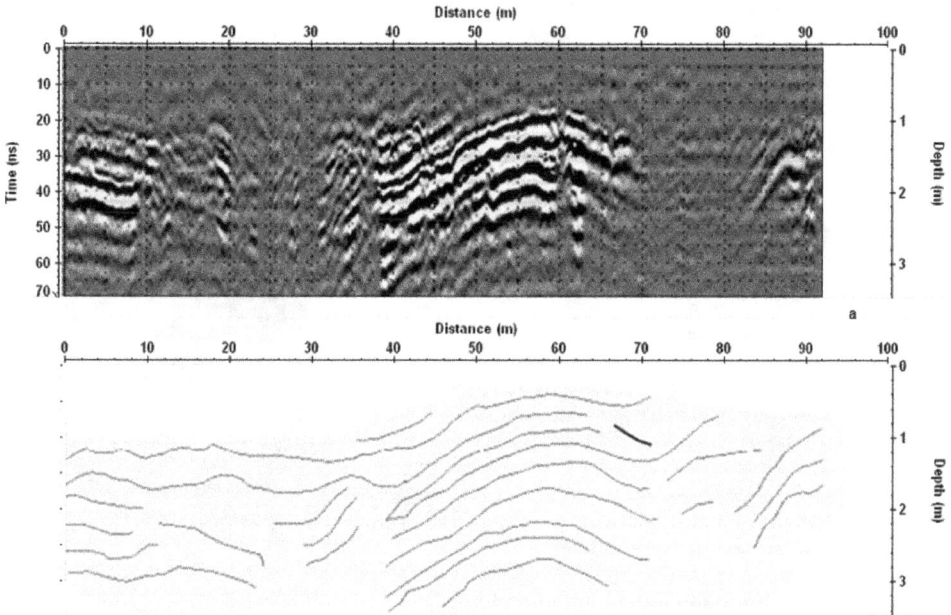

Fig. 15. a) radar section related to the H7 profile acquired in area A; b) geological section reconstructed by radar profile

Finally a pseudo 3D visualization (Figs. 16 and 17) shows the geometrical contact between the two studied formations.

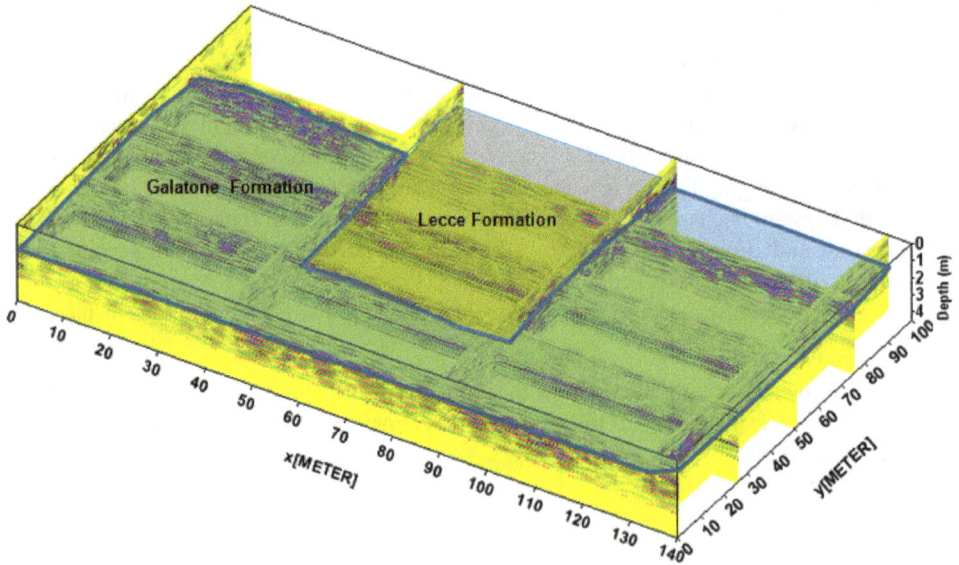

Fig. 16. Area A: pseudo 3D visualization

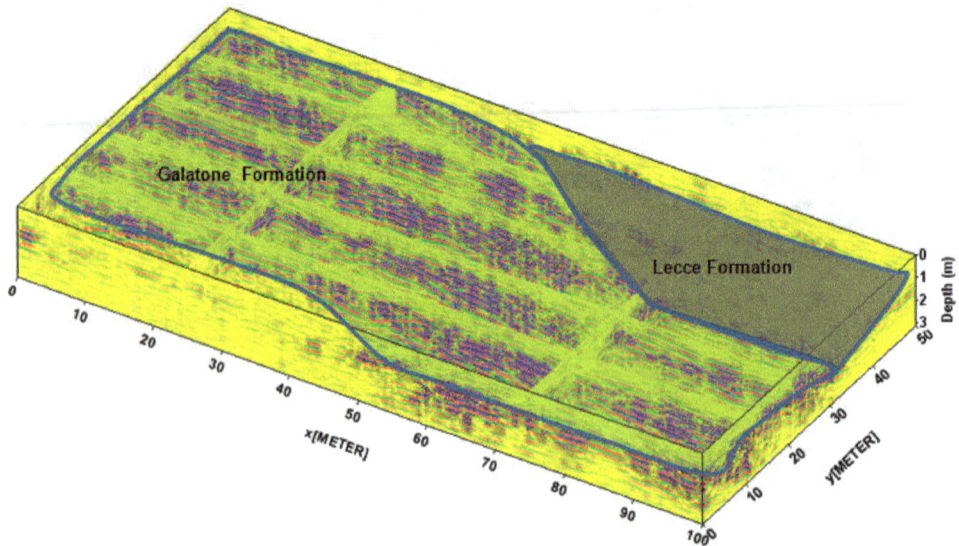

Fig. 17. Area B: pseudo 3D visualization

8. Conclusions

The quality of GPR results obtained in sedimentary rock environments reflect both the lithological nature of sediments its and geometrical relationship together with the care applied to the methodology for data acquisition, processing and interpretation.

In this chapter was described the key steps required to acquired, processing and interpret GPR data in a sedimentary rocks with different lithological characteristic. Data acquired in south Salento were used. Particular importance was placed on ground – truth control obtained from a cut – face. It provide the necessary verification for the interpretation of GPR reflections. In fact the results show that is possible to obtain an improving of the signal to noise ratio using only a traditional filters. The sum of both signals in the time domain, better highlight, at least in some areas, the trend of the stratigraphy and helped to distinguish areas with different electromagnetic characteristics, related to the different lithology. A general good correlation between the radar and the geological section was obtained. The EM wave velocity measurements were very useful for interpretation and characterization of the different lithologies, especially in the areas with low signal to noise ratio. It was able to recognize areas of a succession of limestone and marl and limestone layers, and made up of areas with no obvious surface layer.

This overview outlines an approach suitable for application to future GPR surveys in sedimentary rock environments and should lead to an improvement in the quality of GPR studies, which will aid interpretation of, and comparison between, data sets.

In the successive GPR measurements in the two chosen areas the results obtained on the cut-face were useful to set the field methodology, the antenna frequency, the processing step, and data interpretation. Results shows the characteristics of the sediments of the two studied formations: Galatone Formation is always mildly folded affected by fractures, while the layers of Lecce Formation are horizontal and show geometrical mismatch with the Galatone formation.

With regard to the GPR method it can be concluded that it gives results clearly interpretable when the lithological units is well stratified and presents sharp contrasts of the electromagnetic properties. It allows, through the EM velocity analysis, to solve the evolution of stratification in the first 3m - 4m in depth; to estimate the thickness of the layers; to give the lithological nature of the two formations. Obviously, where the geological formation investigated is characterized by lithological homogeneity with indistinct stratification surfaces, and thus contrasts with weak electromagnetic fields, the results of the GPR method is not easily evaluated in the absence of outcrops that allow calibration of the method.

9. References

Baker, P.L., 1991. Response of ground-penetrating radar to bounding surfaces and lithofacies variations in sand barrier sequences. Explor. Geophys. 22, 19–22.

Barbera C., Bossio A., Mazzei R., Monteforti B., Salvatorini G., 1993. Un flash sul ciclo miocenico del Salento. Soc. Paleont. It. XII convegno Guida alle escursioni, 79-84

Beres, M., Green, A., Huggenberger, P., Horstmeyer, H., 1995. Mapping the architecture of glaciofluvial sediments with three-dimensional georadar. Geology 23, 1087–1090.

Bossio A., Mazzei R., Monteforti B., Salvatorini G., 1992. Notizie preliminari sul Miocene di S.Maria al Bagno-S.Caterina, presso Nardò (Lecce). Paleopelagos, 2, 99-107.

Bossio A., Esu D., Foresi L.M., Girotti O., Iannone A., Luperto Sinni E., Margiotta S., Mazzei R., Monteforti B., Ricchetti G., Salvatorini G., 2000 - Formazione di Galatone, nuovo nome per un'unità litostratigrafica del Salento (Puglia, Italiameridionale). Atti Soc. Toscana Sc. Nat., Mem., Serie A, 105, 151-156.

Bossio, A., Guelfi, F., Mazzei, R., Monteforti, B., Salvatorini, G., Varola, A., 1987. Precisazioni sull'età dei sedimenti pleistocenici di due cave del leccese (San Pietro in Lama e Cutrofiano), Proceedings del Convegno sulle conoscenze geologiche del territorio salentino. Lecce 12 December 1987. Quaderni di Ricerche del Centro Studi Geotecnica e Ingegneria, 11, pp 147-174.

Bossio A., Mazzei R., Monteforti B. & Salvatorini G., 1994. La successione miocenica nell'area tipo delle Calcareniti di Andrano (Puglia, Italia meridionale)", Boll. Soc. Paleont. Italiana, 33, (2), 249-255, Modena.

Bossio A., Esu D., Foresi L.M., Girotti O., Iannone A., Luperto E., Margiotta S., Mazzei R., Monteforti B., Ricchetti G., Salvatorini G., (1998), "Formazione di Galatone, nuovo nome per un'unità litostratigrafia del Salento (Puglia, Italia meridionale)", Atti Soc. Tosc. di Scienze Naturali, Mem., Serie A, 105, 151-156, Pisa.

Bossio A., Foresi L., Margiotta S., Mazzei R., Monteforti B. & Salvatorini G., 1999. Carta geologica del settore nord orientale della Provincia di Lecce; scala 1:25000; settore 7,8,10 scala 1:10000, Università degli Studi di Siena.

Carrozzo M. T., Leucci G., Margiotta S., Negri S., Nuzzo L., 2000 APPLICAZIONE DELLA METODOLOGIA GPR PER LA SOLUZIONE DI PROBLEMI STRATIGRAFICI.Bollettino Geofisico Anno XXIII, Vol 1-2 Gen-Giu 2000, pp 5-16;

Carrozzo M.T., Leucci G., Negri S., Nuzzo L., 2003. GPR SURVEY TO UNDERSTAND THE STRATIGRAPHY AT THE ROMAN SHIPS ARCHAEOLOGICAL SITE (PISA, ITALY). Archaeological Prospection, vol. 10, n. 1, 57-72;

Ciaranfi, N., Pieri, P., Ricchetti, G., 1992. Note alla carta geologica delle Murge del Salento (Puglia centro-meridionale), Mem. Società Geologica Italiana, 106, pp 449-460, Roma.

Colella R., 1994. Rilevamento geologico e analisi litostratigrafia di lembi oligocenici affioranti tra Galatone e S. Maria al Bagno (LE). Tesi di laurea inedita in geologia, Univ. degli Studi di Bari - Facoltà di SMFN - Dip. Di Geol. e Geofisica.

Conyers L. B. and Goodman D., (1997). Ground-penetrating radar – An introduction for archaeologists, AltaMira Press, A Division of Sage Publications, Inc.

D'Alessandro, A., Mastronuzzi, G., Palmentola, G., Sansò, P., 1994. Pleistocene deposits of Salento leccese (Southern Italy): Problematic relationships, Bollettino della Società Paleontologica Italiana, 33 (2), pp 257-263;

Daniels D., Gunton D. J., Scott H.F., 1988, Introduction to subsurface radar. Institution of Electrical Engineers, Proceedings, 135 (F4), 278-320

Davis J. L and Annan A. P., 1989. Ground Penetrating Radar For high resolution mapping of soil and rock stratigraphy. Geophysical Prospecting 37, 531-551.

Del Prete M. and Santagati G., 1972. Depositi oligoalini interposti tra calcari cretacici e Pietra leccese nei dintorni di Lecce.Geol. Appl. e Idrogeol., vol.7

Dix C. M., 1955. Seismic velocities from surface measurement. Geophysics 20, 68-86.

Fruhwirth R.K., Schmoller R. and Oberaigner E.R. 1996. Some aspects of the estimation of electromagnetic wave velocities. Proceedings of the 6th International Conference on Ground Penetrating Radar, Tohoku University, Sendai, Japan, pp. 135–138.

Huisman J.A., Hubbard S.S., Redman J.D. and Annan A.P. 2003. Measuring soil water content with ground penetrating radar: A review. Vadose Zone Journal 2, 476–491.

Lapen, D.R., B.J. Moorman, and J.S. Price. 1996. Using ground-penetrating radar to delineate subsurface features along a wetland catena. Soil Sci. Soc. Am. J. 60:923-931.

Leucci G., Margiotta S., Negri S., 2000. UN CONTRIBUTO PER LA DEFINIZIONE DEI RAPPORTI GEOMETRICI TRA DUE UNITÀ OLIGO-MIOCENICHE DEL SALENTO LECCESE (PUGLIA, ITALIA) MEDIANTE INDAGINI GEOFISICHE CON GEORADAR. Bollettino della Società Geologica Italiana, III fascicolo 2000, pp 703-714;

Leucci G., Margiotta S., Negri S., 2004. GEOLOGICAL AND GEOPHYSICAL INVESTIGATIONS IN KARSTIC ENVIRONMENT (SALICE SALENTINO, LECCE, ITALY). Journal of Environmental and Engineering Geophysics (JEEG), n. 9, 25-34;

Leucci G., 2008. GROUND PENETRATING RADAR:THE ELECTROMAGNETIC SIGNAL ATTENUATION AND MAXIMUM PENETRATION DEPTH, Scholarly Research Exchange: Volume 2008 • Article ID 926091 • doi:10.3814/2008/926091;

Malagodi S.,Orlando L., Piro S. (1996). Approaches to increase resolution of radar signal. 6th International Conference on Ground Penatrating Radar(GPR'96), September 30-October 3, 1996. Sendan, Japan. 283-288.

Margiotta S., 1999. Il contatto Formazione di Galatone – Formazione di Lecce: evidenze stratigrafico – sedimentologiche, Atti Soc. Tosc. di Scienze Naturali, Mem., Serie A, 106, 73-77, Pisa.

Margiotta S. and Ricchetti G., 2002. Stratigrafia dei depositi oligomiocenici del Salento (Puglia), Boll.Soc.Geol.It., 121

Mills, H.H., and M.A. Speece. 1997. Ground-penetrating radar exploration of alluvial fans in the southern Blue Ridge Province, North Carolina. Environ. Eng. Geosci. 3(4):487-499.

Mokma, D.L., R.J. Schaetzl, J.A. Doolittle, and E.P. Johnson. 1990. Ground-penetrating radar study of ortstein continuity in some Michigan haplaquods. Soil Sci. Soc. Am. J. 54:936-938.

Nobes, D.C., Ferguson R.J., and Brierley G.J.. 2001. Ground-penetrating radar and sedimentological analysis of Holocene floodplains: insight from the Tuross valley, New South Wales. Aust. J. Earth Sci. 48:347-355.

van Overmeeren R.A., 1998. Radar facies of unconsolidated sediments in The Netherlands: A radar stratigraphy interpretation method for Hydrogeology. Journal of Applied Geophysics, 40, 1-18.

Palmentola, G., 1987. Geological and geomorphological outlines of the Salento leccese region (Southern Italy), Proceedings del Convegno sulle conoscenze geologiche del territorio salentino. Lecce 12 December 1987. Quaderni di Ricerche del Centro Studi Geotecnica e Ingegneria Lecce, 11, pp 7-23.

Reynolds, J. M., 1998, An Introduction to Applied and Environmental Geophysics: John Wiley & Sons Ltd.

Sandmeier, K. J., 2010, Reflex 6.0 Manual. Sandmeier Software. Zipser Strabe 1, D-76227 (Karlsruhe, Germany).

Sheriff, R. E., and L. P. Geldart (1995), Exploration Seismology, 2nd ed., 592 pp., Cambridge Univ. Press, New York.

Widess, M. B., 1973. How thin is this bed? Geophysics, Vpl. 38, p. 176-1180.

Woodward J., Ashworth P.H., Best J.L., Sambrook Smith G.H., Simpson C.J., 2003. The use and application of GPR in sandy fluvial environments: methodological considerations. In Ground Penetrating Radar in Sediments, Edited by C.S. Bristow and H. M. Jol, Geological Society Special Pubblication, London, 211, 127-142.

Yilmaz O 1987 *Seismic Data Processing* ed E B Neitzel (Tulsa, OK: Society of Exploration Geophysicists)

Section 2

Biostratigraphy

5

The Muhi Quarry:
A Fossil-Lagerstätte from the Mid-Cretaceous
(Albian-Cenomanian) of Hidalgo, Central México

Victor Manuel Bravo Cuevas[1], Katia A. González Rodríguez[1],
Rocío Baños Rodríguez[2] and Citlalli Hernández Guerrero[2]
[1]Area Académica de Biología, Instituto de Ciencias Básicas e Ingeniería,
[2]Licenciatura en Biología, Instituto de Ciencias Básicas e Ingeniería,
Universidad Autónoma del Estado de Hidalgo,
México

1. Introduction

The German word *lagerstätten* (singular, *lagerstätte*) describes sedimentary deposits with an important concentration of fossil material in a variable state of preservation. According to Seilacher (1970), there are two types of *lagerstätte*: Concentration deposits (*Konzentrat-lagerstätte*) and conservation deposits (*Konservat-lagerstätte*). Concentration deposits are originated by sedimentological and biological processes that promote the accumulation of a high density of fossil remains, particularly hard parts such as shells and bones. Conservation deposits are distinguished by uncommon depositional conditions (for example, rapid burial, anoxia, and/or hypersalinity) that favor exceptional preservation of an organism (Seilacher, 1970; Seilacher et al., 1985). Conservation deposits have received more consideration in the literature and in several instances are regarded as "snapshots" of a particular moment in the history of life (Benton & Harper, 2009; Nudds & Selden, 2008). Although concentration deposits are not mainly distinguished by a high-quality preservation of organic remains, its fossil richness provides information suitable to elaborate paleobiological inferences about an ancient community.

A few Mexican marine Cretaceous deposits are distinguished by important concentrations of fossil remains, usually in a high state of preservation. The localities which have been considered as *Fossil-lagerstätten* in México include: Vallecillo quarry, Nuevo León (northern México) of early Turonian age (Blanco-Piñón et. al. 2002); Múzquiz quarries, Coahuila (northern México) of late Turonian-early Coniacian age (Stinnesbeck et al., 2005); Xilitla quarries, San Luis Potosí (central México) of Turonian age (Blanco-Piñon et al., 2006); and Tlayúa quarry, Puebla (central México) of Albian age (Alvarado-Ortega et al., 2007). A sedimentary deposit with comparable characteristics was discovered about 13 years ago in northwestern sector of the state of Hidalgo, central México; the locality is formally known as Muhi Quarry. Paleontological work carried out in the site since its discovery, has allowed recovering an important sample of fossil remains, which represents a diverse marine community of mid-Cretaceous age. Because of the importance of the site, we discuss taphonomic evidence to consider the Muhi Quarry as a *Fossil-lagerstätte*.

2. Historical review

The Muhi Quarry is geographically located at 20°49'21.5" N and 99°15'38.3" W, set in northwestern sector of the state of Hidalgo, central México (Figure 1). The quarry has been worked for more than 30 years for flagstone by the Yánez family inhabitants of the San Pedro town in Zimapán. Before the first visit by personal of the University of Hidalgo in 1998, Sergio and Ignacio Yánez did not understand the presence of fish in the stones. Once they were instructed about the importance of the discovery, they started paying attention on the specimens during the exploitation. Previously they used to sell them together with the slabs for building. At the same time, kids living around the place became hard collectors (Figure 2). The first specimen formally recovered from the quarry corresponds to a beautifully preserved pachyrhrizodontid head. Since that moment, more than 2,200 specimens have been collected including invertebrates and fish.

The real diversity of the quarry is still unknown since new taxa are discovered every year. Although the fauna is not completely described because of the newness of the discovery, it is remarkable the presence of new species and new records for America, including the shrimp *Aeger hidalguensis* (Feldman et al., 2007), which only Cretaceous sister species *Aeger libanensis* Roger, 1946, comes from the Cenomanian of Lebanon (Feldman et al., 2007); and the fishes *Ichthyotringa mexicana* Fielitz & González-Rodríguez, 2008, which sister species were also found in the Cenomanian Lebanese localities of Namoura, Hakel and Hajula; undescribed halecoid fishes previously known from some localities of Western Tethys; agonids earlier found in Cenozoic localities of Europe (González-Rodríguez & Schultze, 2010); and tetraodontiforms (González-Rodríguez et al., 2011).

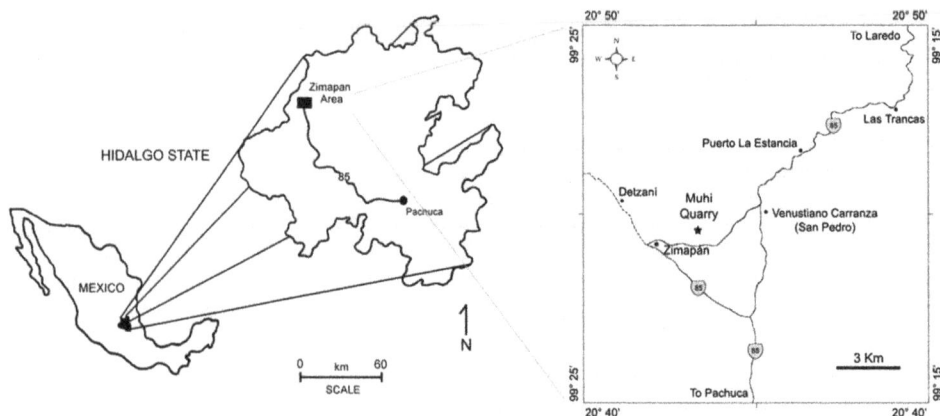

Fig. 1. Index map showing the location of the Muhi Quarry at the Zimapán Area, northwestern sector of Hidalgo, central México.

3. Litostratigraphy, sedimentary environment, and age

The Muhi Quarry is a calcareous rock sequence belonging to the La Negra Facies of the El Doctor Formation, consisting of micritic limestones, bedded and/or nodular cherts, and laminae of barely consolidated mixture of siliciclastic clay and calcium carbonate. The sedimentary sequence shows a lithological variability indicative of an outer sea shelf setting, which received temporal pulses of pelagic waters, and occasionally influx of near-shore waters maybe during storms (Bravo-Cuevas et al., 2009). The age of the Muhi Quarry hinges on its stratigraphic relationships with other Mexican Mesozoic rock units and on the fish fauna that it bears; the deposit is considered of mid-Cretaceous (Albian-Cenomanian) age (Bravo-Cuevas et al., 2009; González-Rodríguez & Bravo-Cuevas, 2005).

Fig. 2. The Muhi Quarry at Zimapán Area, northwestern sector of Hidalgo, central México. A. Panoramic view of the quarry; B. Exploitation of the quarry for flagstone and gravel; C. The Yánez brothers, workers of the quarry; D. Prospecting and collecting of fossil material by kids of the Zimapán area.

4. Material and methods

In order to characterize the fossil assemblage of the Muhi Quarry, we used taphonomic indicators. The fossil assemblage was classified considering life habitat and state of preservation of the record, including autochthonous, parautochthonous or allochthonous concentrations; a mixed assemblage is referred to autochthonous-parautochthonous, parautochthonous-allochthonous, or autochthonous-allochthonous (after Kidwell et al., 1986). In addition, we considered the "time averaging" involved in its formation (Kidwell & Bosence, 1991).

The taphonomic features evaluated comprise: (1) Anatomical completeness, (2) disarticulation, and (3) fragmentation. The deformation of the axial skeleton and closure/openness of jaws were evaluated in fishes. Anatomical completeness (1) refers to the percentage of bones preserved in an individual specimen (Brett & Baird, 1986). We considered three states, as follows: complete (100% of skeletal elements present); partially complete (at least 50% of skeletal elements present); and incomplete (less than 50% of skeletal elements present). Disarticulation (2) was evaluated in crinoids, echinoids, crustaceans, and fishes. This feature refers to the number of hard parts scattered and/or displaced from their original anatomical position (Hill, 1979). Degree of disarticulation was measured as: slightly disarticulated (less than 25% of skeletal elements scattered and/or displaced); moderately disarticulated (25%-50% of skeletal elements scattered and/or displaced); and highly disarticulated (more than 50% of skeletal elements scattered and/or displaced). Thus, articulated specimens mean that all hard parts are in their original anatomical position. Fragmentation (3) refers to the breaking of the skeletal elements, caused by scavengers, predators and/or physical factors (for example, transport produced by currents or wind) (Benton & Harper, 2009; Speyer & Brett, 1988). Three states were recognized: low fragmentation (0-30% of skeletal elements broken), medium fragmentation (30%-70% of skeletal elements broken), and high fragmentation (70-100% of skeletal elements broken) (Brett & Baird, 1986).

According to Bienkowska (2004), deformation of fish axial skeleton includes: vertebral series straight, slightly arched, arched, or S-shaped; this condition is related to a post-mortem effect. Moreover, openness/closure of jaw was used as a relative indicator of level-oxygenation.

5. Fossil concentration of the Muhi Quarry

The fossil concentration of the Muhi Quarry represents a polytypic association defined by an accumulation of invertebrates, vertebrates, and trace fossils. The sample consists of about 2,200 specimens in a variable state of preservation. Invertebrates include ammonites, crustaceans, echinoids, and crinoids. Vertebrate groups comprise cartilaginous fishes and bony fish; remains belonging to an unidentified reptile are also present. Class-level distribution of this record is shown in Figure 3. There are numerous coprolites, probably produced by some fossil groups, such as fish, crustaceans, and/or cephalopods.

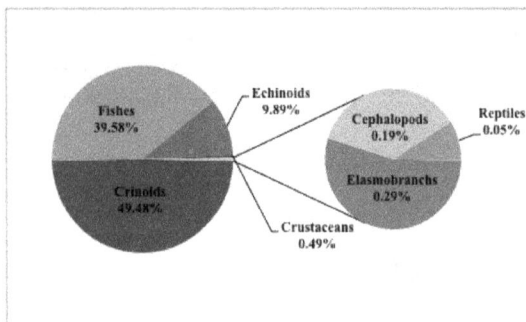

Fig. 3. Class-level distribution of fossil concentration in the Muhi Quarry, mid-Cretaceous (Albian-Cenomanian) of Hidalgo state, central México.

Crinoids are the most abundant group, including about 1,000 un-stalked forms of the Order Comatulida, which are comparable to the Upper Jurassic to Lower Cretaceous genus *Saccocoma* (Turek et al., 1989). Numerous disarticulated echinoid spines (about 200 specimens) are present. Crustaceans and ammonites are less abundant in comparison with other invertebrates of the quarry. Crustaceans include specimens belonging to Aegeridae and Palinuridae families (Feldmann et al., 2007), consisting of about 10 specimens. Ammonites include five poorly preserved specimens tentatively assigned to ?*Mortoniceras* sp. (Bravo-Cuevas et al., 2009; González-Rodríguez & Bravo-Cuevas, 2005).

Fishes are the second most abundant group, consisting of several complete specimens, and numerous skulls, fins, and vertebral series; the sample includes about 800 specimens in a variable state of preservation. The diversity recorded at the moment holds isolated shark teeth and vertebrae, two incomplete bodies of Rajiformes, and 15 Neopterygian families. Majority of the fish groups are under detailed taxonomic study, in addition to an unidentified reptile.

6. Taphonomy: A preliminary approach

6.1 Fishes

The fish assemblage mainly consists of incomplete skeletons highly disarticulated, whereas partially complete skeletons show a variable degree of disarticulation; a low frequency of complete articulated skeletons or slightly disarticulated is observed (Figure 4A). This

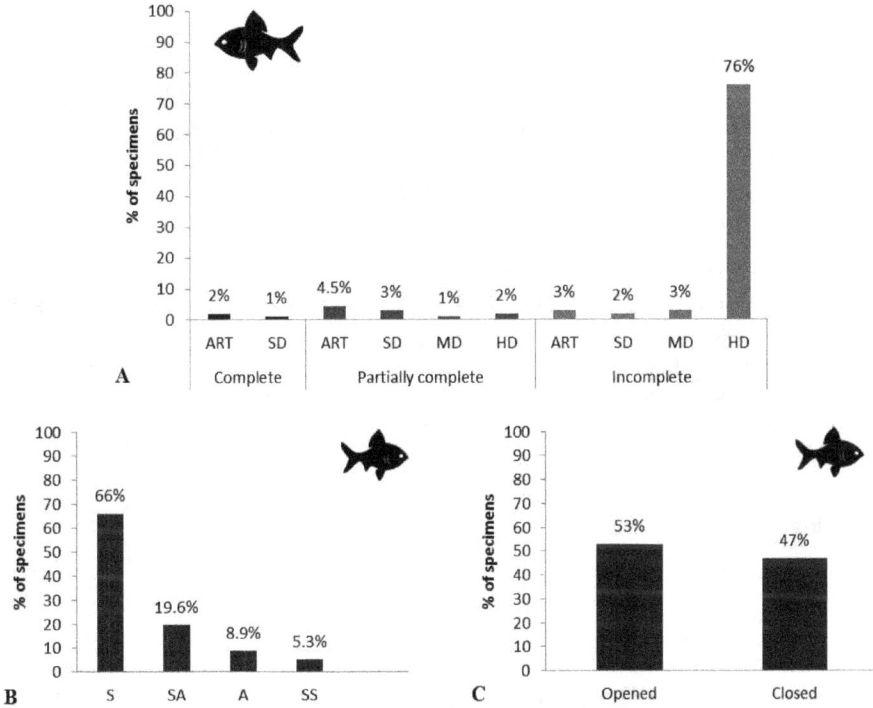

Fig. 4. Frequency distribution of taphonomic features evaluated in bony fish. A) Taphonomic grades of anatomical completeness and disarticulation. B) Deformations of axial skeleton. C) Specimens with jaws opened or closed. Abbreviations: ART= articulated, SD= slightly disarticulated, MD= moderately disarticulated, HD= highly disarticulated, S= straight, SA=slightly arched, A= arched, SS= S-shaped.

pattern points to a slight degree of anatomical completeness and a high state of disarticulation, which suggests that the majority of fish carcasses experienced a full flotation related to water warm conditions (Elder, 1985; Elder & Smith, 1988). In full floating conditions fishes commonly lose dermal skull bones and/or the skull is completely disarticulated from the body (Wilson 1980, 1988a, 1988b); fish thanatocenosis of the Muhi Quarry exhibits this pattern. Moreover, the low degree of anatomical completeness observed, implies a relatively high post-mortem transport, turbulence, decay, and/or bioturbation (McGrew, 1975).

Scavenging frequently occurs in high or sufficient oxygenated waters (Elder, 1985). In particular, the scavenging effect on fish bones produces a pattern of disarticulation that is distinguished by skeletal parts scattered in all directions and without orientation (Smith & Elder, 1985). This condition is rarely observed in the fish skeletons recovered from Muhi Quarry, thus suggesting low or none scavenging activity at the site of deposit, which in turn is indicative of low levels of water oxygenation (Smith & Elder, 1985).

Majority of fish specimens do not show axial skeleton deformation, but some exhibit an arched or S-shaped body (Figure 4B). The bending of the vertebral column is explained as a natural post-mortem effect (Bienkowska, 2004; Jerzmanska, 1960); but, a straight condition of the vertebral column suggests a slow rate of decay which is delayed in conditions of high salt concentrations (Hecht, 1933) and /or low levels of water oxygenation (Schäfer, 1972). Based on these evidences, it is possible that changes in the salinity and oxygenation of the mass water produced the post-mortem effect pattern observed in the Muhi Quarry fishes.

Proportion of fish specimens with an opened and closed jaw is almost equal (Figure 4C, Figure 5H-G). Openness of the jaws indicates sudden death of the fish caused by asphyxiation, poisoning, heat shock or choking up of the gills produced by suspended particles in the water (Ciobanu, 1977, as cited in Bienkowska, 2004; Elder & Smith, 1988; Wilson, 1988b). This suggests that the cause of death of the Muhi Quarry fishes was diverse; it is probable that differential causes of death were related to environmental fluctuations of salinity and oxygenation at the site of deposit.

6.2 Crinoids

Several strata of the Muhi Quarry are covered by comatulids (Figure 6A); its anatomical completeness varies from complete to partially complete individuals, and the majority of the specimens are articulated (Figure 7, Figure 6B-D). Complete disarticulation of recent comatulids occurs relatively fast after a few days, unless they are rapidly buried or removed from the normal cycle of post-mortem disintegration (Blyth Cain, 1968; Liddell, 1975; Meyer, 1971). It is remarkable that majority of the comatulids have their arms coiled, indicating dehydration caused by hypersaline waters (Seilacher et al., 1985). Hence, we propose a sudden death promoted by hypersaline waters, accompanied by a rapid burial in the Muhi Quarry comatulids; but it is also probable that a low-level of oxygenation additionally promoted their death.

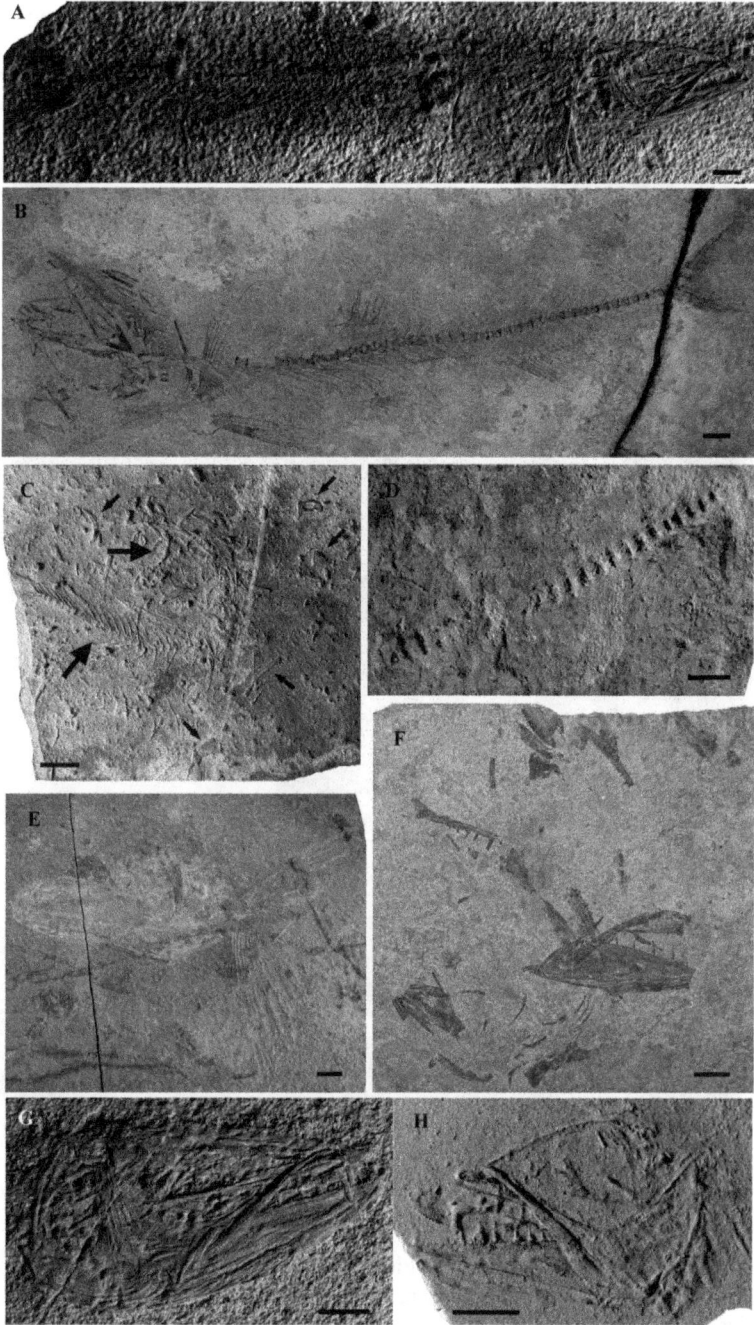

Fig. 5. States of preservation in enchodontids, most abundant fish group from the Muhi Quarry (except D). (A-B) Well-preserved specimens. A) UAHMP-1059, complete and

articulated individual with vertebral series straight. B) UAHMP-608, complete and slightly disarticulated individual with the head rotated upwards. (C-F) Moderate to poorly preserved specimens. C) UAHMP-3166, slab showing a partially complete enchodontid specimen (lower large arrow) with the head bones disarticulated (small arrows), which can be attributed to scavenging activity; a juvenile indeterminate specimen is also preserved in the same slab (upper large arrow). D) UAHMP-2968, isolated vertebral series of unidentified specimen. E) UAHMP-2202, skull slightly disarticulated. F) UAHMP-2203, skull bones highly disarticulated. (G-H) Openness/Closure of jaws. G) UAHMP-2170, specimen with jaws closed. H) UAHMP-784, specimen with jaws opened. Scale bars equal 1 cm.

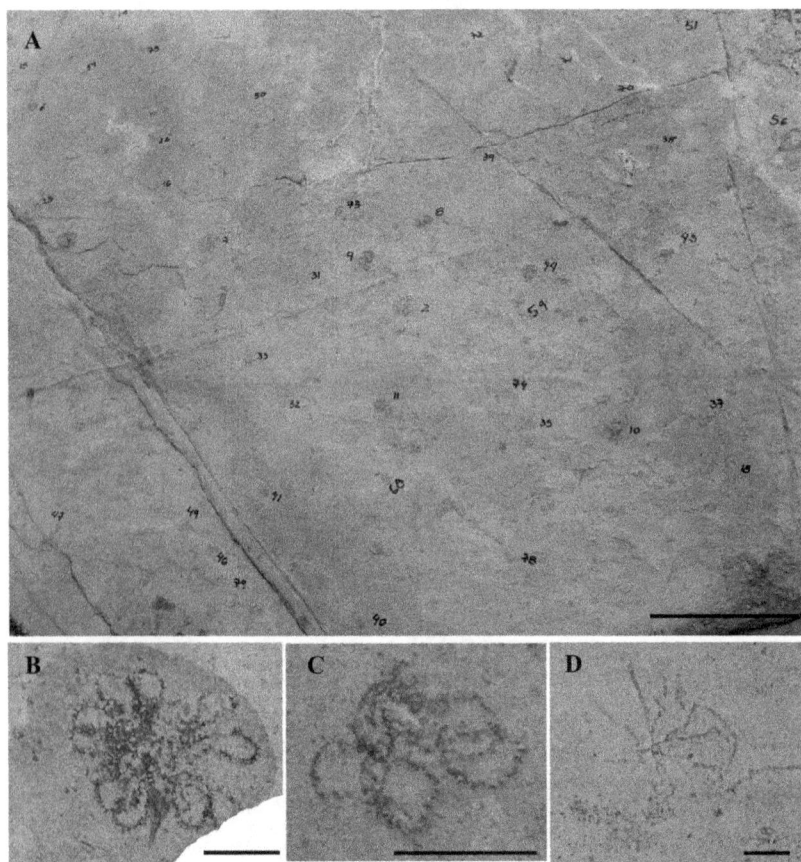

Fig. 6. States of preservation in the comatulids of the Muhi Quarry. A) Complete stratum covered by comatulids in a fine state of preservation (Scale bar equals 5 cm). (B-D) Well-preserved specimens. B) UAHMP-2892, complete and articulated individual. C) UAHMP-3099, partially complete and articulated individual. D) UAHMP-3685, incomplete and slightly disarticulated individual. Scale bars in B-D equal 1 cm

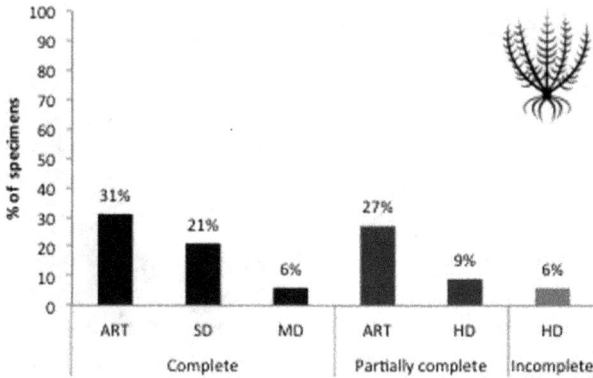

Fig. 7. Frequency distribution of taphonomic features evaluated in crinoids, including anatomical completeness and disarticulation. Abbreviations are as in Figure 4.

6.3 Other invertebrates

Crustaceans belonging to Aegeridae are well preserved (Figure 8A-B); majority of the specimens are complete or partially complete individuals, a small proportion are disarticulated and/or broken (Figure 9). The presence of non-fragmented fragile skeletal elements (such as the lightly sclerotized exoskeleton of crustaceans) is indicative of a low-energy environment (Brett & Baird, 1986). It has been shown that scavengers, bacteriological decay, and infaunal bioturbation rapidly destroyed fresh carcasses of decapods; therefore, they occasionally persist more than one to two weeks, whereas in a setting without any physical disturbance, disintegration occurs after one week (Plotnick, 1986). Hence, the presence of well-preserved crustaceans in the Muhi Quarry suggests a rapid burial, and non-physical damage of their carcasses. Only one specimen belonging to *Palinurus* is poorly preserved; a similar condition also is observed in ammonites referable to ?*Mortoniceras* sp. (Figure 8C-D).

On the other hand, echinoids are represented by numerous isolated spines (Figure 8E-F). This is a common condition in fossil echinoids preservation, due to spines are separated from the body almost immediately after their death; implying post-mortem transport. Echinoid spines are heavy enough for avoiding individual or collective transport over long distances; thus, they are usually found near or at the original habitat of the individual (Schäfer, 1972).

6.4 General taphonomic consideration

The preliminary taphonomic evaluation performed in the macrofossils of the Muhi Quarry, suggests that fluctuations of oxygenation-level and salinity occurred at the site of deposit, related to temporal lowering of dissolved oxygen and to a salinity increment in the mass water. Based on the depositional environment proposed by Bravo-Cuevas et al. (2009), it is probably that these stagnant conditions were controlled by alternated influx of neritic waters and open oceanic waters, further occasionally influx of coastal waters. Additional detailed taphonomic and geochemical studies will lead to precise and corroborate this contention.

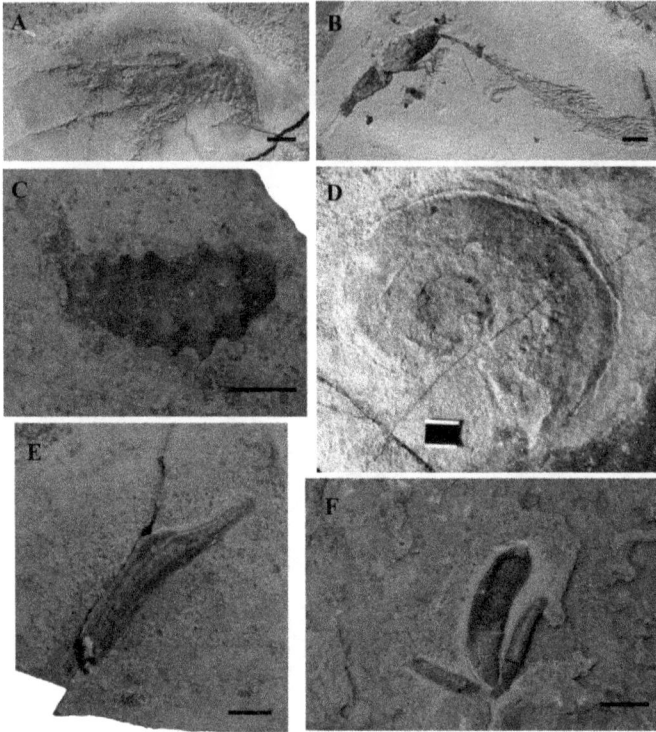

Fig. 8. States of preservation in other groups of invertebrates from the Muhi Quarry. (A-B) Aegeridae well preserved specimens. A) UAHMP-711, complete and articulated individual. B) UAHMP-824, complete and slightly disarticulated individual. (C-D) Ammonoid specimens poorly preserved. C) UAHMP-1465, isolated fragment of shell. D) UAHMP- 2032, mold of an ammonite referable to ?*Mortoniceras* sp. (E) UAHMP-702, isolated and partially articulated and (F) UAHMP-741, partially articulated spines of echinoids. Scale bars equal 1 cm.

7. Characterization and classification of the fossil assemblage

The fossil assemblage from the Muhi Quarry is mainly composed of nektonic animals, such as bony fish, sharks, crustaceans belonging to *Aeger* (Clarkson, 1998), and ammonites referable to ?*Mortoniceras* sp. [this genus probably had a deep-nektonic mode of life (Reboulet et al., 2005)]; floating organisms like the comatulids are also present. It is reasonable to expect that several members of these groups may potentially live at or near the site of deposit; the occurrence of fishes in different ontogenetic stages (for example, enchodontids and berycoids) (Fielitz & González-Rodríguez, 2010; González-Rodríguez, & Fielitz, 2009) is consistent with this argument. Specimens from the Muhi Quarry show a variable degree of disarticulation, which is related to a differential movement of the specimen from its original source; nevertheless, it does not imply that they are out of its life habitat. Both considerations are related to a parautochthonous association (Kidwell et al., 1986; Martin, 1999). Additionally, it has been observed that carbonate sediments normally consist of the autochthonous or parautochthonous remains of organisms living at or near the site of the deposit (Maiklem, 1968).

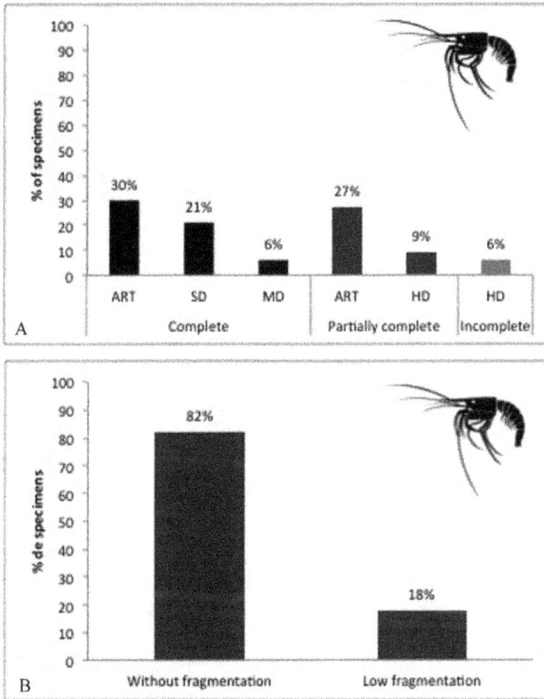

Fig. 9. Frequency distribution of taphonomic features evaluated in crustaceans. A) Taphonomic grades of anatomical completeness and disarticulation. B) Taphonomic grades of fragmentation. Abbreviations are as in Figure 4.

The record includes fossil material belonging to lepisosteids and to the spiny lobster *Palinurus*. Extant lepisosteid species are primarily freshwater fishes that occasionally swim into coastal marine or brackish environments, whereas fossil species are known from freshwater deposits or are thought to have been temporary immigrants from nearby freshwater tributaries (Grande, 2010). On the other hand, *Palinurus* is an epibenthic form typical of rocky, muddy and/or sandy bottoms (Alvarez et al., 1996). Considering the site of deposit that represents the Muhi Quarry, both groups occur in a foreign life habitat; furthermore, specimens referable to these taxa exhibit a low degree of preservation. Thus, they would be considered as allochthonous forms. Given this, the association of the Muhi Quarry is representative of a mixed assemblage, integrated mainly by parautochthonous forms and in a lesser degree by autochthonous forms, which is referred as a parautochthonous-allochthonous association (following Kidwell et al., 1986).

The fossil assemblage from the Muhi Quarry shows an effect of time-averaging. The presence of skeletal elements at the site of deposit with a differential durability (teeth, bones, and chitinous exoskeletons), which persist it at or near the sediment surface for variable periods of time, evidences an accumulation of specimens that not necessarily were alive at the same time (Flessa, 2008). Taphonomic evidence may also indicate time-averaging; particularly, specimens belonging to enchodontids, the most abundant group of fishes in the quarry, show a mixture of states of preservation, indicating some degree of time-averaging. By the same token, variation in preservation among taxa is also indicative of time averaging; the co-occurrence of fish skeletons in a variable state of preservation, with delicate articulated chitinous exoskeletons belonging to crustaceans supports this argument. This variability in preservation within and among taxa leads to classified the fossil assemblage as a *within-habitat assemblage*, suggesting a time-averaging over time intervals ranging from years to thousand years (Flessa, 2008); specifically in marine fossil assemblages it would be expected at least centuries of extent of time-averaging (Kidwell & Bosence, 1991; Martin, 1999).

8. The Muhi Quarry: A lagerstätte

There exist various *Fossil-Lagerstätten* in México, characterized by diverse fauna of invertebrates and vertebrates, coming from different ages and several paleoenvironments: The Tlayúa quarry of Albian age is a locality frequently compared with Solnhofen because of the great diversity of fauna and the excellent state of preservation, including details of muscles, gills, digestive tracts, and stomach contents with fishes making up 70%–80% of the macrofossils deposited in a restricted basin with marine and freshwater influence (Alvarado-Ortega et al., 2007; Applegate et al., 2006); the El Chango and El Espinal quarries of Albian-Cenomanian age (Alvarado-Ortega et al., 2009) with plant remains, mollusks, crustaceans, insects and different fish taxa. Vega et al. (2006) suggested that these deposits were accumulated within a shallow lagoon or estuary with occasional freshwater influence; the Vallecillo quarry of early Turonian age (Blanco-Piñón et al., 2002) where the fauna consisting of invertebrates, sharks, neopterygians, and latimerioids was deposited on an open shelf, which is supported by the pelagic assemblage and the absence of submarine barriers in the region around Vallecillo, Nuevo León (Ifrim et al., 2005, 2010); the El Rosario quarries of late Turonian-early Coniacian age containing plant remains, invertebrates and vertebrates with anatomical details of soft tissues preserved and some specimens conserved in 3D, that were deposited in an open marine shelf environment, at least 100 km away from coast line (Stinnesbeck et al., 2005); and the Xilitla quarries of Turonian age containing invertebrates and fishes deposited in an open shelf environment with low energy influx (Blanco-Piñón et al., 2006). Although there are some similarities among supraspecific fish taxa within Muhi Quarry and other Late Cretaceous localities of México, at species level the Muhi fish are different, most of them representing new species. Moreover the paleoenvironment condition where the biota was deposited is dissimilar in most cases; although the proposed paleoenvironment for the Muhi quarry seems to be comparable to those of El Rosario and Xilitla quarries, the lithology, and paleobiota are quite different. We have not found inoceramids that are common in both localities and either plant remains.

The Muhi Quarry fauna has been previously compared with some Cenomanian fossil *lagerstätten* of the eastern Tethys such as Namoura, Hakel and Haula in Lebanon, Jebel Tselfat, Kem Kem beds and Daura in Morocco, Ein-Yabrud in Jerusalem, English Chalk in England, and Comen in Slovenia (Bravo-Cuevas et al., 2009; González-Rodríguez & Bravo-Cuevas, 2005), but there is a closer similarity with the lower Cenomanian fish fauna of Hakel and Haula in Lebanon which paleobiota also includes crustaceans and land plants. In these localities the fish beds consist of thin-bedded, siliceous limestone alternating irregularly with more massive limestone, nodules and lenses of impure chert occur occasionally throughout the beds (Patterson, 1967). The fish beds were deposited in small basins, 250 m across, which Hückel (1970) interpreted as sinkholes formed by tectonic activity on the contemporary seafloor, at the outer margin of the continental shelf (Forey et al., 2003). The main difference among the Muhi Quarry and Hakel and Hajula quarries is the presence of land plants and terrestrial fauna in the Lebanese localities. This condition does not occur in the Mexican outcrop, suggesting that the Muhi fauna was deposited offshore.

9. Conclusion

The fossil concentration of Muhi Quarry represents a mixed assemblage integrated by a parautochthonous-allochthonous association; furthermore, it is evidenced that the fossil assemblage may be considered as a *within-habitat assemblage*. A preliminary taphonomic approach suggests that at site of deposit, occurred conditions of hypersalinity and/or low-level oxygenation, favoring an important accumulation of organic remains and its eventual differential preservation; further detailed taphonomic and geochemical studies will lead to precise and corroborate this argument. Nevertheless, stagnant conditions are commonly associated to the formation of a *Fossil-lagerstätte*. The Muhi Quarry is comparable in preservational and environmental conditions to other mid-Cretaceous Mexican quarries considered as *lagerstätten*, such as El Rosario and Xilitla, northern and central México, as well as to the Hakel and Hajula quarries of eastern Tethys, although the fish fauna seem to be endemic for the Atlantic Ocean and the Paleo- Caribbean province.

10. Acknowledgment

We thank to Sergio and Ignacio Yánez, workers of the Muhi quarry for collecting fossil material since 1999. Also we want thank the Biologists Jorge Alberto González and Jaime Priego Vargas for editing the figures of these report.

11. References

Alvarado-Ortega, J., Ovalles-Damián, E. & Blanco-Piñón, A. (2009). The Fossil Fishes from the Sierra Madre Formation, Ocozocoautla, Chiapas, Southern Mexico. *Palaeontologia Electronica*, Vol. 12, No.2, (August 2009), pp.1-22, ISSN 1094-8074

Alvarado-Ortega, J., Espinosa-Arrubarrena, L., Blanco, A., Vega, F. J., Benammi, M. & Briggs, D. E. G. (2007). Exceptional preservation of soft tissues in cretaceous fishes from the Tlayúa Quarry, Central Mexico. *Palaios*, Vol.22, No.6, (December 2007), pp. 682–685, ISSN 0883-1351

Álvarez, J., Villalobos, L. & Lira, E. (1996). Decápodos, In: *Biodiversidad, Taxonomía y Biogeografía de artrópodos mexicanos: hacia una síntesis de su conocimiento*, Llorente-Bousquets J., García-Aldrete, A. N. & González-Soriano, E., pp. 106-116, CONABIO/UNAM, ISBN 968-36-4857-6, México

Applegate, S. P., Espinosa-Arrubarrena, L., Alvarado-Ortega, J. & Benammi, M. (2006). Revision on Recent Investigations in the Tlayúa quarry. In: *Studies on Mexican Paleontology*, Vega, F. J., Nyborg, T.G., Perrilliat, M.C., Montellano-Ballesteros, M., Cevallos-Ferriz, S.R.S. & Quiroz-Barroso, S.A., pp. 276-304, Series Topics on Geobiology, Vol. 24, Kluwer Academic Publishers B. V., ISBN 1-4020-3882-8, Netherlands

Benton, M. J. & Harper, D. A. T. (2009). *Introduction to Paleobiology and the fossil record* (First edition), Wiley-Blackwell, ISBN 978-1-4051-4157-4, Singapore

Bienkowska, M. (2004). Taphonomy of ichthyofauna from an Oligocene sequence (Tylawa Limestones horizon) of the Outer Carpathians, Poland Geological. *Quarterly Geology*, Vol.48, No.2, pp. 181–192, ISSN 16417291

Blanco, A., Duque-Botero, F. & Alvarado-Ortega, J. (2006). Lower Turonian Fossil Lagerstätten in Mexico: their relationship to OAE-2. *Geological Society of America, Abstracts with Programs*, Vol.38, No.7, p. 148, ISSN 0016-7592

Blanco-Piñon, A., Frey, E., Stinnesbeck, W. & López-Oliva, J. G. 2002. Late Cretaceous (Turonian) fish assemblage from Vallecillo, northeastern Mexico. *Neues Jahrbuch für Geologie und paläontologie, Abhandlungen,* Vol. 225, No.1, pp. 39-54, ISSN 0077-7749

Blyth Cain, J. D. (1968). Aspects of the depositional environment and paleoecology of crinoidal limestones. *Scottish Journal of Geology*, Vol.4, No. 3, (September, 1968), pp. 191–208, ISSN 0036-9276

Bravo-Cuevas, V. M., González-Rodríguez, K. A. & Esquivel-Macías, C. (2009). Advances on stratigraphy and paleontology of the Muhi Quarry from the Mid-Cretaceous (Albian-Cenomanian) of Hidalgo, Central Mexico. *Boletín de la Sociedad Geológica Mexicana*, Vol.69, No.2, (September 2009), pp. 155-165, ISSN 1405-3322

Brett, C. E. & Baird, G. C. (1986). Comparative taphonomy: a key to paleoenvironmental interpretation based on fossil preservation. *Palaios*, Vol.1, No.3, (June 1986), pp. 207-227, ISSN 0883-1351

Clarkson, E. N. K. (1998). Invertebrate paleontology and evolution (Fourth edition), Blackwell Publishing, ISBN 978-0-632-05238-7, Singapore

Elder, R. L. (1985). Principles of aquatic Taphonomy with examples from the fossil record. Thesis, University of Michigan, Ann Arbor, Michigan, pp. 1-336

Elder, R. L & Smith, G. R. (1988). Fish Taphonomy and environmental inference in paleolimnology. *Palaeogeography, Palaeoclimatology, Palaeoecology*, Vol.62, No-1-4, (January, 1988), pp. 577-592, ISSN 0031-0182

Feldman, R., Vega, F. J., Martínez-López, L., González-Rodríguez, K., González-León, O. & Fernández-Barajas, M.R. (2007). Crustacea from the Muhi Quarry (Albian-Cenomanian), and a review of Aptian Mecochiridae (Astacidea) from Mexico. *Annals of Carnegie Museum*, Vol.76, No.3, (December 2007), pp. 135-144, ISSN 0097-4463

Fielitz, C. & González-Rodríguez, K. A. (2010) A new species of *Enchodus* (Aulopiformes: Enchodontidae) from the Cretaceous (Albian to Cenomanian) of Zimapán, Hidalgo, Mexico. *Journal of Vertebrate Paleontology*, Vol.30, No.5, (September, 2010), pp. 1343-1351, ISSN 0272-4634

Fielitz, C. & González-Rodríguez, K. A. (2008). A new species of *Ichthyotringa* from the El Doctor Formation (Cretaceous), Hidalgo, México, In: *Mesozoic Fishes 4–Homology and Phylogeny*, Arratia, G., Schultze, H. P. & Wilson, M. V. H., pp. 373-388, Verlag Dr. Pfeil, ISBN 978-3-89937-080-5, Munchen Germany

Flessa, K. W.(2008). Time-averaging, In: *Palaeobiology II*, Briggs, D. E. G. & Crowther, P. R., pp. 292-296, Blackwell Publishing, ISBN 0-632-05147-7, United Kingdom

Forey, L. P., Yi, L., Patterson, C. & Davies, C. E. (2003). Fossil fishes from the Cenomanian (Upper Cretaceous) of Namoura, Lebanon. *Journal of Systematic Palaeontology*, Vol. 1, No. 4, (December 2003), pp. 227-330, ISSN 1477-2019

González-Rodríguez, K., Schultze, H. P. & Arratia, G. (2011). Unexpected appearance of advanced neoteleosts in the Cretaceous and the controversy between fossil record and molecular clock. *IV Congreso Latinoamericano de Paleontología de Vertebrados*, San Juan, Argentina (September, 2011)

González-Rodríguez, K. & Schultze, H. P. (2010). A fossil agonid (Actinopterygii, Teleostei, Percomorphacea) from the Albian-Cenomanian of México. In *Fifth International*

Meeting on Mesozoic Fishes – Global Diversity and Evolution, Abstract book and field guides. González-Rodríguez & G. Arratia, pp. 50, Ciencia al Día, Vol. 19, Universidad Autónoma del Estado de Hidalgo, (August 2010), ISBN 9786074821192

González-Rodríguez, K. & Fielitz, C. (2009). Los peces fósiles, In: *Los fósiles del estado de Hidalgo*, González-Rodríguez, K. Cuevas-Cardona, C. & Castillo-Cerón, J., pp. 65-78, Universidad Autónoma del Estado de Hidalgo, ISBN 978-607-482-047-8

González-Rodríguez, K. & Bravo-Cuevas, V. M. (2005). Potencial fosilífero de la Cantera Muhi (Formación El Doctor: Albiano-Cenomaniano) de la región de Zimapán, Estado de Hidalgo. *PALEOS Antiguo*, Vol.1, No.1, (October 2005), pp. 27-42, ISSN 1870-7009

Grande, L. (2010). An empirical synthetic pattern study of gars (Lepisosteiformes) and closely related species, based mostly on skeletal anatomy. The resurrection of Holostei. *American Society of Ichthyologists and Herpetologists. Special Publication 6. Supplementary Issue of Copeia*, Vol. 10, No. 2A, (October 2010), pp. 1-871, ISSN 0045-8511

Hecht, F. (1933). Der Verbleib der organischen Substanz der Tierre bei meerischer Einbettung. *Senckenbergiana*, Vol.15, No.3-4, pp. 165-249, ISSN: 0037-2110

Hückel, U. 1970. Die Fischschiefer von Haquel und Hjoula in der Oberkreide des Libanon. *Neues Jahrbuch für Geologie und Paläontologie. Abhandlungen*, Vol. 135, pp. 113-149, ISSN 0077-7749

Ifrim, C., Stinnesbeck, W. & Frey, E. (2005). Upper Cretaceous (Cenomanian-Turonian and Turonian-Coniacian) open marine Plattenkalk-deposits in NE Mexico. Abstracts book. *4th International Symposium on Lithographic Limestone and Plattenkalk*, Eichstät, Germany, pp. 15-16

Ifrim, C., Giersch, S., González-González, A. H., Stinnesbeck, W., Frey, E. & López-Oliva, J. G. (2010). The Turonian platy limestone at Vallecillo, Nuevo León, México, and its fishes. In: *Fifth International Meeting on Mesozoic Fishes – Global Diversity and Evolution*. Abstract book & field guides, González-Rodríguez, K. & Arratia, G., pp. 119-134, Ciencia al Día, Vol. 19, Universidad Autónoma del Estado de Hidalgo, (August 2010), ISBN 9786074821192

Jerzmańska, A. (1960). Ichtiofauna łupków jasielskich z Sobniowa. *Acta Palaeontologica Polonica*, Vol.5, No.4, pp. 367–412, ISSN 0567-7920

Kidwell, S. M. & Bosence, D. W. J. (1991). Taphonomy and time-averaging of marine shelly faunas, In: *Taphonomy: Releasing the data locked in the fossil record*, Allison, P. A. & Briggs, D. E. G., pp. 116-209, Plenum Press, ISBN 0-306-43876-3, New York, USA

Kidwell, S. M., Fürisch, F. T. & Aigner, T. (1986). Conceptual framework for the analysis and classification of fossil concentrations. *Palaios*, Vol.1, No.3, (June 1986), pp. 228-238, ISSN 0883-1351

Liddell, W.D. (1975). Recent crinoid biostratinomy. *Geological Society of America Abstracts with Programs*, Vol.4, p. 1169, ISSN 0016-7592

Maiklem, W. R. (1968). Some hydraulic properties of bioclastic carbonate grains. *Sedimentology*, Vol.10, No.2, (March, 1968), pp. 101-109, ISSN 1365-3091

Martin, R. E. (1999). Taphonomy: A process approach (First edition), Cambridge University Press, ISBN 0-521-59171-6, Cambridge, United Kingdom

McGrew, P. O. (1975). Taphonomy of Eocene fish from Fossil Basin, Wyoming. *Fieldiana Geology*, Vol.33, No.14, pp. 257-270, ISSN 0096-2651

Meyer, D. L. (1971). Post-mortem disarticulation of Recent crinoids and ophiuroids under natural conditions. *Geological Society of America Abstracts with programs*, Vol.3, p. 645, ISSN 0016-7592

Nudds, J. & Selden, P. (2008). Fossil–Lagerstätten. *Geology Today*, Vol.24, No.4, (July-August 2008), pp. 153-158, ISSN 1365-2451

Patterson, C. (1967). New Cretaceous berycoid fishes from the Lebanon: *Bulletin of the British Museum (Natural History) Geology*, Vol. 14, No.3, pp. 69-109, ISSN 0968-0462

Plotnick, R. E. (1986). Taphonomy of a modern shrimp: implications for the arthropod fossil record. *Palaios*, Vol.1, No.3, (June 1986), pp. 286-293, ISSN 0883-1351

Reboulet, S., Giraud F. & Proux, O. (2005). Ammonoid abundance variations related to changes in trophic conditions across the Oceanic Anoxic Event 1d (Latest Albian, SE France). *Palaios*, Vol.20, No.2, (April, 2005), pp. 121-141, ISSN 0883-1351

Roger J., (1946). Les invertébrés des couches à poissons du Crétacé supérieur du Liban. *Mémoires de la Société Géologique de France*, Paris, Vol. 23, pp. 1-92, ISSN 0037-9409

Schäfer, W. (1972). Ecology and palaeoecology of marine environments (First edition), Chicago University Press, ISBN 0-05-002127-3, Chicago, USA

Seilacher, A. (1970). Begriff und Bedeutung der Fossil-Lagerstätten. *Neues Jahrbuch für geologie und Paläontologie- Monatshefte*, 1970, No.1, pp. 34-39, ISSN 0028-3630

Seilacher, A., Reif, W. E., Westphal, F., Riding, R., Clarkson, E. N. K. & Whittington, H. B. (1985). Sedimentological, Ecological and Temporal Patterns of Fossil Lagerstätten. *Philosophical Transactions of the Royal Society of London, Series B, Biological Sciences*, Vol.311, No.1148, (October 1985), pp. 5-24, ISSN 0080-4622

Smith, G. R. & Elder, R. L. (1985). Environmental interpretation of burial and preservation of Clarkia fishes, In: *Late Cenozoic history of the Pacific Northwest*, Smiley, C. J., pp. 85-93, American Association for the Advancement of Science Pacific Division, ISBN 0-934394-06-7, San Francisco, California, USA

Speyer, S.E. & Brett, C.E. (1988). Taphofacies models for epeiric sea environments: Middle Paleozoic examples. *Palaeogeography, Palaeoclimatology, Palaeoecology*, Vol.63, No.1-3, (February, 1988), pp. 225-262, ISSN 0031-0182

Stinnesbeck, W., Ifrim, C., Schmidt, H., Rindfleisch, A., Buchy, M. C., Frey, E., González-González, A., Vega, F., Cavin, L., Keller, G. & Smith, K. T. (2005). A new lithographic limestone deposit in the Upper Cretaceous Austin Group at El Rosario, county of Múzquiz, Coahuila, northeastern Mexico. *Revista Mexicana de Ciencias Geológicas*, Vol. 22, No. 3, pp. 401-418, (December 2005), ISSN 1026-8774

Turek, V., Marek, J. & Benes, J. (1989). *Fossils of the World* (edited by Brown, J.) (First edition), Arch Cape Press, ISBN 0-517-67904-3, New York, USA

Vega, F.J., García-Barrera, P., Perrilliat, M. C., Coutiño, M.A. & Mariño-Pérez, R. (2006). El Espinal, a new plattenkalk facies locality from the Lower Cretaceous Sierra Madre Formation, Chiapas, southeastern Mexico. *Revista Mexicana de Ciencias Geológicas*, Vol. 23, No. 3, pp. 323-333. (December 2006), ISSN 1026-8774

Wilson, M.V.H. (1988a). Paleoscene #9: Taphonomic processes: information loss and information gain. *Geoscience Canada*, Vol.15, No.2, pp. 131-148, ISSN 0315-0941

Wilson, M.V.H. (1988b). Reconstruction of ancient lake environments using both autochthonous and allochthonous fossils. *Palaeogeography, Palaeoclimatology, Palaeoecology*, Vol.62, No.1-4, (January, 1988), pp. 609-623, ISSN 0031-0182,

Wilson, M.V.H. (1980). Eocene lake environments: depth and distance-from-shore variation in fish, insect, and plant assemblages. *Palaeogeography, Palaeoclimatology, Palaeoecology*, Vol. 32, pp. 21-44, ISSN 0031-0182

6

The Paleogene Dinoflagellate Cyst and Nannoplankton Biostratigraphy of the Caspian Depression

Olga Vasilyeva[1] and Vladimir Musatov[2]
[1]Institute of Geology and Geochemistry Ural Branch RAS,
[2]Lower Volga Institute of Geology and Geophysics,
Russia

1. Introduction

The Pricaspian Depression in the southeast of the East European Platform is one of the deepest depressions formed over the Baikal folded basement. The sedimentary cover consists of four structural levels corresponding to the major stages of its formation. The Riphean and the Vendian beds constitute the first structural level. The second one comprises deposits from the Devonian to the Upper Triassic. This includes the Lower Permian plastic salt-bearing beds. The post-salt deposits of the second structural level are up to 4-5 km thick. The third structural level of the sedimentary cover is composed of the Triassic-Oligocene beds, with the total thickness as high as 2.5 km; in some sections, the Paleogene sediments are over 1000 m thick. The principal accumulations of oil, gas and potassium salts are confined to the second and the third structural levels of the Pricaspian Depression.

For lack of any core drilling, no biostratigraphic exploration of the sedimentary cover, the Paleogene beds included, has been made in the region since the middle of the past century. The biostratigraphic method has been substantially improved during that period: zonal scales from nannoplankton (Aubri, 1986; Martini, 1974; Okada & Bukri, 1980; Perch-Nielsen, 1985; Varol, 1998) and foraminifers (Berggren et al., 1995) have been refined, a standard zonal scale has been developed from the dinocysts (Luterbacher et al., 2004; Powell, 1992), that have never been studied in that region before.

The latest micropaleontologic examinations of the Paleogene beds from the Northern Pricaspian have shown the sediments to comprise organic-walled microphytoplankton (dinoflagellate cysts) and calcareous nannoplankton that potentiate detailed division and dating of Paleogene sequences (Vasilyeva & Musatov, 2010a). Examination of dinoflagellate cysts and nannoplankton from the Central Pricaspian is promising in terms of zonal scales direct comparisons from those groups, correlations with the adjacent areas and West Siberia; the latter region is known for the Paleogene beds to be represented exclusively by siliceous terrigenous bodies. Furthermore, dinoflagellate cyst occurrences and analyses of palynologic remains distribution over the water area of the Pricaspian paleobasin are interesting in terms of studying fluctuations of the sea level, reconstructing the paleoecologic environment, evolution of the extensive marine basin in the Cenozoic, formation of local residual reservoirs upon lowering of the world ocean level. The present paper aims at (1) biostratigraphic division of the Paleogene section in the Central Pricaspian Region (the Elton key well) and dating the regional lithostratons; (2) correlating of the beds from the Central and the Northern Pricaspian Regions; (3) summarizing and making direct comparisons of the zonal scales from dinocysts and nannoplankton in the Pricaspian Region; (4) diagnosing fluctuations of the sea level by means of analyzing paleoecologic characteristics of the phytoplankton associations.

2. Geological setting

During the Cenozoic, the Pricaspian Depression belonged to the northern margin of the Peri-Tethys region and used to be a major epicontinental basin connected with both, the Paleo-Atlantic and the seas of northwestern Europe and with Paleo-Arctic through the Turgaj strait and the West Siberian sea basin. An extensive region of the Pricaspian Depression used to plunge in the Cenozoic, with the northern Uzeni-Utvinskaya structural zone remaining more stable (Zhuravlev, 1970). The structure of the Pricaspian Depression is peculiar for vast occurrences of geomorphologic features of the Permian salt-dome tectonics. Salt domes, up to ten kilometers in height and across, lie fairly close over the depression and create substantially dissected relief of the surface of sedimentation (Zhuravlev, 1970). Development of major positive geomorphologic structures was sure to affect the processes of sedimentogenesis and the hydrologic regime of the basin in the Cenozoic epoch. The most complete sections of the Paleogene have been penetrated in the inter-dome areas of the Volga and Ural central interfluve (Grachev et al., 1971; Pechenkina & Kholodilina, 1971). During the Cenozoic, the basin used to remain deep there. For that matter, the lithologic sequences there are described as being monotonous, with dominating fine clayey gaize-like rocks; carbonate interlayers occur in the lowermost and in the upper part of the section.

3. Pilot stratigraphic study

Complex investigations of the numerous wells drilled in the 50-70-ties of the last century failed to provide any uniform understanding of the region's suite stratigraphy. To make stratigraphic divisions of the Paleogene sections, either the North Caucasian scheme, based on high similarity of the faunal complexes (Pechenkina & Kholodilina, 1971; Razmyslova & Nikitina, 1975), or isochronous foraminifer strata with geographic names were used (Grachev et al., 1971). The insufficiently elaborate foraminifer scale of that period did not allow to date the Paleogene sequences unambiguously. Nevertheless, principal stratigraphic levels of the subseries and the corresponding faunal complexes have been outlined.

To assess the complexity of a section lithologic division, we made use of V.G. Grachev's lithostratigraphic chart (Grachev et al., 1971). A series of lithostrata may be traced in the Paleogene from the central part of the Pricaspian Depression. The Danian is represented by a clayey-marl sequence of the Algai formation, with a complex of foraminifers *Acarinina inconstans, Globoconusa* sp. ("the Furmanovskian" layers according to V.G. Grachev) and by the Cygan formation (gray calcareous clays with marl interlayers, saturated with glauconite) with foraminifers *Globorotalia angulata*. The Upper Paleocene is composed of dark gray, low calcareous gaize-like clays and gaizes with mollusk fauna. This sequence corresponds to the lower member of V.G. Grachev's Kaztalov layers, with foraminifers *Spiroplectammina spectabilis* (Grachev et al., 1971). The subseries upper part corresponds to the clay sequence - argillite-like, gaize-like, calcareous, with marl and sandstone interlayers; this is distinguished as the upper member of the Kaztalov layers. It comprises siliceous microbiota and foraminifers *Acarinina subsphaerica* (Grachev et al., 1971). In the Lower Eocene, the Bostandyk formation is recognized (clays - aleuritic, low calcareous, with glauconite) with complexes of foraminifers *Globorotalia subbotinae* and *G. marginodentata*. The Eocene middle subseries is represented by the Kopterek formation, with the foraminifer fauna of *Truncorotalia aragonensis* and *Acarinina crassaformis* (Grachev et al., 1971). The nannoplankton pilot study in the Volga-Pricaspian Region has been made by V.A. Musatov (Musatov, 1996).

4. Material and methods

4.1 Lithologic structure of the Paleogene section from the Elton key well

The Elton key well was drilled in 1960 in the administrative territory of Kazakhstan, between the lakes of Elton and Aralsor, 50 km east of Elton (the Aralsor field) (Fig. 1). The section has penetrated the Paleogene beds from the Danian to the Ypresian (thickness of 165 m) within the dome arch, that's why the lithostrata are relatively thin, probably partially eroded. The Paleogene deposits lie discordantly over the light gray Maastrichtian pelitomorphic marls (839–785 m). From the bottom upwards, the Paleogene section is represented by the following lithostratons.

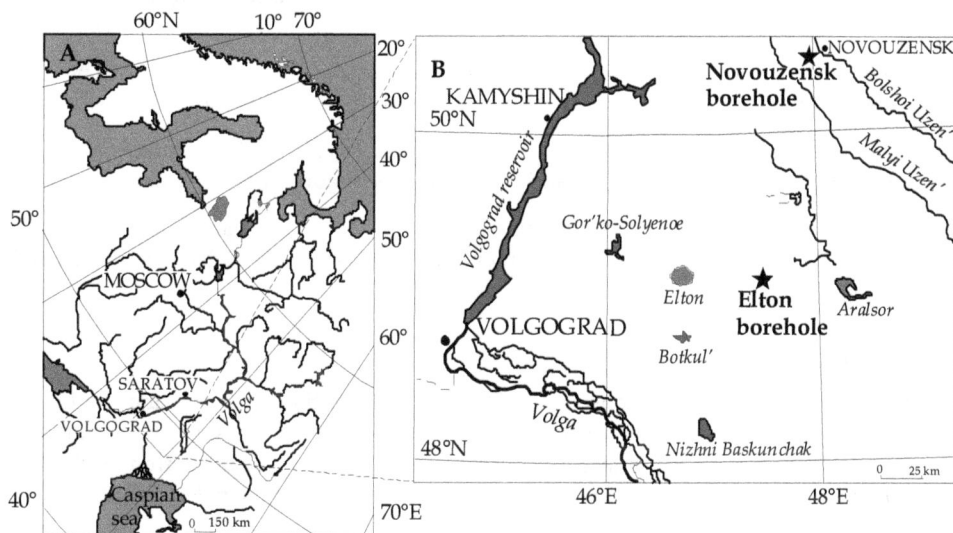

Fig. 1. Geographic location of the region studied (A), the Novouzenk key well and the Elton well (B).

4.1.1 The Danian. The Cygan formation

785–763 m – clays: gray, calcareous, occasionally sandy, laminated, with glauconite. A marl interlayer (0.1 m) lies in the lower part of the interval. In the upper portion, the clays are bioturbated, with macrofauna remains and fish scales.

763–749 m – clays: light gray, greenish, calcareous, sandy; in the upper part – bioturbated, with fish scales and macrofauna remains. The content of aleurite material increases in the interval lower part. Marl, about 1 m thick, occurs in the base of the interval.

4.1.2 Selandian–Thanetian. The Kaztalov formation

749–711 m – clays: dark gray, almost black, dense, gaize-like, silty, micaceous, occasionally low calcareous, intensely bioturbated, in the upper part – with macrofauna, fish scales and bones. The rock carbonate and sand contents increase in the suite lower interval.

711–695 m – clays: gray, dense, non-calcareous, slightly gaize-like.

695–690 m – alternating interlayers of glauconite sandstones (0.3–1.5 m thick), clays – dark gray, gaize-like, sandy, low calcareous (0.3–1.3 m thick) and marls – sandy, finely laminated (0.5 m thick). Clay interlayers contain vegetable remains and fish scales.

4.1.3 The Ypresian. The Bostandyk formation

690–680 m – clays: dark gray, black, gaize-like, sandy, occasionally calcareous, micaceous, contain vegetable detritus and fish scales. In the basement, the clays are laminated, with lenses and sprinkles of aleurite material.

680–669 m – alternating clays – dark gray, black, dense, shaly and sandy, with macrofauna fragments and fish scales. The rocks are low calcareous.

669–660 m – clays: dark gray, black, sandy, occasionally calcareous, with lenses and nests of sandy-aleurite material. Microfauna, macrofauna detritus and fish scale occur.

660–630 m – clays: black, dense, micaceous, calcareous, occasionally finely laminated, in the basement – with patches and nests of sandy material, macrofauna and fish scales.

630–620 m – clays: dark gray, black, gaize-like, homogenous, in the upper part of the interval – sandy. In the uppermost of the interval, the clays contain vegetable detritus and microfauna.

4.2 Sampling

The material for the operation consisted of 70 samples from the Paleogene and the Upper Cretaceous sections, selected for nannoplankton and dinocyst analyses. By the moment of sampling, the core was 70-50% intact. The average of 4-8 samples was taken from each 10-m interval, thus, the sections micropaleontologic description was fairly detailed (Fig. 2).

4.3 Palynologic maceration

The sample palynologic processing was carried out in the laboratory at the West Siberian Institute for Oil and Gas Problems (Tyumen) in accordance with the standard procedures accepted at the Russian Academy of Sciences (Petrova, 1986). (1) A 150g rock sample was treated with 10% hydrochloric acid to dissolve calcium carbonates, with subsequent washing of the residue to neutral reaction. (2) The residue was treated with tetrasodium pyrophosphate ($Na_4P_2O_7 \times 10H_2O$) for dispersion of clay and washed several times for elimination of argillaceous material. (3) The residue was centrifuged with heavy liquid (K_2CdI_4) (specific gravity of 2.25) to isolate palynologic remains. (4) The macerate was treated with 10% hydrofluoric acid, with subsequent flushing to remove the fluorosilicate compounds. The glycerine macerate was examined under the Karl Zeiss Jena microscope, with incident light and magnification of x400, x600.

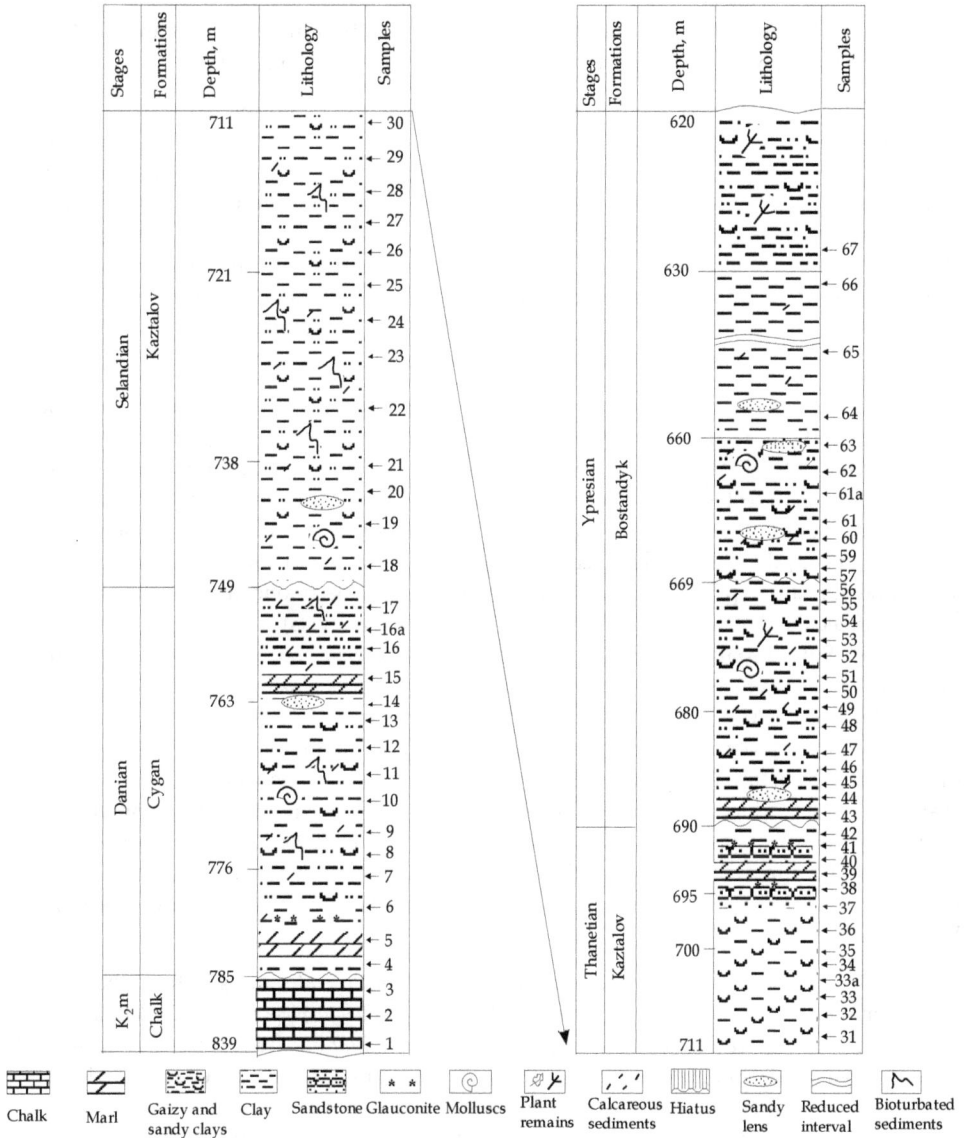

Fig. 2. The Elton well lithologic structure and sampling for dinocysts and nannoplankton

4.4 Palynologic diagnostics and investigation techniques

Species diagnostics of dinoflagellate cysts was performed in accord with the dinocyst catalogue summarized by Fensome & Williams (2004). The section biostratigraphic division according to dinocysts was accomplished in accord with the standard zonal scale accepted for the Paleogene of Northwestern Europe (Luterbacher et al., 2004). For the Maastrichtian, several alternative zonal scales have been developed from dinocysts (Kirsch, 1991; Marheinecke, 1992; Schiøler & Wilson, 1993; Slimani, 2001). The sequences of the zonal species in the Elton section allow to compare them to the chart by P. Schiøler and G. Wilson, established for the Maastrichtian from Denmark (Schiøler & Wilson, 1993).

The results thus obtained were analyzed as follows. (1) Counting all the palynologic fragments, i.e. microphytoplankton (a), pollen and spores (b) (up to 200–250 specimens if possible), to determine relative contents of palynologic remains of ecologically heterogeneous geneses and to diagnose the proximal-distal signals of associations in accord with the methods proposed by some workers (Brinkhuis, 1994; Powell et al., 1996; Stover & Hardenbol, 1994; Wall et al., 1977). (2) Composition analyses of the microphytoplankton complex: the number of dinocyst species (a), the number of species of dinocyst gonyaulacoid morphotypes (b), perodinioid ones (c), prasinophytes and acritarchs (d) were used to determine the basin transgressive-regressive impulses.

The possibility of interpreting the species composition and the diversity of phytoplankton complexes as the reflection of sea level fluctuations in the Pricaspian Region is provided by two principal provisions. (1) According to Haq et al. (1988), the most pronounced features of a rhythm are manifested on the inner and on the middle shelf. Those are associated with substantial depth fluctuations, with accumulation of differentiated sediments, shelf exposure during the stage of low sea level (SB1 boundary type), changing contours of the shore line. Drastic depth fluctuations and the changes of the photic layer water mass were sure to affect the organic-walled plankton (dinoflagellate cysts) composition in the inner shelf zone. Substantial lowering of the sea level is interpreted as a stress situation for plankton paleophytocoenoses; this is manifested in decreasing species diversity of associations (Dale, 1996; Pross, Schmiedl, 2002). (2) Current studies of the Eocene deposits have shown the compositions of organic-walled phytoplankton on the shelf to have differentiated structure (Brinkhuis, 1994). In open marine basins, gonyaulacoid dinoflagellate cysts show the tendency to genera distribution from the inner neritic zone to open oceanic settings (Brinkhuis, 1994; Crouch & Brinkhuis, 2005; Dale, 1996; Pross & Schmiedl, 2002), demonstrating proximal-distal signals. In case of transgressive movement of the water masses in an epicontinental basin, the share of taxa of the outer neritic zone grows in the littoral parts.

An extensive group of gonyaulacoid cysts represents neritic water masses (from the inner to the outer ones): *Spiniferites, Achomosphaera, Fibrocysta, Cleistosphaeridium, Cordosphaeridium*. Representatives of *Areoligera/Glaphyrocysta, Cordosphaeridium* are frequently associated with transgressive phases in the neritic zone (Crouch & Brinkhuis, 2005; Guasti et al., 2006; Habib et al., 1992; Iakovleva et al., 2001; Powell et al., 1996; Toricellii et al., 2006). Of the gonyaulacoid cyst category, *Impagidinium* and *Nematosphaeropsis/Cannosphaeropsis* genera are clearly distinguished; those occur in the shelf outer zone and oceanic settings and are characteristic of oligotrophic conditions (Brinkhuis, 1994; Dale, 1996; Sluijs et al., 2005). A group with opposite paleoecologic features is represented by peridinioid dinocysts. In the Paleogene sections from West Europe, this group of organic plankton is peculiar for the shore-nearest settings associated with lagoons and bays, i.e. zones with the environment of advanced eutrophic properties. Similar paleoecologic adaptations in peridiniods are generally related with metabolism types, aptitude to heterotrophic and mixotrophic nutrition (Powell et al., 1996). It should be added that peridinioid dinocysts constitute the core of organic-walled plankton associations in the Paleogene interior (epicontinental) seas of Eurasia (Andreyeva-Grigorovich et al., 2011), particularly in the Pricaspian basin (Vasilyeva & Musatov, 2010a, b). The transgressive-regressive phases there are peculiar for dynamics of dinocyst species diversity and changing taxonomic compositions of principal paleoecologic groups of organic-walled phytoplankton.

4.5 Nannoplankton

Calcareous nannoplakton has been revealed in all the intervals represented by marls and calcareous rocks. No coccolithophorids have been recognized in the middle part of the Kaztalov formation. Nannoplankton was studied from the preparation made of rock powder alcohol suspension, without any enrichment. Biostratigraphic division of most of the section was made according to Martini scale (Martini, 1971); in the upper section, Okada and Bukry zonal scale (Okada & Bukry, 1980) was used. Fractional zonal divisions, similar to those in the scale developed by Varol (1998), are not always acquirable due to obviously poor species compositions of the nannoplankton complexes or total absence of nannoplankton from the section intervals. All the biozones in the section have been recognized either from the first appearance or from vanishing of the index species, according to the standard definition of zonal divisions (Martini, 1971; Okada & Bukry, 1980). The species were determined under the AxioPlan 2 (Karl Zeiss) microscope.

5. The section biostratigraphic division

5.1 Dinocysts

Dinoflagellate cysts have been revealed practically from all over the section; those are represented by diversely composed complexes of various saturations. The most productive and rich associations have been revealed from the Maastrichtian, Danian and Ypresian section intervals. Stratigraphic occurrences of some dinocyst taxa, zone intervals and the most significant biostratigraphic events are shown in Fig. 3.

5.1.1 *Triblastula utinensis* zone

The dinocyst complex from the Maastrichtian marls (839–785 m interval) is described from FO *Triblastula utinensis, Isabelidinium cooksoniae* and the lack of *Alterbidinium acutulum*, which allows to distinguish the *Triblastula utinensis* zone and its upper subzone *Cannosphaeropsis utinensis* (Schiøler, Wilson, 1993) in that interval of the section. The zone is dated as the upper part of the Lower Maastrichtian – the lowermost of the Upper Maastrichtian (Schiøler & Wilson, 1993).

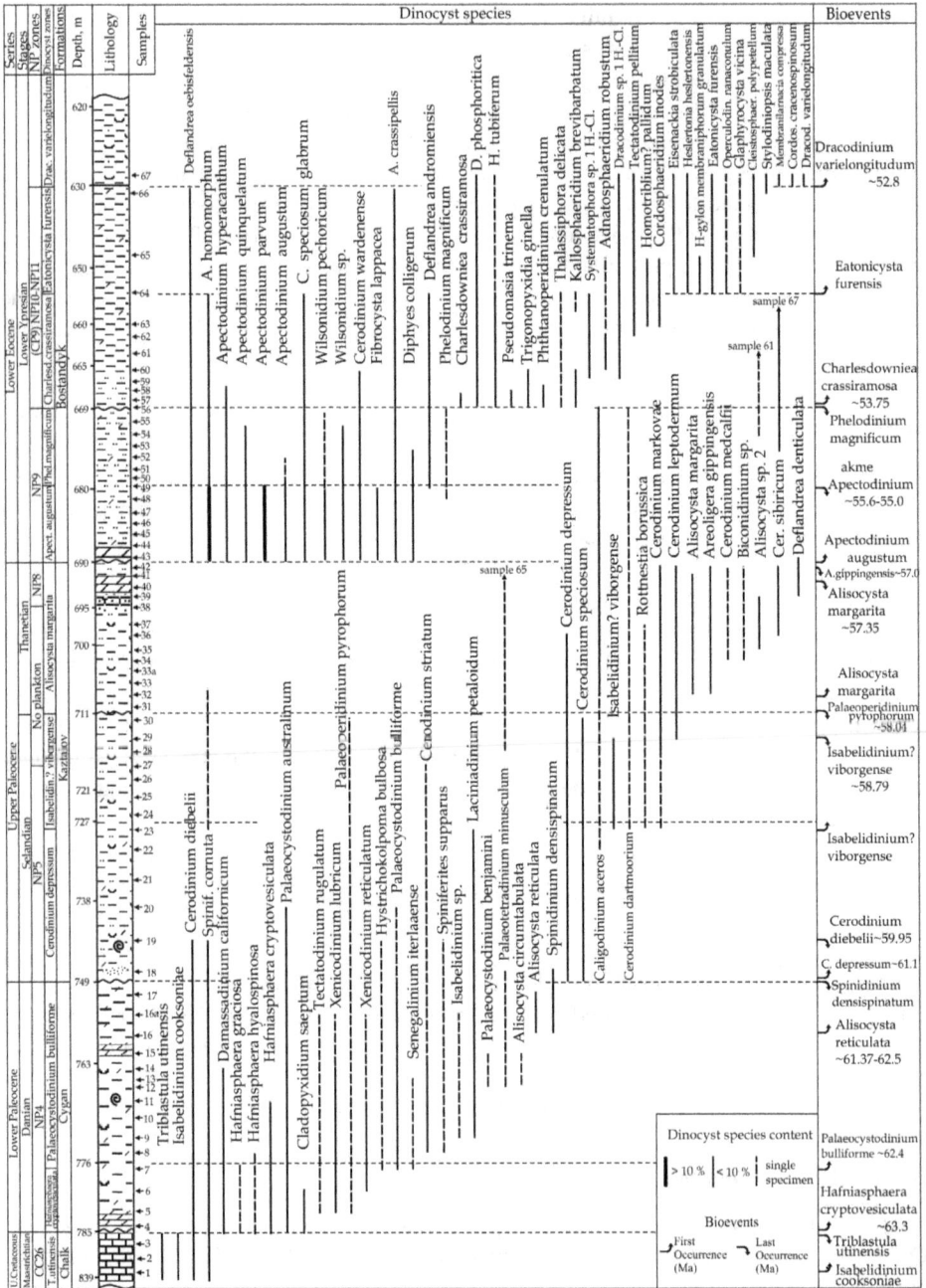

Fig. 3. Stratigraphical occurrences of certain dinocyst and acritarch species in the Elton section. Most important dinocyst events (first occurrence, last occurrence, akme) and their ages are shown (from Luterbacher et al., 2004; Williams et al., 2004). For lithological legend see Fig. 2.

5.1.2 *Hafniasphaera cryptovesiculata* zone

The 785–776 m interval is referred to the *Hafniasphaera cryptovesiculata* (Hansen, 1977) zone from FO *Hafniasphaera cryptovesiculata, H. graciosa*. The biostraton has been originally recognized by J. Hansen in the Danian limestones from Denmark; it correspondsʙ to the upper part of the *Damassadinium californicum* (*Danea mutabilis*) zone (Hansen, 1977). The D2a zone may be directly correlated with the nannoplankton scale and corresponds to the nannoplankton zones NP3 (part)–NP4 (part) (Costa & Manum, 1988; Luterbacher et al., 2004; Powell, 1992;). Characteristic taxa from the Pricaspian complex: *Hafniasphaera hyalospinosa, Palaeocystodinium australinum, Tectatodinium rugulatum, Xenicodinium lubricum*. Availability of *Palaeocystodinium australinum* suggests that the biozone corresponds to the upper part of the D2a biochron from the standard zonal scale (Luterbacher et al., 2004) and is referred to the Late Danian.

5.1.3 *Palaeocystodinium bulliforme* zone

The 778–749 m interval is referred to the D2b *Palaeocystodinium bulliforme* zone from the appearance of the zonal species (Luterbacher et al., 2004). *Senegalinium iterlaaense, Hystrichokolpoma bulbosa, Cerodinium striatum, Laciniadinium petaloidum* appear in the interval basement. Within the interval (samp.12), FO *Palaeotetradinium minusculum, Alisocysta circumtabulata, Palaeocystodinium benjamini* is recognized. In the upper part of the interval (samp.16a), FO *Alisocysta reticulata*, LO *P. benjamini, A. circumtabulata* is pronounced. In the top of the interval, LO *Tectatodinium rugulatum, Xenicodinium lubricum, X. reticulatum, A. reticulata* occurs, appearance of small forms of *Alterbidinium* spp. is recorded, inclusive of *A. dilwynense*. The biozone corresponds to the following biochrons: *Cerodinium striatum* in J. Powell scale for northwestern Europe (Powell, 1992), DP2 (Ekofisk Fm, Maureen Fm) of the zonal chart for the North Sea basin (Mudge & Bujak, 2001), a part of the D2 (Costa & Manum, 1988); it correlates with a part of the NP4 nannoplankton zone (Luterbacher et al., 2004). The zone is referred to the Late Danian.

5.1.4 *Cerodinium depressum* zone

The 749–730 m interval corresponds to the zone D3b *Cerodinium depressum* (Heilmann-Clausen, 1985; Luterbacher et al., 2004), zone *Cerodinium speciosum* (Powell, 1992) from FO *Cerodinium depressum, C. speciosum* and the lack of *Damassadinium californicum*. The interval is described from LO *Hystrichokolpoma bulbosa, Cerodinium diebelii, Spiniferites supparus* (samp.19), *Palaeocystodinium australinum, P. bulliforme* (samp.20), *Laciniadinium petaloidum* (samp.21). Solitary *Spinidinium densispinatum* occur in the interval basement. According to Heilmann-Clausen (1985), the zone is indirectly correlatable with the nannoplankton zones NP4 (part)–NP5 (part). Characteristic species in the Elton section comprise *Palaeotetradinium minusculum, Palaeocystodinium australinum, P. lidiae, Membranosphaera maastrichtica, Laciniadinium petaloidum*. The *Cerodinium depressum* zone is referred to the Selandian.

5.1.5 *Isabelidinium*? viborgense zone

The 730–711 m interval is distinguished as the *Isabelidinium*? *viborgense* zone, as an interval from FO to LO *I.*? *viborgense*. The taxon has been first recognized in the upper part of the Kerteminde formation, within the mass of the overlying non-carbonate Selandian clays from the section of the Viborg 1 well (Viborg zone 2 –Viborg zone 3) (Heilmann-Clausen, 1985). The bioevent *I.? viborgense* is manifested in the upper part of the zone *Cerodinium speciosum* (Powell, 1992), *Cerodinium depressum* (Luterbacher et al., 2004), traced in the North Sea basin (Heilmann-Clausen, 1994), in the sections from West Siberia (Amon et al., 2003), Povolzhye-Pricaspian region (Aleksandrova, 2001; Vasilyeva & Musatov, 2010a). The *Isabelidinium? viborgense* species was proposed by A. Köthe (2003) in Germany as an additional marker of the *Cerodinium speciosum* zone. In the Northern Pricaspian, the interval of that species occurrence is limited by the NP5 nannoplankton zone (Vasilyeva & Musatov, 2010a). LO *Palaeoperidinium pyrophorum, Cerodinium striatum, C. speciosum* is observed in the upper part of the zone in the Elton section. The *I.? viborgense* zone is referred to the Selandian (the middle part).

5.1.6 *Alisocysta margarita* zone

The 711–690 m interval corresponds to the *Alisocysta margarita* biochron (Luterbacher et al., 2004). The biozone is described from FO *Alisocysta margarita, Areoligera gippingensis* in the interval basement and from the lack of the Upper Selandian zonal markers: *P. pyrophorum, P. australinum, C. striatum*. The Elton association is represented by diverse *Cerodinium: C. medcalfii, C. markovae, C. depressum, C. leptodermum, C. sibiricum*, as well as by *D. denticulata* and numerous *Fromea laevigata*. The zone is referred to the Upper Selandian – Thanetian; it correlates with the NP6–NP8 nannoplankton zones (Luterbacher et al., 2004). The lack of the Selandian zonal species makes it possible to compare the biozone with its Thanetian part (Heilmann-Clausen, 1994).

5.1.7 *Apectodinium augustum* zone

The Bostandyk formation is represented by productive dinoflagellate complexes of diverse species compositions, with numerous *Apectodinium* species in the lower part and extremely rare wetzelielloideans in the upper part of the formation. Similar distribution of wetzelielloids is not typical for Russia's interior boreal basins, but is characteristic of the Crimea-Caucasian region and of the Ukraine (Andreyeva-Grigorovich, 1991). Biozones in the upper part of the Bostandyk formation are recognized as close analogues of the standard zones in accord with the correlative taxa (Costa & Manum, 1988; De Coninck 1990; Heilmann-Clausen & Costa, 1989).

The 690-680 m interval refers to the standard D5a *Apectodinium augustum* zone (Costa & Manum, 1988; Luterbacher et al., 2004) from the appearance of the zonal taxon *A. augustum*, acme of the *Apectodinium* species. The contents of the *Apectodinium homomorphum, A. quinquelatum, A. parvum, A. hyperacanthum* species in this interval are up to 70% high, drastically reducing above the depth of 679 m. *A. homomorphum* dominates. *Wilsonidium pechoricum* is present. The D5a zone corresponds to the IETM interval (Bujak & Brinkhuis, 1998; Crouch et al., 2001; Crouch & Brinkhuis, 2005; Egger et al., 2003), corresponds to the Initial Eocene and correlates with the upper part of the NP9 nannoplankton zone (Luterbacher et al., 2004).

5.1.8 *Phelodinium magnificum* zone

The 680-669 m interval of the Bostandyk formation corresponds to the D5b *Phelodinium magnificum* zone (Luterbacher et al., 2004). The zone is specified as an interval above the top of the D5a *Apectodinium augustum* zone to the appearance of the earliest species of *Wetzeliella* spp. (Heilmann-Clausen & Costa, 1989). In the Elton section, the biozone is described from occasional species of *Apectodinium*, inclusive of *A. augustum*, also *Cerodinium wardenense*, *Phelodinium magnificum, Deflandrea andromiensis*. No acme of *Deflandrea oebisfeldensis* is observed. The D5b *Phelodinium magnificum* zone correlates with the NP10 nannoplankton zone and is referred to the Initial Ypresian.

5.1.9 *Charlesdowniea crassiramosa* zone

The 669–655 m interval is distinguished as the D6b *Charlesdowniea crassiramosa* zone from FO *Ch. crassiramosa, Defalndrea phosphoritica*. The appearance of those species is known from the Early Ypresian D6b zone in the sections from Britain (London Clay) and northwestern Germany (Wursterheide research well) (Costa & Manum, 1988; Heilmann-Clausen & Costa, 1989). Availability of dinocysts *Trigonopyxidia ginella, Adnatosphaeridium robustum, Kallosphaeridium brevibarbatum,* acritarchs *Pseudomasia trinema* reflects common properties with the *Phthanoperidinium crenulatum* zone (middle part of the Orchies Formation in the stratotype Ypresian section from Belgium) (De Coninck, 1988). The *Charlesdowniea crassiramosa* biozone corresponds to the D6b *Wetzeliella meckelfeldensis* standard zone (Luterbacher et al., 2004) and is referred to the Early Ypresian.

5.1.10 *Eatonicysta furensis* zone

The 655–630 m interval is recognized as the *Eatonicysta furensis* zone. It is described from FO *E. furensis, Eisenackia strobiculata, Heslertonia heslertonensis, Hystrichostrogylon membraniforum granulatum, Operculodinium nanaconulum, Glapyrocysta vicina* in the interval basement. There are no wetzelielloideans within the zone. The biozone is a close analogue of the D7a (*Eatonicysta ursulae – Dracodinium solidum*) zone from the section of the Wursterheide well in Germany (Heilmann-Clausen & Costa, 1989), *Eatonicysta ursulae* zone (Roubaix formation) from the Kallo section in Belgium (De Coninck, 1988).

5.1.11 *Dracodinium varielongitudum* zone

The 630–620 m interval is recognized as the *Dracodinium varielongitudum* zone from the zonal species appearance (solitary occurrence), the presence of *Cordosphaeridium cracenospinosum, Stylodiniopsis maculatum, Membranilarnacia compressa* (Heilmann-Clausen & Costa, 1989). The biozone corresponds to the standard D7c *Dracodinium varielongitudum* biochron and is referred to the Middle Ypresian (Luterbacher et al., 2004). Some new dinocyst species from the Pricaspian Depression sections are systematically described in previous studies (Andreyeva-Grigorovich et al., 2011; Vasilyeva, 2011).

5.2 Nannoplankton

Calcareous nannoplankton complexes have been revealed practically throughout the section, with the exception of some intervals. They are fairly representative in terms of species composition, which provides reliable identification of zonal divisions. The most productive and rich associations have been revealed in the Maastrichtian, Danian, Selandian and the Early Ypresian section intervals.

5.2.1 *Nefrolithus frequens* (CC26) zone

The nannoplankton complex from the Maastrichtian marls (839–785 m) is peculiar for fairly high species diversity, inclusive of *Lithraphidites quadratus* and *Nefrolithus frequens*, which makes it possible to refer the beds to the Late Maastrichtian (zone CC26) (Perch-Nielsen, 1985; Sissing, 1977).

5.2.2 *Ellipsolithus macellus* zone

The 785–763 m interval is referred to the lower part of the *Ellipsolithus macellus* (*Coccolithus robustus*) zone (Martini, 1971) from the presence of solitary *Ellipsolithus macellus, Coccolithus robustus* and numerous *Prinsius martini* in the complex. The assemblage is principally represented by numerous *Coccolithus pelagicus, C. cavus, C. subpertusus, Cruciplacolithus tenuis, Cr. primus, Chiasmolithus danicus, Placozygus sigmoides*. Species *Lanternithus duocavus, Biantolithus sparsus, Neochiastozygus saepes, N. eosaepes* are substantially less frequent. At the level of 776 m, sparse *Chiasmolithus edentulus* appear, at the level of 770 m - rare *Ellipsolithus distichus*. Minor amounts of Cretaceous species occur throughout the section. On the basis of the species composition, this section part may be supposed to correspond to the NNTp4D-F subzones of Varol zonal scale for the North Sea (Varol, 1998).

The 763–749 m interval. At the level of ~763, fairly numerous *Neochiastozygus perfectus* and solitary small *Fasciculithus* sp. appear. Stratigraphically higher, at the level of 750, *Fasciculithus ulii, F. janii, F. magnus* appear. Besides, *Braarudosphaera bigelowii* occur quite frequently. The principal composition of the complex remains the unchanged. It should be noted that at that level all the coccoliths become substantially larger. The presence of the above mentioned species makes it possible to outline the probable level of the *Fasciculithus* first radiation at 763 m. This section interval may be compared to the upper half of the NP4 zone according to Martini (1971), or with zones NNTp5B-NNTp8 (?) specified by Varol (1998).

5.2.3 *Fasciculithus tympaniformis* zone

The 749–737 m interval is peculiar for the appearance of *Fasciculithus tympaniformis, F. ulii, F. janii, F. pileatus, Sphenolithus primus, Chiasmolithus bidens, Ch. consuetus*; in the upper part of the section *N. protenus* appear. The complex principal composition does not change as compared to the previous one, but nannoplankton productivity reduces substantially, and at the level of 738 m it disappears. This part of the section correlates reliably with the zones NP5 by Martini (1971), CP4 by Okada & Bukry (1980), NNTp9-10 by Varol (1998). It should be noted that *Sphenolithus primus* appears in the Elton key well substantially later than in the more northerly sections (Novouzensk key well, Ozinki section). No nannoplankton has been revealed in the 737–702 m interval.

5.2.4 *Heliolithus riedelii* zone

The 702–690 m interval is peculiar for appearing *Heliolithus riedelii*, drastic increase of *Fasciculithus tympaniformis* content, rare *Discoaster falcatus, D. nobilis, D.megastypus, Heliolithus kleinpellii, Chiasmolithus frequens, Neochiastozygus junctus* appear. It should be noted that in the upper part of the interval all the coccoliths are very large, overgrown, teratic forms are frequently encountered, many of them have additional calcite knobs. This is especially characteristic *Braarudosphaera bigelowii, Sphenolithus* spp. Nannoplankton productivite, however, is insignificant, the species diversity stays at the same level as in the lower intervals: 15–20 species. This interval correlates reliably with the NP8 and CP7 zones.

5.2.5 *Discoaster multiradiatus* zone

The 690–669 m interval is peculiar for appearing *Discoaster multiradiatus, Discoaster salisburgensis,* numerous *Romboaster cuspis, R. bitrifida, R. spineus,* the content of *Neochiastozygus junctus* increases, *Campylosphaera eodela* occur infrequently. Relative to the previous interval, the nannoplakton productivity has increased substantially, but the species diversity has hardly changed. In terms of the complex species composition, this part of the section may be reliably referred to the upper part of the NP9 zone or to the CP8b subzone.

5.2.6 *Discoaster diastypus* zone

The 669–630 m interval. From the depth of 669 m, upwards in the section, gradual degrease of the nannoplankton productivity is recorded. The complexes become depleted in terms of both, species and amounts. Characteristic species *Discoaster diastypus* appears, solitary *Neococcolithus dubius,* many Paleocene species disappear, inclusive of *Heliolithus riedelii, Placozygus sigmoides, Cruciplacolithus tenuis, Ellipsolithus macellus* (occasional specimens occur only in the basement). No *Tribrachiatus bramlettei, T. contortus* have been encountered in the complex, probably due to substantial impoverishment. At the same time, no *Discoaster lodoensis* have been revealed, characteristic of the NP12 (CP10) zone. This makes it possible to compare this part of the section with the CP9 zone, without dividing it into subzones. In the 630–620 m interval, the nanoplankton complex is quite poor, thus, no zonal affiliation may be determined.

6. Rhythm successions and paleoecologic interpretations

6.1 The Danian

Beds from the Danian section are represented by well differentiated rocks comprising marl interlayers. Though boundaries of certain rhythm phases (erosion horizons, maximum flooding surface) are not clearly distinguished, the intervals recognized by the authors differ from each other, correlate positively enough with definite rhythm stages and are confirmed by characteristics of the plankton associations.

6.1.1 The 785–763 m interval

Rocks from the interval lower part (785–776 m) are represented by gray calcareous clays with patches of glauconite sand comprising a marl interlayer (0.1 m thick). The upper part of the interval (776–763 m) is composed of gaize-like sandy clays with fish scales, macrofauna remains, numerous fine burrows of deposit feeders.

In the interval base, drastic increase is observed in species diversity of dinoflagellate cysts represented mainly by gonyaulacoid forms (Fig. 4). Those are dominated by representatives of the *Fibrocysta, Fibradinium, Xenicodinium, Tectatodinium* genera – the taxa characteristic of open marine settings from the inner to the outer part of neritic zone. The occurrence of a solitary specimen of *Impagidinium* (samp. 5) is most probably accounted for by transportation from the basin deeper outer part during the transgressive phase of the rhythm. The taxa from that association (*Spiniferites, Hystrichokolpoma, Achomosphaera, Areoligera, Hafniasphaera*) refer to cosmopolitan proximochorate cysts frequent in the well-stirred, non-stratified water masses from the neritic zone. Subsequent stepwise reduction of dinocyst species diversity (DSD) is manifested in the upper part of the interval (776–763 m). Sparse peridinioid cysts represented by the *Palaeocystodinium* species, more rarely by *Cerodinium, Palaeotetradinium, Spinidinium, Alterbidinium, Senegalinim*, demonstrate increases in the lower (samp. 6) and in the upper (samp.12) parts of the interval, but they are forced out from a major part by dinoflagellate gonyaulacoid forms. Generally, from two to four peridiniod species are present in a complex of organic-walled plankton.

Terrestrial palynomorphs are of subordinate importance in macerates. Those consist mostly of triaperturate pollen (stemma *Postnormapolles*), comparable in its organization type to modern wind-pollinated angiosperms. Such pollen grains are transported at long ranges by air streams, which doesn't disagree with interpretation of the plankton complex characterizing the open sea settings in the inner shelf conditions. Sparse prasinophyte algae and acritarchs (*Fromea laevigata*, species *Leiosphaeridia, Comatosphaeridium, Cyclopsiella, Paralecaniella, Paucilobomorpha*) indicate proximity of littoral facies.

Similar distribution of the organic plankton taxa and the interval lithologic structure may be interpreted as the highstanding system tract (HST) in the inner shelf conditions. Successive appearance of the *Hafniasphaera cryptovesiculata* (63.3 Ma), *Palaeocystodinium bulliforme* (62.2–62.4 Ma) dinocysts and *Coccolithus robustus, Ellipsolithus macellus* (63.2 Ma), *E. distichus* nannoplankton makes it possible to compare this interval with the Da-3 Danian rhythm (Hardenbol et al., 1998).

6.1.2 The 763–749 m interval

In the lower part the rocks are represented by a marl layer, about 1.0 m thick, in the upper part – by grayish green calcareous clays with burrows of deposit feeders, macrofauna remains, fish scales. The peak of dinocyst species diversity coincides with FO *Alisocysta reticulata* and the coccolithophorid first radiation including solitary small *Fasciculithus* spp. The dinocyst association is represented by roughly equal taxon participation of gonyaulacoid (species *Achomosphaera, Fibradinium, Alisocysta, Membranosphaera, Damassadinium, Cordosphaeridium, Spiniferella, Thalassiphora*) and peridinioid (species *Laciniadinium, Palaeocystodinium, Alterbidinium, Palaeotetradinium, Cerodinium*) cysts. Both groups show brief DSD peaks slightly above the marl interlayer. Terrestrial palynomorphs are of subordinate importance; they are represented by small angiosperm forms (*Triporopollenites* spp., *Aquillapollenites* spp.) and by conifers. The interval lithologic structure and dinocyst species distribution correspond to the HST stage that corresponds to the next Danian rhythm Da-4 (Hardenbol et al., 1998).

6.2 The Selandian

The 749–711 m interval, attributed to the Selandian, is represented by an extremely poorly differentiated clayey gaize-like sequence. Some fragments of this part of the section are represented by slightly higher calcareous rocks comprising sparse nannoplankton, or by

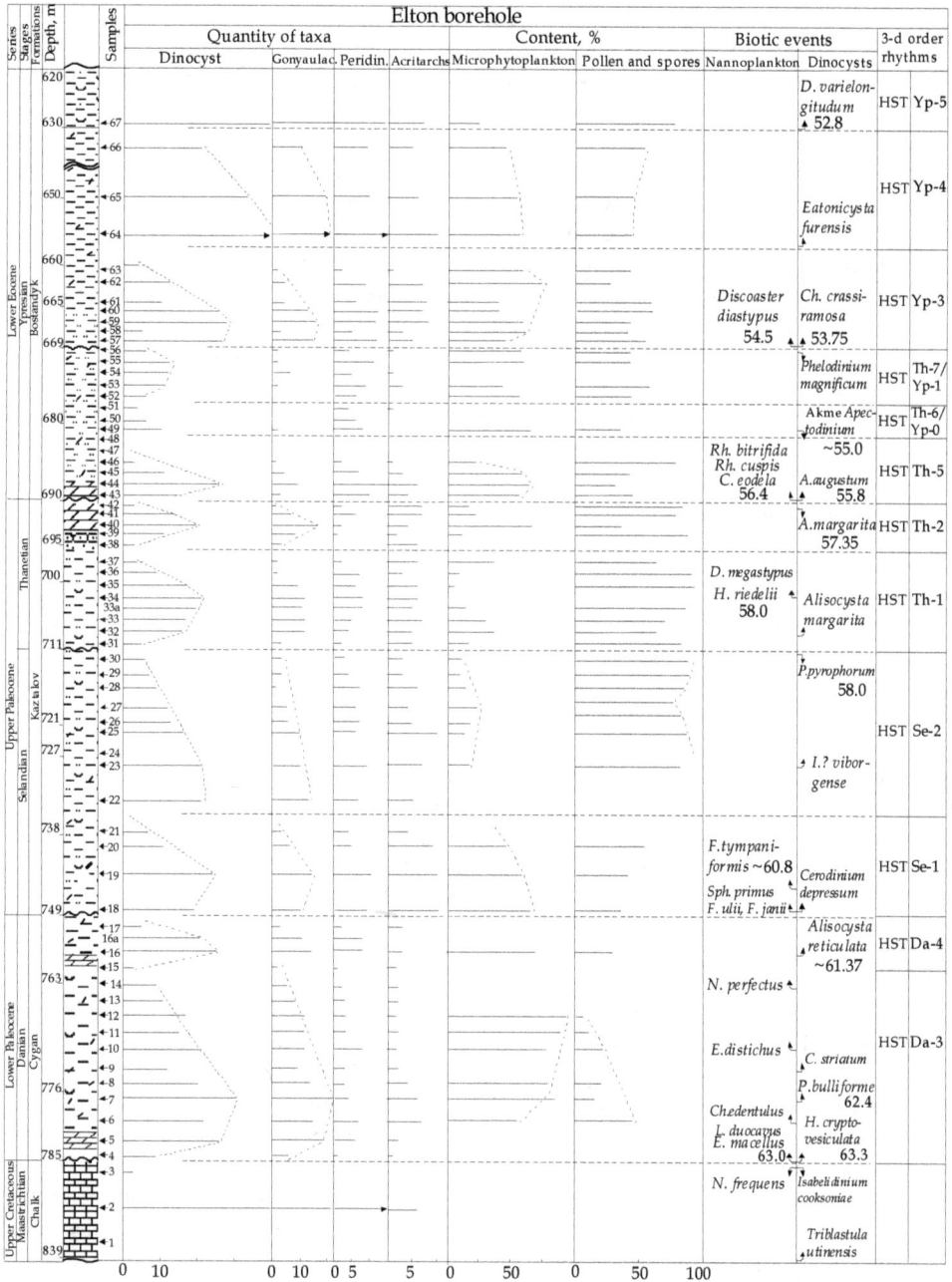

Fig. 4. Dynamic change in the palynomorphs assemblage structures and their interpretation according to Hardenbol 3-d order rhythms scale (Hardenbol, 1994; Hardenbol et al., 1998)

higher sandy clays, or by rock areas with intense bioturbation. The features of lithologic structure induce one to rely mostly on micropaleontologic evidence for rhythm recognition. Terrestrial palynomorph are absolutely dominant among all the palynologic remains (Fig. 4). They are represented by angiosperm pollen of the stemma *Normapolles* (*Trudopollis* spp., *Oculopollis* spp., *Pompeckjoidaepollenites* spp.), *Postnormapolles* (*Triatriopollenites* spp., *Subtriporopollenites* spp.) and by spore plants (*Gleicheniidites* spp., *Laevigatosporites* spp., *Cicatricosisporites* spp., *Stereisporites* spp.), with subordinate participation of Gymnosperms (*Pinaceae, Taxodiaceae*). Organic-walled phytoplankton comprises substantial amounts of non-marine plankton groups (prasinophytes and acritarchs). Prasinophyta species diversity is comparable to those of peridiniod and gonyaulacoid dinoflagellates. On the whole, dinocyst species composition is very scanty. Palynofacies composition and structure in this interval of the section are indicative of shallow-water littoral settings with low aeration of water masses. Gradual decline of connections with the open ocean is also indicated by the complexes of carbonate nannoplankton; gradual impoverishment of associations is observed starting from the 749 m level. Two distinct successive trends of DSD have been revealed in this part of the section.

6.2.1 The 749–738 m interval

The rocks are represented by dark gray, almost black, gaize-like, dense clays with rare macrofauna, microfauna and fish scales. Clays from the interval basement are higher calcareous and enriched in sandy and aleuritic materials. In this part of the interval, increase (24 taxa) and decrease (down to 7 species) of the DSD index are observed (Fig.4). Dinocysts are represented by gonyaulacoid (genera *Fibradinium, Membranosphaera, Spiniferites*) and peridinioid (genera *Palaeotetradinium, Palaeocystodinium, Cerodinium*) forms in practically equal shares. Green algae comprise *Fromea laevigata, Paucilobimorpha apiculata, Palambages morulosa, Paralecaniella identata, Botryococcus* spp., *Veryhachium* spp., *Leiosphaeridia* spp., *Diacrocanthidium spp.* and demonstrate dynamics similar to that in dinocysts. The first appearances of *Cerodinium depressum* dinocysts and *Fasciculithus tympaniformis* nannofossils make it possible to compare the 749–738 m interval with the Selandian first rhythm Se-1 (Hardenbol, 1994) represented by the HST stage in shallow-water littoral settings.

6.2.2 The 738–711 m interval

This section interval is composed of gray, dense, gaize-like clays. The rocks become progressively less calciferous upwards in the section. The top 6 m are composed of non-carbonate rocks with numerous deposit-feeder burrows. This part of the section is peculiar for consistent impoverishment of the dinocyst species composition (from 23 to 6 taxa), absolute dominations of *Fromea laevigata* acritarchs against the background of numerous prasinophyte algae. Palynomorphs of terrestrial origin quantitatively dominant. It is only in the lower part that impoverished complex of the *Fasciculithus tympaniformis* zone plankton occurs, coccoliths disappear entirely above the 735 m level. The interval is marked with *Isabelidinium? viborgense* dinocysts (samp. 23) appearance and *Palaeoperidinium pyrophorum* (LO in samp. 30) disappearance in the roof. We believe the interval to correspond to the Selandian second rhythm Se-2 (Hardenbol, 1994) and to correlate with the HST stage in shallow-water conditions.

6.3 Thanetian

The Thanetian beds are represented by two lithological distinct sequences: the lower, the clayey one, and the higher, the sandy-marly mass. The micropaleontologic features are similarly distinct.

6.3.1 The 711–695 m interval

The sediments are composed of gray, dense, gaize-like, non-calcareous clays with deposit-feeder burrows. The interval is close to the previous (738–711 m) in lithologic composition. A clear peak of species diversity in organic-walled phytoplankton (5–21–4 taxa) is manifested in this interval. It is formed by peridinioid (genera *Cerodinium, Palaeotetradinium, Senegalinium*), chorate and marginate (genera *Cordosphaeridium, Spiniferites, Achomosphaera, Alisocysta, Spiniferella, Areoligera, Caligodinium*) cysts. Diversity of prasinophyte algae and domination of terrestrial palynomorphs suggest that the association has been formed in littoral settings. Pollen is represented by various angiosperm taxon of the stemma *Postnormapolles*, to a lesser extent - by *Normapolles* and conifers. *Alisocysta margarita, Areoligera gippingensis* dinocyst and *Heliolithus riedelii* nannoplankton appearance (at the 705 m level) make it possible to refer the 711–695 m interval to the Thanetian initial rhythm (Hardenbol et al., 1998).

6.3.2 The 695–690 m interval

Drastic change of lithologic composition is observed in this part of the section. The rocks are represented by alternating interlayers of glauconite sandstones (0.75–1.5 m thick), marls (0.2–0.3 m thick) and calcareous sandy clays (0.3–1.3 m). The sediment composition is suggestive of highly dynamic sedimentation settings which has affected the phytoplankton composition, as well. The DSD impulse in the interval has been formed mostly by gonyaulacoid cysts represented by proximochorate forms – genera *Alisocysta, Spiniferites, Cordosphaeridium, Achomosphaera, Melitasphaeridium* and *Caligodinium, Impletosphaeridium*. Occurrences of those taxa are characteristic of inner neritic water masses. In the interval this association is completed with sparse peridinioids *Cerodinium markovae, C. leptodermum, Deflandrea denticulata, Lentinia wetzelii* – typical species from the shore-nearest shelf part. The observed impulse of dinocyst diversity occurs within acritarch facies, similarly to the situation in the previous interval. *Fromea laevigata* abundance disappears only in samples 39 and 40; acritarchs are substituted with chorate dinocysts there demonstrating a negative "mirror" peak. Palynomorphs of terrestrial genesis are still numerous and represented by diverse high-order taxa. The species composition of the calcareous nannoplancton complexes stays at a low level, in the range of 12-15 species; productivity increases substantially in some marl layers. This section fragment lies within the *Alisocysta margarita* (dinocysts) and *Heliolithus riedelii* (nannoplankton) zones and corresponds to the Th-2 rhythm (Hardenbol et al., 1998).

6.4 Ypresian

The Ypresian beds in the Elton section are represented by a monotonous clay mass, therefore, lithologic indices of changing rhythms are practically unpronounced. Paleoecologic changes in compositions of the plankton associations, however, are manifested most vividly in this part of the section.

6.4.1 The 690–680 m interval

The sediments in the interval are represented by low calcareous, sandy, indistinctly banded clays with lenticles of aleurite material. Clays comprise plant detritus and fish scales. There is an interlayer of dark gray marl in the basement (0.25 m thick) – with thin intercalations of aleurite-sandy material of quartz-glauconite composition.

The interval is manifested as a sharp, brief impulse of diversity in a dinocyst assemblages, mostly of wetzelielloids of the *Apectodinium* genus, inclusive of *A. augustum*, *A. parvum*, *A. quinqellatum*, and *Wilsonidium* spp. (Fig. 4). *A. homomorphum* dominates among the *Apectodinium* species. Relative content of the *Apectodinium* spp. cysts in this association is up to 70% high. The wetzelielloid group is supplemented with other peridinioid dinoflagellates, namely *Deflandrea*, inclusive of *D. andromiensis*, *Cerodinium*, *Palaeotetradinium*. Gonyaulacoids are diverse, but sparse. This group of taxa from the inner neritic zone is composed of cosmopolitan species *Achomosphaera*, *Hystrichokolpoma*, *Oligosphaeridium* and fibrose forms with thickened walles: *Kallosphaeridium*, *Tectatodinium*, *Caligodinium*, *Fibrocysta*, *Hafniasphaera*. Acritarchs and prasinophytes are solitary in this association; *Paralecaniella*, *Leiosphaeridia*, *Pterospermella*, *Paucilobimorpha* are permanently diagnosed. Organic-walled plankton dominated in the interval, while terrestrial palynomorphs are of subordinate importance; they are represented by fine triaperture and tricolporate pollen (*Subtriporopollenites* spp., *Triporopollenites* spp., *Tricolporopollenites* spp.). Pollen of the Stemma *Normapolles* and cryptogams are rare.

Paleophytocoenosis (690–680 m interval) developed in the conditions of inner neritic, fairly well aerated water masses. The lithologic sequence most probably reflects the HST stage within the Th-5 rhythm (Hardenbol & al., 1998), which is determined from FO *Apectodinium augustum*, acme of the *Apectodinium* spp. (dinocysts) and fairly rich nannoplankton of the CP8b subzone. The nannoplankton productivity is high enough, but its species diversity does not exceed 15–20 species. All the coccolithophores are well preserved, there are no teratic or overgrown forms. The distinct acme of *Apectodinium* spp. reflects the event of the Initial Eocene Thermal Maximum (IETM), globally correlatable about 55 million year ago (Bujak & Brinkhuis, 1998; Crouch et al., 2001; Egger et al., 2003; Steurbaut et al., 2003).

6.4.2 The 680–669 m interval

The rocks are represented by alternating low calcareous, dark gray, sandy clays and denser shaly black clays. Macro- and microfauna fragments and fish scales occur in the sediments. Associations of organic-walled microphytoplankton demonstrate two short DSD impulses and characterize different rhythms. They can't be described in detail due to material scantiness. The dinocyst complex from the interval lower part (samp. 49–51), however, belongs to the *Apectodinium augustum* zone and is represented exclusively by wetzelielloids – *Apectodinium* species, inclusive of *A. augustum*, and *Wilsonidium* spp. The upper part of the interval (samp. 52–56) reflects a mixed complex of peridiniod dinoflagellates, inclusive of the genera *Deflandrea*, *Cerodinium*, *Apectodinium*, *Wilsonidium*, *Phelodinium*, *Palaeotetradinium*. Species *Deflandrea andromiensis* and *D. oebisfeldensis* dominate there. Gonyaulacoids are rare; they are represented by *Diphyes colligerum*, *Achomosphaera crassipellis*, *Spiniferites ramosus*. Poor preservation of palynologic material should be noted, as well as abundance of coalified debris and numerous fragments of amorphous organic matter. The nannoplankton complex is substantially impoverished (8–12 species), its productivity reduces. Based on palynologic features and on the occurrence frequency of marker dinocysts, one may suggest that the 680–669 m imterval reflects fragments of two successive rhythms of the Initial Ypresian: Yp-0 and Yp-1 (Hardenbol et al., 1998; Powell et al., 1996).

6.4.3 The 669–660 m interval

The entire upper part of the Elton section, starting from the depth of 669 m, is represented by substantially different plankton characteristics. Well saturated and representative dinocyst complexes are peculiar for predominance of gonyaulacoid morphotypes. Moreover, the wetzelielloids that make the basis for biostratigraphic division of the Ypresian beds from Russia's epicontinental section are extremely rare or absent. At the same time, dynamics of species diversity in dinocyst associations is most distinctly manifested in this part of the section.

The 669–660 m interval is represented by dark gray, almost black, low calcareous, sandy clays with microfauna, macrofauna detritus and fish scales. Sand material occurs as lenses and superfine laminae. Sharp increase and gradual decrease of the DSD-criterion is observed in this part of the section. Dinoflagellate associations are represented by various gonyaulacoid morphotypes, mostly by chorate cysts. Representatives of *Achomosphaera*, *Spiniferites*, *Cordosphaeridium*, *Hystrichosphaeridium* occur most frequently; those are cosmopolitan taxa typical of neritic water masses. In the group of peridinioids, *Deflandrea* are the most characteristic and diverse ones. Wetzellieloids are represented by solitary specimens of *Charlesdowniea crassiramosa*, *Apectodinium homomorphum*, *A. hyperacanthum*, *Dracodinium* sp. 1 Heil.-Claus. Acritarchs and prasinophytes are rare. Among those, *Paralecaniella*, *Paucilobimorpha*, *Leiosphaeridia*, *Cymatiosphaera* occur most consistently. The relative content of organic-walled phytoplankton in this interval is generally above the amount of terrestrial palynomorphs. All the palynologic remains are peculiar for poorly preserved material. Many dinocysts are represented by fragments, the periphragm structure is damaged, frayed. Large amounts of amorphous organics and coalified debris are recorded. The nannoplankton species diversity gradually reduces upwards in the section – to 4-8 species. Finds of (FO) *Charlesdoniea crassiramosa*, *Discoaster diastypus* (nannoplankton) make it possible to correlate the 669–660 m interval with the first major transgression of the Lower Ypresian Yp-3 (HST stage) and the zone from the *Wetzeliella meckelfeldensis* dinocysts.

6.4.4 The 660–630 m interval

This interval of the Elton section is composed of black, dense, indistinctly laminated, low calcareous clays with minor admixture of sandy-aleurite material as patches and lenses. Microfauna, macrofauna fragments and fish scales occur in the sediments. The sharpest and the highest DSD peak is observed in this part of the section (Fig. 4). The composition of the organic-walled microphytoplankton complex is rather unexpected. On the one hand, high taxonomic diversity is observed in chorate dinocysts. Among those, representatives of the genera *Achomosphaera*, *Spiniferites*, *Hystrichsphaeridium*, *Cordosphaeridium* are the most common ones. Sporadically, genera *Operculodinium*, *Homotriblium*, *Fibrocysta*, *Cleistosphaeridium*, *Membranosphaera* occur, i.e. typical representatives of neritic zone. Solitary *Impagidinium dispertitum* occur, reflecting the influence of deeper settings. On the other hand, the peridinioid part of the phytoplankton spectrum is represented only by representatives of dominating *Deflandrea* and *Cerodinium*. There are practically no wetzelielloid dinocysts present.

The most of *Deflandrea* specimens are found as fragments – endocysts. Periphragm in the preserved forms is thinned, perforated or covered with granules, spines. The preservation state of the palynologic material reflects unfavorable chemical conditions during sedimentation and, probably, the initial stage of organic matter destruction. Poor preservation is characteristic of terrestrial palynomorphs, as well. Besides, macerates contain large numbers of foraminifer inner capsules, scolecodont remains, fragments of amorphous organics, detrital material. Constant presence of *Botryococcus, Palambages, Leiosphaeridia* is characteristic of those palynofacies. Such structures of taphocoenosis and palynofacies, material preservation state and sediment compositions suggest that the paleophytocoenosis has been formed in the conditions of limited water exchange and disturbed circulation within the water column, which was, probably, caused by partial isolation of that basin part. Rise of the sea level and inflow of transgressive water masses used to provide phytocoenosis (diverse gonyaulacoid dinocysts) short-term taxonomic renovation, but closed settings favored formation of reducing environment, stagnation, development of deflandreoids and prasinophyte alga complexes representing the autochthonous biota there. The calcareous nannoplankton complexes are extremely poor, just with 5-6 species, inclusive of *Discoaster diastypus*. It is the inflow of marker dinocysts (*Eatonicysta furensis, Stylodiniopsis maculatum*) that allows correlation of the 660–630 m interval with the *Dracodinium simile* zone and the consistent Lower Ypresian rhythm Yp-4.

6.4.5 The 630–620 m interval

The sediments in the interval are represented by dark gray, uniform, low arenaceous, gaize-like clays. In some site the clays are layered, irregularly pyritized, comprise fauna detritus and plant remains.

Organic-walled phytoplankton has been examined only in one sample, since just a small core fragment from that part of the section was preserved. Association of high species diversity has been revealed compared to the complex from the top of the previous interval. The phytoplankton is represented by mixed ecologic groups. Diverse neritic gonyaulacoid forms occur. Those generally comprise representatives of genera *Achomosphaera, Cordosphaeridium, Cleistosphaeridium, Membranilarnacia, Tectatodinium. Impagidinium* and *Eatonicysta* are observed. The ecologic group from littoral settings consists of peridinioid forms of the genera *Dracodinium, Wetzeliella, Deflandrea, Senegalinium, Apectodinium*. This is supplemented with abundant plankton of prasinophyte algae and acritarchs – *Pterospermella, Paralecaniella, Cymatiosphaera, Cyclopsiella, Botryococcus*. Microforaminifer capsules, scolecodonts and fragments of amorphous organics occur substantially less frequently. Terrigenous palynomorphs represented by pollen of the Stemma *Postnormapolles*, tricolporate grains (Angiosperms), taxodians (Gymnosperms) and fern spores quantitatively dominate over plankton.

The DSD peak and appearance of *Dracodinium varielongitudum* dinocysts (52.8 Ma) allow to refer the interval to the next rhythm of the Ypresian, Yp-5 (Hardenbol et al., 1998). Peridinioid diverse composition and satisfactory material preservation state suggest that the paleoecologic conditions of this algocoenosis formation were substantially improved in that time interval. The species composition and productivity of nannoplankton, however, has not changed as compared to the previous interval.

7. Discussion

7.1 Palynologic data interpretation

Examination of the Elton section has shown that notwithstanding some uncertainty and probable faults in counting the taxa of the species rank (different preservation of palynologic material, taxonomically understudied phytoplankton groups, divisiveness in species diagnostics of paleontologic species), making use of the criterion of dinocyst species diversity (DSD) allows clear structuring of associations in the Paleogene beds from the Pricaspian basin. The dinocyst species diversity always reduces to minimum values in the intervals with apparent lithologic evidences of hiatus. In the section intervals associated with development of calcareous facies the DSD impulses are confined to interlayers of marls and marly clays. We believe the dynamics of plankton species diversity in the Pricaspian region to reflect rhythmic fluctuations of the basin depth and the changes of the water-mass volume within the photic zone.

The necessary criterion for rhythm substantiation consists in the lithologic features of the rhythm sequential stages: determination of the lowstanding sea tract (LST), transgressive system tract (TST), highstanding basin phase (HST), shelf margin wedge (SMW), rhythm dividing surfaces and limits (Haq et al., 1988). Since the Elton section is situated in the deepest plunged part of the depression and is represented mostly by monotonous clay sequences, clear lithologic criteria of the rhythm stages are unrecognizable or partially diagnosed (in the Danian beds). Therefore, a necessity arises to compare the sections from the Elton and from the Novouzensk key well drilled in the northern, shallower shelf zone of the Pricaspian Depression (Vasilyeva & Musatov, 2010b) (discussed below).

Distribution of coalified debris (phytoclasts) – fine black and coarse brown, fossilized cuticle fragments (single-layered epidermis of higher plants) and tracheides (cells of plant conducting system) is definitely oriented in the specified intervals and may provide additional information for rhythm diagnostics. We did not count those microfossil groups or microforaminifer organic capsules or fragments of amorphous organic matter. We recorded obvious accumulations of organic-origin remains in individual levels and samples.

High contents of pollen and spores (palynomorphs of terrestrial origin) throughout the Paleogene section from the Central Pricaspian happened to be fairly unexpected while palynologic data analyses. The Elton well is over 1000 km away from the supposed shoreline of the paleobasin. Large share of all pollen and spore groups in macerates may be accounted for by the presence of major islands developed over positive geomorphologic constructions. Let us remark here, that according to the data of geophysical exploration (Zhuravlev, 1970), the entire area of the Pricaspian Depression is covered with salt-dome structures, inclusive of those up to 10 km high and across. Most probably, the islands were represented by salt domes, similar to the Elton and Baskunchak ones. An extensive system of submerged domes and islands in the shelf of the Pricaspian basin used to constitute geomorphologic constructions similar to barrier reefs. In this case, phytocoenosis of organic-walled plankton peculiar for high diversity and the presence of acritarchs, prasinophyte algae and peridinean dinocysts, looks quite accountable, as well. The depths of the basin with such paleophytocoenosis do not generally exceed 50 m.

7.2 Correlating the sections from the Elton and the Novouzensk key wells

The two section correlation allows preliminary estimation of transgressive rhythm and corresponding sediment occurrences in the shelf of the Pricaspian paleobasin (Fig. 5). Adequately detailed paleontologic sampling enables us to describe the regional paleontologic sedimentation settings and their changes in the Cenozoic.

7.2.1 The Danian

The Danian beds from the Elton and the Novouzensk wells are poorly correlatable and may be rather regarded as complementary to each other. The marl sequence from the Lower Danian Algai formation is represented in the Novouzensk section, but is absent from the Elton well. The Algai formation marl pelitomorphic sediments (up to 40 m thick) are common in the Central Pricaspian region, but are locally absent from many salt-dome arches. The Algai formation is described from nannoplankton of the zones NP2 *Cruciplacolithus tenuis* and NP3 *Chiasmolithus danicus*; additional sampling from the Novouzensk well indicates that dinocysts from the D1 with *Xenicodinium lubricum* are also present.

The Upper Danian is represented by calcareous clays from the Cygan formation. In the Central Pricaspian region, its thickness ranges from 20 to 140 m, increasing in the interdome downfolds. The Cygan formation is 36 m thick in the Elton well. The phytoplankton paleontologic analysis shows the formation to have been formed within two successive rhythms of the Upper Danian – Da-3 и Da-4. In the Elton section, sediments corresponding to the Da-3 rhythm are represented by black, calcareous, sandy clays (HST); they correspond to the nannoplankton zone NP4 *Ellipsolithus macillus* (*Coccolithus robustus*) and sequential appearances of the *Hafniasphaera cryptovesiculata* and *Palaeocystodinium bulliforme* dinocysts (zones D2a and D2b). Only the Da-4 rhythm beds may be traced in both wells. In the submerged part of the Pricaspian region, they are composed of greenish gray calcareous clays (14 m thick) with a marl interlayer in the basement; in the northern marginal part (Novouzensk well) – of black sandy clays (36 m thick). Sediments of the Da-4 rhythm are peculiar for the appearance of dinocysts *Alisocysta reticulata* (FO and LO), *Spinidinium densispinatum* (FO), *Alterbidinium circulum* (FO). Supposedly, the level of fasciculite first radiation (small *Fasciculithus* sp.) begins to show there. This level is not traceable in the North Pricaspian region, probably due to the development of an island barrier of salt domes and hampered water exchange. In the Danian, the Pricaspian region used to be an open warm-water sea basin inhabited by rich complexes of heat-loving biota (foraminifers, brachiopods, sea urchins, nannoplankton, dinocysts). The tendency to shoaling became obvious by the end of the Danian.

7.2.2 Selandian and Thanetian

Selandian and Thanetian beds are consistently traceable in both wells. They are represented by low calcareous, gaize-like clays of mostly terrigenous genesis. Two Selandian rhythms correlate consistently in the Central and in the North Pricaspian regions. In the Central Pricaspian, the Se-1 and Se-2 beds are lithologically manifested as a uniform sequence of gray, gaize-like clays of Kaztalov formation. In the basin northern marginal part – as differentiated bodies of the Lower Syzran and the Upper Syzran subformations separated with an erosion horizon. The appearance sequence of *Cerodinium depressum* (FO) dinocysts

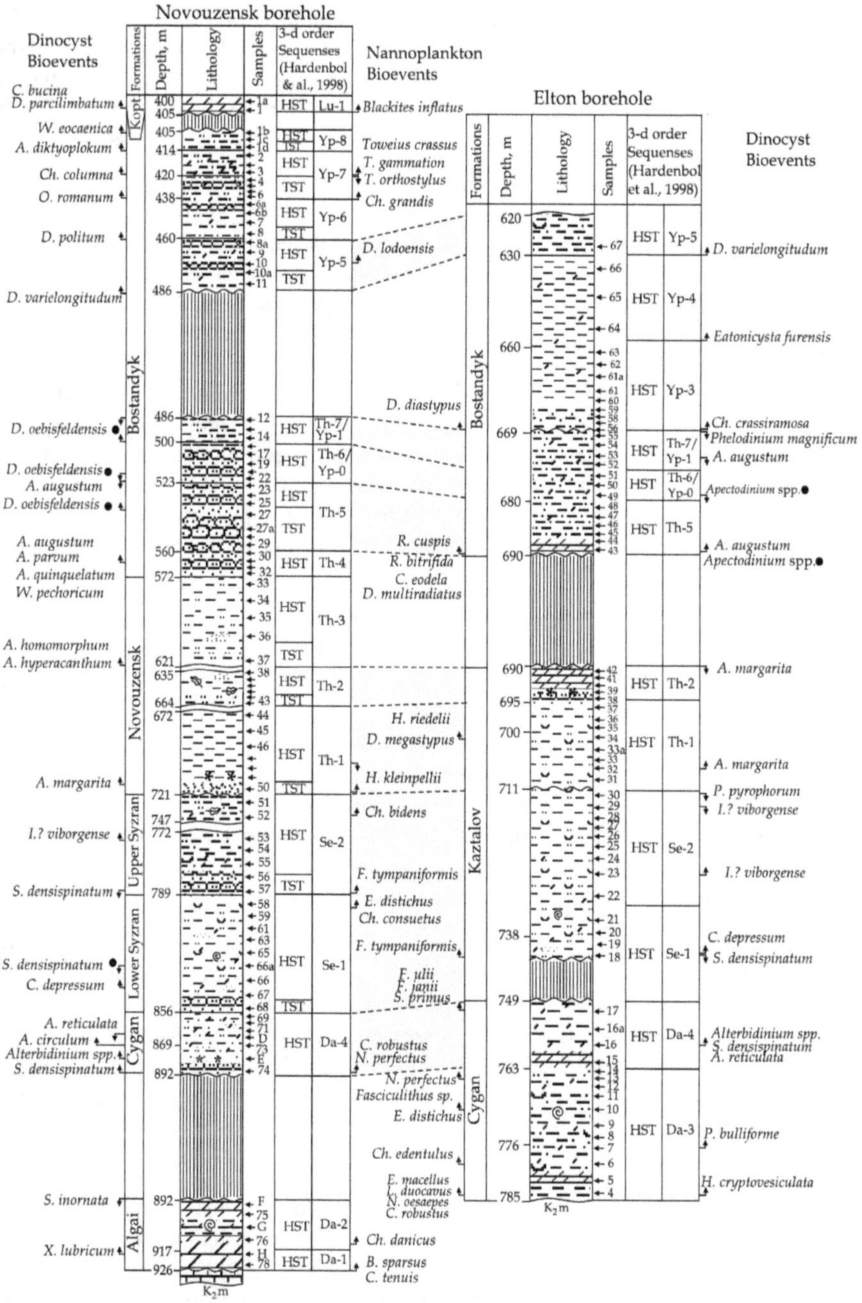

Fig. 5. Correlation of the rhythms in the sections from the Northern and Central Caspian Depression (the Novouzensk key well and the Elton well) based on micropaleontological data. Legend as in Fig. 2.

characterizes the Se-1 rhythm, while FO and LO *Isabelidinium? viborgense*, LO *Palaeoperidinium pyrophorum* - the Se-2 rhythm. The stratigraphic gap between the Danian and the Selandian is clearly manifested in the Elton well. The successive change of dinocyst assemblages there is most comparable with the Danian-Selandian transition in the Viborg 1 section, Denmark (Heilmann-Clausen, 1985).

In the Elton section, the basement of the Kaztalov formation coincides with the appearance of a series of nannoplankton species of the genus *Fasciculithus* (*F. pileatus*, *F. ulii*, *F. janii,*), i.e. with the clearly manifested second radiation of fascicullites. The zonal species *F. tympaniformis* appears several meters higher. This biotic event marks the basement of the Selandian stage. In the Novouzensk section, the nannoplankton complexes are poorer and *Fasciculithus tympaniformis* is recorded only in the basement of the Upper Syzran subformation (Se-2 rhythm). During the Selandian, the sea basin remained warm, shallow-water, and was affected by substantial drift of siliceous material from the continent; that was probably associated with general sinking of the continental margin and widening of the water area. Note, that substantial part of the Volga Uplands is composed of a thick sequence of gaizes and siliceous clays. A vast system of islands, shoals and deeper straits favored formation of organic-walled biota phytocoenoses with peridinioid dinocysts and prasinophyte algae. Nannoplankton associations are poor and in some intervals (Kaztalov formation) have not been revealed at all.

The Lower Thanetian is represented by two lithologically differentiated sequences corresponding to two Thanetian rhythms: Th-1 and Th-2. The lower clay sequence (the upper part of the Kaztalov formation from the Elton section and the lower part of the Novouzensk formation in the Novouzensk section) is described from the appearance of *Alisocysta margarita* dinocysts. It corresponds to the Thanetian first rhythm manifested in the shallow-water settings of the inner neritic zone. Organic plankton associations are represented by peridinioids with numerous prasinophytes and acritarchs. Nannoplankton is absent from most of the beds. Some finds of *Heliolithus kleinpellii* coccoliths in the basement of the Novouzensk formation and *Heliolithus riedelii* in the uppermost of the Kaztalov formation make it possible to date that rhythm quite clearly.

The upper sandy sequence corresponds to the Thanetian second rhythm Th-2. This is manifested by formation of highly dynamic beds in the top of the Kaztalov suite in the Central Pricaspian region and black sandy clays in the North Pricaspian region. Paleontologically, it corresponds to the *Alisocysta margarita* zone (dinocysts) and NP8 *Heliolithus riedeii* (nannoplankton). The Upper Thanetian beds (Th-3 rhythm) are represented in the North Pricaspian region, but are missing from the Elton section.

7.2.3 Ypresian

Lithostratigraphic sequences of the initial Ypresian are fairly consistent in the Pricaspian Depression. In the Central Pricaspian region, they are represented by sandy-clayey, low calcareous sediments from the Bostandyk formation (Th-5, Th-6/Yp-0, Th-7/Yp-1), in the northern marginal part – by a thick sandstone sequence with subordinate clay interlayers. It is remarkable, that two sections separated by 150 km differ so drastically in the contents of organic-walled phytoplankton at that stratigraphic level. Both associations refer to the *Apectodinium augustum* zone. But while acme of the *Apectodinium* species is observed in the depression central, submerged part, in the littoral zone, the Early Eocene climate maximum (IETM) is manifested by acme of *Deflandrea oebisfeldensis*, with subordinate participation of *Apectodinium* spp. We believe this to be associated with depth differences and diverse adaptations of wetzelielloid and deflander dinocysts to the physical-chemical properties of the environment within the inner shelf. Note, that in the basin northern marginal part, no nannoplankton has been revealed at that level; in the central part it is represented by diverse-species complex of the CP8b zone.

Certain significant events occur in the Early Ypresian. In the Central Pricaspian region, the Lower Ypresian is represented by a thick sequence of black, shaly, low calcareous clays formed during the Yp-3 and Yp-4 rhythms. There are no corresponding beds in the Novouzensk well. The Early Ypresian associations of organic-walled phytoplankton and calcareous nannoplankton in the Elton section reflect the features of unfavorable paleoecologic settings, stagnant, reducing environments. One may suppose that the stratigraphic gap revealed in the Novouzensk section was not of local, but of regional character. If the northern marginal part of the depression is accepted to have experienced tectonic uplift, the basin inner, submerged parts happened to become partially isolated; thus, suppressed plankton associations were formed (dinocysts and nannoplankton). Most probably, open interconnections with the Tethys and normal marine sedimentation were restored in the Middle Ypresian, during the Yp-5 rhythm, peculiar for appearance of *Dracodinium varielongitudum* dinocysts in both, northern and central parts of the Pricaspian Depression.

Subsequent biostratigraphic sequences of nannoplankton and dinocysts, inclusive of the Lower Lutetian, have been revealed only in the Novouzensk well. Sandy-clayey non-calcareous rocks occur in the Elton section above the 620 m depth. The core is in bad condition, highly curtailed and unsampled.

We believe that the Novouzensk and the Elton sections complement each other and may constitute the basis for biozonal division of the Pricaspian Depression Paleogene from dinocysts and nannoplankton (Fig. 6). The proposed chart covers the interval from the Lower Danian to the Lower Lutetian and is based on direct correlations of dinocysts and nannoplankton.

8. Conclusion

Examination of the Elton key well has shown that practically entire Paleogene section from the Central Pricaspian region (from the Upper Danian to the Middle Ypresian) is described from productive complexes of dinocysts and nannoplankton. Zonal nannofossil complexes have not been revealed only in a portion of the Upper Selandian and in the lowermost of the Thanetian. The Cygan formation from the basement of the section is referred to the Upper Danian.

It is represented by nannoplankton of the zone NP4 *Ellipsolithus macellus* (*Coccolithus robustus*); in its upper part, small *Fasciculithus* spp. appear. Successive appearances of dinocysts (FO) *Hafniasphaera cryptovesiculata, Palaeocystodinium bulliforme, Alisocysta reticulata* are peculiar for the Cygan formation. The Kaztalov formation is represented by nannoplankton of the zones NP5 *Fasciculithus tympaniformis* (in the lower part) and NP8 *Heliolithus riedelii* (in the top) and corresponds to the Selandian and a part of the Thanetian.

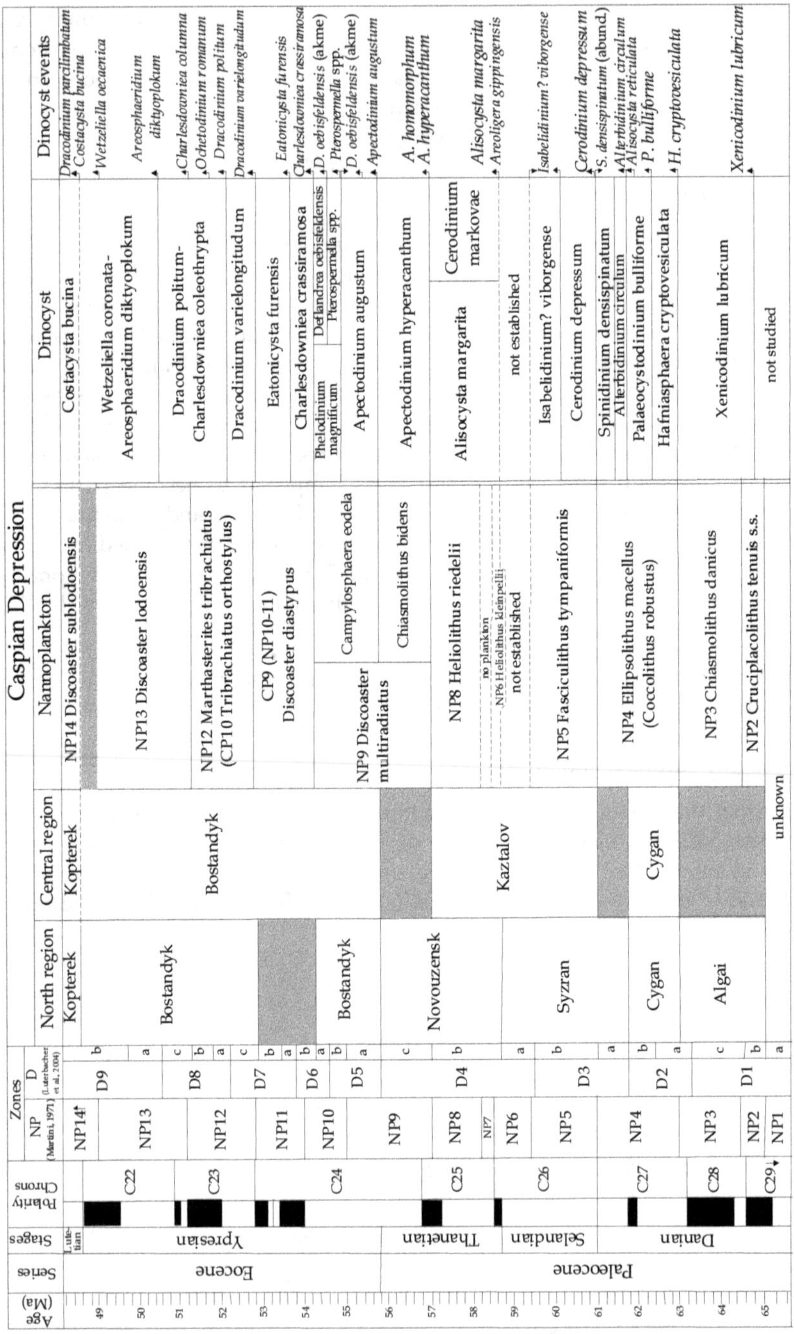

Fig. 6. The chart of the Danian-Lower Lutetian zonal division from nannoplankton and dinoflagellate cysts in the Caspian Depression

First occurence of a number of fasciculites (*F. ulii, F. janii*), that are predecessors of the zonal species - *F. tympaniformis,* markes the base of the Kaztalov formation. Sequences of dinocysts FO *Cerodinium depressum,* FO and LO *Isabelidinium? viborgense,* LO *Palaeoperidinium pyrophorum,* FO *Alisocysta margarita* represent the Kaztalov formation. This is separated from the underlying and from the overlying sequences with a gap. The Bostandyk formation from the top of the Elton section is referred to the Ypresian. The Bostandyk formation is characterized from successive dinocyst events, inclusive of FO *Apectodinium augustum,* acme *Apectodinium* spp., LO *Phelodinium magnificum,* FO *Charlesdowniea crassiramosa,* FO *Eatonicysta furensis,* FO *Dracodinium varielongitudum.* Nannoplankton of the CP8b, CP9 and CP10 zones has been revealed in that formation. Section correlations in the North and Central Pricaspian regions (from the Novouzensk and the Elton key wells) make it possible to present a chart for biostratigraphic division of the Paleogene section from dinocysts and nannoplankton.

Comparison of the organic-walled microphytoplankton complexes from the region shows the dinocyst species diversity to grow substantially in the Central Pricaspian relative to the basin northern marginal part. At that, the share of gonyaulacoid dinoflagellates within the plankton composition increases appreciably. This group is represented mostly by cosmopolitan chorate forms (*Achomosphaera, Spiniferites, Cordosphaeridium, Fibrocysta*), supplemented by species *Xenicodinium, Tectatodinium, Kallosphaeridium, Impletosphaeridium.* Distribution of peridionioids within the Pricaspian inner shelf demonstrates fairly well defined trends during the Ypresian. The wetzelielloid group is inclined to occur in more open marine facies, while deflandroids occur ubiquitously and are preserved in disturbed marine environment determined by restricted sea circulation. Diversity of prasinophyte algae and acritarchs within the phytoplankton denotes formation of local shallow-water, freshened, low-current settings.

Paleoecologic analyses of the organic-walled microphytoplankton complexes and correlations of the two thoroughly dated sections make it possible to reveal cyclicity corresponding to the third-order rhythms and to trace the dynamics of the Pricaspian basin evolution in the Early Cenozoic. Both sections belong to the inner shelf zone, but the Elton section is represented by the deeper facies, farther remote from the shoreline. Breaks in both sections generally correspond to dropout of one or two definite rhythms, which is caused by local tectonic movements.

9. Appendix – List of micro- and nannoplakton

A. Dinoflagellate cysts

Achomosphaera crassipellis *Deflandre & Cookson, 1955) Stover & Evitt, 1978*
Adnatosphaeridium robustum *(Morgenroth, 1966) De Coninck, 1975*
Alisocysta circumtabulata *(Drugg, 1967) Stover et Evitt, 1978*
Alisocysta reticulata *Damassa, 1979*
Alisocysta margarita *(Harland, 1979) Harland, 1979*
Alisocysta *sp. 2 Heilmann-Clausen 1985*
Senegalinium ?dilwynense *(Cookson & Eisenack, 1965) Stover & Evitt, 1978*
Alterbidinium acutulum *(Wilson, 1967) Lentin & Williams, 1985*
Alterbidinium *spp. (including* A. prominense *Vassilyeva,* Alterbidinium compactum *Vassilyeva in Andreyeva-Grigorovich et al., 2011)*

Apectodinium homomorphum *(Deflanre & Cookson, 1955) Lentin & Williams, 1977*
Apectodinium hyperacanthum *(Cookson & Eisenack, 1965) Lentin & Williams, 1977*
Apectodinium quiquelatum *(Williams & Down, 1966) Costa & Downie, 1979*
Apectodinium parvum *(Alberti, 1961) Lentin & Williams, 1977*
Apectodinium augustum *(Harland, 1979) Lentin & Williams, 1981*
Areoligera gippingensis *Jolly, 1992*
Biconidinium *sp.*
Caligodinium aceras *(Manum & Cookson, 1964) Lentin & William,s 1973*
Cannosphaeropsis utunensis *O. Wetzel, 1933*
Cerodinium depressum *(Morgenroth, 1966) Lentin & Williams, 1987*
Cerodinium diebelii *(Alberti 1959) Lentin & Williams 1987*
Cerodinium leptodermum *(Vozzhennikova 1963) Lentin & Williams 1987*
Cerodinium markovae *(Vozzhennikova 1967) Lentin & Williams 1987*
Cerodinium medcalfii *(Stover 1974) Lentin & Williams 1987*
Cerodinium sibiricum *(Vozzhennikova 1967) Lentin & Vozzhennikova 1990*
Cerodinium striatum *(Drugg 1967) Lentin & Williams 1987*
Cerodinium speciosum *(Alberti 1959) Lentin & Williams 1987*
Cerodinium speciosum *subsp. glabrum (Gocht 1969) Lentin & Williams 1987*
Cerodinium dartmoorium *(Cookson & Eisenack 1965) Lentin & Williams 1987*
Cerodinium wardenense *(Williams& Downie 1966) Lentin & Williams 1987*
Charlesdowniea crassiramosa *(Williams & Downie 1966) Lentin & Vozzhennikova 1989*
Cladopyxidium saeptum *(Morgenroth 1968) Stover & Evitt 1978*
Cordosphaeridium ?cracenospinosum *Davey & Williams 1966*
Cordosphaeridium inodes *(Klumpp 1953) Eisenack 1963*
Cleistosphaeridium polypetellum *(Islam 1983) Islam 1993*
Damassadinium californicum *(Drugg 1967) Fensom et al., 1993*
Deflandrea andromiensis *Vozzhennikova 1967*
Deflandrea denticulata *Alberti 1959*
Deflandrea oebisfeldensis *Alberti 1959*
Deflandrea phosphoritica *Eisenack 1938*
Diphyes colligerum *(Deflandre & Cookson 1955) Cookson 1965*
Dracodinium varielongitudum *(Williams et Downie 1966) Costa et Downie 1979*
Dracodinium *sp. 1 Heilmann-Clausen in Heilmann-Clausen & Costa 1989*
Eatonicysta furensis *Stover & Evitt 1978*
Eisenackia ?strobiculata *Morgenroth 1977*
Fibrocysta lappacea *(Drugg 1970) Stover & Evitt 1978*
Glaphyrocysta ?vicina *(Eaton 1976) Stover et Evitt 1978*
Hafniasphaera cryptovesiculata *Hansen 1977*
Hafniasphaera graciosa *Hansen 1977*
Hafniasphaera hyalospinosa *Hansen 1977*
Heslertonia heslertonensis *(Neale & Sarjeant 1962) Sarjeant 1966*
Homotriblium tenuispinosum *Davey & Williams 1966*
Hystrichokolpoma bulbosum *(Ehrenberg 1938) Morgeroth 1968*
Hystrichostrogylon membraniphorum subsp. granulatum *Heilmann-Clausen in Heilmann-Clausen & Costa 1989*
Hystrichosphaeridium tubiferum *(Ehrenberg 1938) Wetzel 1933*

Isabelidinium cooksoniae *(Alberti 1959) Lentin & Williams 1977*
Isabelidinium? viborgense *Heilmann-Clausen 1985*
Impagidinium dispertitum *(Cookson & Eisenack 1965) Stover & Evitt 1978*
Kallosphaeridium brevi*barbatum De Coninck 1969*
Laciniadinium petaloidum *Vassilyeva in Andreyeva-Grigorovich et al., 2011*
Lentinia ?wetzelii *Morgenroth 1966*
Membranilarnacia compressa *Bujak 1994*
Membranilarnacia glabra *Agelopoulus 1967*
Membranosphaera maastrichtica *Samoilovich in Samoilovich et Mtchedlishvili 1961*
Melitasphaeridium *sp.*
Operculodinium nanaconulum *Islam 1983*
Oligosphaeridium *sp.*
Palaeoperidinium pyrophorum *(Ehrenberg 1938) Sarjeant 1967*
Palaeocystodinium australinum *(Cookson 1965) Lentin et Williams 1976*
Palaeocystodinium bulliforme *Ioannides 1986*
Palaeocystodinium benjamini *Drugg 1967*
Palaeocystodinium golzowense *Alberti 1961*
Palaeocystodinium lidiae *(Gorka 1963) Davey 1969*
Palaeotetradinium minusculum *(Alberti 1961) Stover et Evitt 1978*
Phthanoperidinium cren*ulatum (De Coninck 1975) Lentin & Williams 1977*
Rottnestia borussica *(Eisenack 1954) Cookson & Eisenack 1961*
Senegalinium iterlaaense *Nohr-Hansen & Heilmann-Clausen 2001*
Spinidinium densispinatum *Stanley 1965*
Spiniferites supparus *(Grugg 1967) Sarjeant 1970*
Systematophora *sp. 1 Heilmann-Clausen*
Stylodiniopsis maculatum *Eisenack 1954*
Tectatodinium rugulatum *(Hansen 1977) McMinn 1988*
Tectatodinium pellitum *Wall 1967*
Triblastula utinensis *O. Wetzel 1933*
Trigonopyxidia ginella *(Cookson & Eisenack 1960) Down & Sarjeant 1965*
Thalassiphora pelagica *(Eisenack 1954) Eisenack & Gocht 1960*
Thalassiphora delicata *Williams & Downie 1966*
Wilsonidium pechoricum *Iakovleva & Heilmann-Clausen 2010*
Wilsonidium *sp.*
Xenicodinium lubricum *Hansen 1977*

B. Prasinophyta and Acritarcha

Botryococcus *sp.*
Comasphaeridium *sp. Staplin et al., 1965*
Cyclopsiella *sp. Drugg & Loeblich Jr.1967*
Diacrocanthidium *sp. Deflandre & Foucher 1967*
Fromea ?laevigata *(Drugg 1967) Stover & Evitt 1978*
Leiosphaeridia *sp. Eisenack 1958*
Paralecaniella identata *(Deflandre & Cookson 1955) Cookson & Eisenack 1970*
Palambages morulosa *O. Wetzel 1961*
Paucilobimorpha ?apiculata *(Cookson & Eisenack 1962) Prössl 1994*

Pseudomasia trinema *De Coninck 1969*
Veryhachium *sp.*

C. Nannoplankton

Biantholithus sparsus *Bramlette & Martini, 1964*
Braarudosphaera bigelowii *(Gran & Braarud, 1935) Deflandre, 1947*
Campylosphaera eodela *Bukry & Percival, 1971*
Chiasmolithus (Sullivania) danicus *(Brotzen, 1959) ex van Heck & Perch-Nielsen, 1987*
Chiasmolithus bidens *(Bramlette & Sullivan, 1961) Hay & Mohler, 1967*
Chiasmolithus (Sullivania) consuetus *(Bramlette & Sullivan) Hay & Mohler, 1967*
Coccolithus (Ericsonia) cavus *Hay & Mohler, 1967*
Coccolithus robustus *(Bramlette & Sullivan, 1961) Wise et al., 2002*
Coccolithus pelagicus *(Wallich, 1877) Schiller, 1930*
Coccolithus subpertusus *(Hay & Mohler, 1967) Wei & Pospichal, 1991*
Cruciplacolithus tenuis *(Stradner, 1961) Hay & Mohler, 1967*
Cruciplacolithus primus *Perch-Nielsen, 1977*
Cruciplacolithus frequens *(Perch-Nielsen, 1977) Romein, 1979*
Discoaster araneus *Bukry, 1971*
Discoaster diastypus *Bramlette & Sullivan, 1961*
Discoaster multiradiatus *Bramlette & Riedel, 1954*
Discoaster megastypus *(Bramlette & Sullivan, 1961) Bukry, 1973*
Discoaster nobilis *Martini, 1961*
Ellipsollithus distichus *(Bramlette & Sullivan, 1961) Sullivan, 1964*
Ellipsolithus macellus *(Bramlette & Sullivan, 1961) Sullivan, 1964*
Fasciculithus tonii *Perch-Nielsen, 1971*
Fasciculithus tympaniformis *Hay & Mohler, 1967*
Fasciculithus janii *Perch-Nielsen, 1971*
Heliolithus riedelii *Bramlette & Sullivan, 1961*
Markalius inversus *(Deflandre in Deflandre & Fert, 1954) Bramlette & Martini, 1964*
Neochiatozygus eosaepes *Perch-Nielsen, 1981*
Neochiastozygus junctus *(Bramlette & Sullivan, 1961) Perch-Nielsen, 1971*
Neochiastozygus perfectus *Perch-Nielsen, 1971*
Neochiastozygus saepes *Perch-Nielsen, 1971*
Neococcolithes dubius *(Deflandre in Deflandre & Fert, 1954) Black, 1967*
Placozygus sigmoides *(Bramlette & Sullivan, 1961) Romein, 1979*
Prinsius martinii *(Perch-Nielsen, 1969) Haq, 1971*
Rhomboaster bitrifida *Romein, 1979*
Rhomboaster cuspis *Bramlette & Sullivan, 1961*
Rhomboaster spineus *(Shafik & Stradner, 1971) Perch-Nielsen, 1984*
Sphenolithus primus *Perch-Nielsen, 1971*
Lithraphidites quadratus *Bramlette & Martini 1964*

10. Acknowledgments

We are grateful to managers of OAO "Saratovneftegas" who gave us the opportunity to study and collect core samples from the Novouzensk and Elton key wells.

11. References

Aleksandrova, G.N. (2001). Palynological Characteristics of Paleocene Deposits in the Lower Volga Region (Borehole 28, town of Dubovka). *Stratigraphy and Geological Correlation*, Vol. 9, No. 6, pp. 529-602, ISSN 0869-5938

Amon, E.O., Vasilyeva, O.N. & Zhelezko, V.I. (2003). Stratigraphy of the Talitsa Horizon (Paleocene) in the Central Trans-Urals. *Stratigraphy and Geological Correlation*, Vol. 11, No 3, pp. 278-292, ISSN 0869-5938

Andreyeva-Grigorovich, A.S. (1991). Phytoplankton (Dinocysts and Nannoplankton) Zonal Stratigraphy of the Paleogene of southern USSR (thesis). Kiev (Ukraine); Geological Institute, 47 p (in Russian)

Andreyeva-Grigorovich, A.S., Zaporozhec, N.I., Shevchenko T.V., Aleksandrova G.N., Vasilyeva O.N., Iakovleva A.I., Stotland, A.B. & Savitskaya, N.A. (2011). *Atlas of Paleogene Dinocysts of Ukraine, Russia and adjacent countries*, Naukova dumka, ISBN 978-966-00-1044-4, Kiev, Ukraine (in Russian)

Aubri, M.-P. (1986). Paleogene calcareous nannoplankton biostratigraphy of Northwestern Europe. *Palaeogeography, Palaeoclimatology, Palaeoecology*, Vol. 55., pp. 267–334, ISSN 0031-0182

Berggren, W.A., Kent, D.V., Swiher, C.C. & Aubry, M.-P. (1995). A revised Cenozoic chronology and chronostratigraphy, In Berggren, W.A., et al. (Eds.), *Geochronology, Time Scales, and Global Stratigraphic Correlation*, SEPM (Society for Sedimentary Geology) Special Publication, Vol.54, ISBN n.d., pp. 129-212

Brinkhuis, H. (1994). Late Eocene to early Oligocene dinoflagellate cysts from the Priabonian type-area (northeast Italy): biostratigraphy and palaeoenvironmental interpretation. *Palaeogeography, Palaeoclimatology, Palaeoecology*, Vol.107, pp. 121–163, ISSN 0031-0182

Bujak, J.P. & Brinkhuis, H. (1998). Global warming and dinocysts changes across the Paleocene/Eocene boundary, In: *Late Paleocene–Early Eocene biotic and climatic events in the marine and terrestrial records*, M.-P. Aubry et al. (Eds.), pp. 277–295, New York: Columbia Univ. Press, ISBN n.d

Costa, L.A. & Manum, S.B. (1988). The description of the interregional zonation of the Paleogene (D1-D15) and Miocene (D16-D20). *Geologisches Jahrbuch*, Reihe A, Helt 100, pp. 321-330, ISSN 0341-6399

Crouch, E. & Brinkhuis, H. (2005). Environmental change across the Paleocene–Eocene transition from eastern New Zealand: a marine palynological approach. *Marine Micropaleontology*, Vol.56, pp. 138–160, ISSN 0377-8398

Crouch, E.M., Heilmann-Clausen, C., Brinkhuis, H., Morgans, H.E.G., Rogers, K.M., Egger, H., & Schmitz, B. (2001). Global dinoflagellate event associated with the late Paleocene Thermal Maximum. *Geology*, April 2001, Vol.29, No.4, pp. 315–318, ISSN 0091-7613

Dale, B. (1996). Dinoflagellate cysts ecology: modeling and geological applications, In: *Palynology: principles and applications*, J. Jansonius, & D.C. McGregor. (Eds.), pp. 1249–1276, American Association of Stratigraphic Palynologists Foundation, Vol.3, ISBN 9-931871-03-4, USA

De Coninck, J. (1988). Ypresian organic-walled phytoplankton in the Belgian Basin and adjacent areas. *Bulletin de la Societe Belge de Geologie*, Vol.97, No.3–4, pp. 287–319, ISSN 0-772-9464

Egger, H., Fenner, J., Heilmann-Clausen, C., Rögl, F., Sachsenhofer, R. & Schmitz, B. (2003). Paleoproductivity of the northwestern Tethyan margin (Anthering section, Austria) across the Paleocene–Eocene transition, In: *Causes and consequences of globally warm climates in the Early Paleogene*, S.L. Wing et al. (Eds.), pp. 133–146, Geological Society of America Special Paper, Vol.369, ISSN 0072-1077

Fensome, R.A. & Williams, G.L. (2004). The Lentin et Williams Index of Fossil Dinoflagellates. *American Association of Stratigraphic Palynologists*, Contributions Series, No.42, 909 p, ISSN 0160-8843

Grachev, N.V., Zhizhchenko, B.P., Kolyhkalova, L.A. & Kholodilina, T.S. (1971). Paleogene sediments in the central Volga and Ural rivers basin, In: *Cainozoic Stratigraphy and Paleogeography of the south USSR gas and oil-bearing regions*, B.P. Zhizhchenko, V.A. Ivanova (Eds.), VNIIGas Publications, No. 31/39–32/40, pp. 36-45, Nedra, Moscow (in Russian)

Guasti, E., Speijer, R., Brinkhuis, H., Smit, J. & Steurbaut, E. (2006). Paleoenviromental change at the Danian-Selandian transition in Tunisia: Foraminifera, organic-walled dinoflagellate cyst and calcareous nannofossil records. *Marine Micropaleontology*, Vol.55, pp. 1–17, ISSN 0377-8398

Habib, D., Moshcovitz, S. & Kramer, C. (1992). Dinoflagellate and calcareous nannofossil response to sea-level change in Cretaceous–Tertiary boundary sections. *Geology*, Vol. 20, pp. 165–168, ISSN 0264-8172

Hansen, J.M. (1977). Dinoflagellate stratigraphy and echinoid distribution in Upper Maastrichtian and Danian deposits from Denmark. *Bulletin Geological Society Denmark*, Vol. 26, August 1977, pp. 1-26, ISSN 0011-6297

Haq, B.U., Hardenbol, J. & Vail, P.R. (1988). Mesozoic and Cenozoic chronostratigraphy and cycles of sea-level change. In: *Sea-level changes: an integrated approach*, Society of Economic Paleontologists and Mineralogists, Special Pubication No.42, pp. 71-108, Tulsa, Oklahoma (USA), ISBN 0-918985

Hardenbol, J. (1994). Sequence stratigraphic calibration of Paleocene and Lower Eocene continental margin deposits in the NW Europe and US Gulf Coast with the oceanic chronostratigraphic record, *Meeting Procceedings "Stratigraphy of the Paleocene"*, ISBN n.d., GFF, Vol. 116, pp. 49-51

Hardenbol, J., Thierry, J., Farley, M.B., Jacquin T., Graciansky P.-C. & Vail P.R. (1998). Mesozoic and Cenozoic sequence chronostratigraphic framework of European basins, In: *Mesozoic and Cenozoic sequence stratigraphy of European Basins*, P.-C. Graciansky, J. Hardenbol, T. Jacquin, P.R. Vail (Eds.) Society of Economic Paleontologists and Mineralogists, Special Publication No.60, Tulsa, Oklahoma, USA, ISBN 0-918985

Heilmann-Clausen, C. & Costa, L.I. (1989). Dinoflagellate Zonation of Uppermost Paleocene? to Lower Miocene in the Wursterheide Research Well, NW Germany, *Geologisches Jahrbuch*, A 111, pp. 431-521, ISSN 0341-6399

Heilmann-Clausen, C. (1985). *Dinoflagellate stratigraphy of the uppermost Danian to Ypresian in the Viborg 1 borehole, central Jylland, Denmark*. Danmarks Geologiske Undersøgelse. S. A. No.7, pp. 1-69, ISBN n.d

Heilmann-Clausen, C. (1994). Review of Paleocene dinoflagellates from the North Sea region. *Meeting Proceedings "Stratigraphy of the Paleocene"*, GFF, Vol. 116, pp. 51-53, ISBN n.d

Iakovleva, A.I., Brinkhuis, H., & Cavagnetto, C. (2001). Late Paleocene–Early Eocene dinoflagellate cysts from the Turgay Strait, Kazakhstan: correlation across ancient seaways. *Palaeogeogaphy, Palaeoclimatology, Palaeoecology*, Vol. 172, pp. 243–268, ISSN 0031-0182

Kirsch, K.-H. (1991). *Dinoflagellatenzysten aus der Obekreide des Helveticums und Nordultrahelveticums von Oberbayern*, Munchner Geowissenchaftliche, Abh.22, 306 p, ISBN n.d., München

Köthe, A. (2003). Paleogene Dinoflagellates from northwest Germany – Biostratigraphy and Paleoenvironment. *Geological Jahrbuch*, A 118, pp. 3-111, ISSN 0341-6399

Luterbacher, H.P., Ali, J.R., Brinkhuis, H., Gradstein, F.M., Hooker, J.J., Monechi, S., Ogg, J.G., Powell, J., Rohl, U., Sanfilippo, A. & Schmitz, B. (2004). The Paleogene Period, In F.M. Gradstein, J.G. Ogg, Smith A. (eds.) A Geologic Time Scale. Cambridge (UK): Cambridge University Press, pp. 384-408, ISBN n.d.

Marheinecke, U. (1992). Monographie der Dinozysten, Achritarcha und Chlorophyta des Maastrichtium von Hemmoor (Hierdersachsen). *Palaeontographica*, Abt. B, Vol.227, Lfg. 1-6, August 1992, pp.1-173, ISSN 1846-1933

Martini, E. (1971). Standard Tertiary and Quaternary calcareous nannoplankton zonation. *Proceedings of the Second Planktonic Conference*, Farinacci A. (Ed.), Roma, pp. 739-785, ISBN n.d

Mudge, D.C. & Bujak, J.P. (2001). Biostratigraphic evidence for evolving palaeoenvironments in the Lower Paleogene of the Faeroe-Shetland Basin. *Marine and Petroleum Geoogy*, Vol.18, pp. 577-590, ISSN 0264-8172

Musatov, V. A. (1996). *The Paleogene Nannoplankton Biostratigraphy of the Lower Volga Region (thesis)*, Saratov (Russian Federation): Lower Volga Institute of Geology and Geophysics, 25 p (in Russian)

Okada, H. & Bukry, D. (1980). Supplementary modification and introduction of code numbers to the low-latitude coccolith biostratigraphic zonation (Bukry, 1973, 1975). *Marine Micropaleontoogy*, Vol.5, pp. 321-325, ISSN 037-8398

Pechenkina, A.P. & Kholodilina, T.S. (1971). Some data on Paleogene deposits stratigraphy from Volga and Ural rivers basin based on Foraminiferas, In: *Cainozoic Stratigraphy and Paleogeography of the south USSR gas and oil-bearing regions*, B.P. Zhizhchenko, V.A. Ivanova (Eds.), VNIIGas Publications, No. 31/39-32/40, pp. 25-36, Nedra, Moscow (in Russian)

Perch-Nielsen, K. (1985). Cenozoic calcareous nannofossils. In: *Plankton Stratigraphy*, H. Bolli, J. Saunders, K. Perch-Nielsen (Eds.), pp. 427-554, Cambridge University Press (UK), ISBN 0 521 23576 6

Petrova, I.A. (1986). *Recommendations for methods of palynological preparation of sediments*. VSEGEI Publications, Leningrad, 77 p (in Russian)

Powell, A.J. (1992). Dinoflagellate cysts of the Tertiary System, In: *A stratigraphic index of Dinoflagellate cysts*, A.J. Powell (Ed.), pp. 155-251, London: British Micropaleontol. Soc. Ser., Published in 1992 by Chapman & Hall, ISBN 0412-362805, London (UK)

Powell, A.J., Brinkhuis, H. & Bujak, J.P. (1996). Upper Paleocene-Lower Eocene dinoflagellate cyst sequence biostratigraphy of southeast England. *Correlation of the Early Paleogene in Northwest Europe*, R.W.O'B. Knox, R.M. Corfield, R.E. Dunay (Eds.), Geological Society Special Publication, London. Vol.101, pp. 145-183, ISBN 0305-8719

Pross, J. & Schmiedl, G. (2002). Early Oligocene dinoflagellate cysts from the Upper Rhine Graben (SW Germany): paleoenvironmental and paleoclimatic implications. *Marine Micropaleontology*, Vol.45, pp. 1-24, ISSN 0377-8398

Razmyslova, S.S. & Nikitina, J.P. (1975). The Caspian Depression, In: *Stratigraphy of USSR. Paleogene system*, D.V. Nalivkin (Ed.), pp.204-219, VSEGEI Publications, Nedra, Moscow (in Russian)

Schiøller, P. & Wilson, G.J. (1993). Maastrichtian dinoflagellate zonation in Dan Field, Danish North Sea. *Review of Palaeobotany and Palynology*, Vol.78, pp. 321-351, ISSN 0034-667

Sissingh, W. (1977). Biostratigraphy of Cretaceous calcareous nannoplankton. Geologie en Mijnbouw, Vol.56, No.1, pp.37-65, ISSN 0016-7746

Slimani, H. (2001). Les kystes dinoflagellésdu Campanian au Danien dans la region de Maastricht (Belgique, Pays-Bas) et de Turnhout (Belgique): biozonation et corrélation avec d'autres regions en Europe occidentale. *Geologica et Paleontologica*, Vol.35, August 2001, pp. 161-201, ISSN 0344-659X

Sluijs, A., Pross, J. & Brinkhuis, H. (2005). From greenhouse to icehouse: organic-walled dinoflagellate cysts as paleoenvironmental indicators in the Paleogene. *Earth-Science Reviews*, Vol.68, pp. 281–315, ISSN 0012-8252

Steurbaut, E., Magioncalda, R., Dupuis, C., van Simaeus, S., Roche, E & Roche M.-I.E. (2003). Palynology, paleoenvironments, and organic carbon isotope evolution in lagoonal Paleocene-Eocene boundary settings in North Belgium, In: *Causes and consequences of globally warm climates in the Early Paleogene*, S.L. Wing, P.Gingerich, B. Schmitz, E. Thomas (Eds.). Geological Society of America, Special Paper No.369, pp. 291–317, ISSN 0072-1077

Stover, L.E. & Hardenbol, J. (1994). Dinoflagellates and depositional sequences in the lower Oligocene (Rupelian) Boom clay formation, Belgium. *Bulletin de la Societe Belge de Geologie*, Vol. 102, No.1–2, pp. 2–77, ISSN 0379-1807

Torricelli, S., Knezaurek, G. & Biffi, U. (2006). Sequence biostratigraphy and palaeoenvironmental reconstruction in the Early Eocene Figols Group of the Tremp-Graus Basin (south-central Pyrenees, Spain). *Palaeogeography, Palaeoclimatology, Palaeoecology*, Vol.232, pp. 1–35, ISSN 031-0182

Varol, O. (1998). Palaeocene calcareous nannofossil biostratigraphy, *Proceedings of the International Nannofossil Association Conference (Nannofossils and their applications)*, pp. 267-310, ISBN 0-7458-0237-0, London (UK), 1987

Vasilyeva, O.N. (2011). New wetzielloidae dinocyst species (Dinophyceae) from Eocene deposits of Eastern Pricaspian. *IGG UrB RAS Publications*, Vol.158, pp. 14-20, ISBN 978-5-94332-090-3, Available from http://www.igg.uran.ru/Publications/Tr158_2011/Tr158_Contents.htm (in Russian)

Vasilyeva, O.N. & Musatov, V.A. (2010a). Paleogene Biostratigraphy of the North Circum-Caspian Region (Implication of the Dinocysts and Nannoplankton from the Novouzensk reference borehole), Part 1: Age Substantional and Correlation of Deposits. *Stratigraphy and Geological correlation*, Vol.18, No.1 (January-February 2010), pp. 83-104, ISSN 0869-5938

Vasilyeva, O.N. & Musatov, V.A. (2010b). Paleogene Biostratigraphy of the North Caspian Region based on Dinocysts and Nannofossils from the Novouzensk borehole. Article 2: Biotic events and Paeoecological Settings. *Stratigraphy and Geological correlation*, Vol.18, No.2, (March-April 2010), pp. 177-199, ISSN 0869-5938

Wall, D., Dale, B., Lohmann, G.P. & Smith, W.K. (1977). The environmental and climatic distribution of dinoflagellate cysts in the North and South Atlantic and adjacent seas. *Marine Micropaleontology*, Vol. 30. pp. 319–343, ISSN 0377-8398

Williams, G.L., Brinkhuis, H., Pearce, M.A., Fensome, R.A. & Weegink, J.W. (2004). Southern Ocean and global dinoflagellate cyst events compared: Index events for Late Cretaceous–Neogene. In: *Proceedings of the Ocean Drilling Program, Scientific Results*, Vol.189, N.F. Exon, J.P. Kennett, M.J. Malone (Eds.), Available from http://www-odp.tamu.edu/publications/189_SR/107/107.htm

Zhuravlev, V.S. (1970). The Caspian Depression, In: *Geology of USSR, Vol. XXI, Western Kazakhstan*, Part I (Geological description), Book 2, A.V. Sidorenko (Ed.). pp. 108-165, Nedra, Moscow (in Russian)

7

Pliocene Mediterranean Foraminiferal Biostratigraphy: A Synthesis and Application to the Paleoenvironmental Evolution of Northwestern Italy

Donata Violanti
Earth Science Department, Turin University,
Italy

1. Introduction

In this chapter biostratigraphic concepts and methods are applied to the Pliocene foraminiferal assemblages of the central Piedmont (Northwestern Italy), with examples of correlations at regional and Mediterranean scale. The order of discussion of topics is as follows: foraminifers and biostratigraphy and ecobiostratigraphy methods, a review of previous definitions of the Pliocene series and its stages, the Mediterranean foraminiferal biozonation, brief description of the Northwestern Italy Pliocene succession and of Piedmont selected sections, correlations and applications to the paleoenvironmental and tectonic history of the area. Locations of the stratotype sections, of cited sites, and of the studied sections are reported in Fig. 1.

Foraminifers are one of the most widespread marine protozoans, inhabiting both the water colomn with planktonic forms and bottom sediments from the inner neritic zone to bathyal depths with very diversified taxa. Planktonic species, known from the Jurassic, are characterized by a very high evolutive rate and provide the biostratigraphic markers in pelagic, deep sea sediments of Upper Mesozoic and Cenozoic. Appearance or occurrence, as well as disappearance of a taxon represent biostratigraphic data.

Moreover, other data could be obtained by environmentally controlled parameters, such as frequency peaks or coiling changes of a taxon occurring nearly simultaneously in the same area. In the Mediterranean Pliocene, ecobiostratigraphical data, recognized all over the region, are the delayied re-entry of some planktonic (*Globorotalia scitula*) and benthic species (*Cibicidoides robertsonianus, Siphonina reticulata*) during the lower Early Pliocene, slightly later than the basin infilling after the Mediterranean Salinity Crisis. Ecobiostratigraphy based correlations are obviously more dependent from environmental parameters than biostratigraphical events, but are useful to improve the stratigraphic resolution almost at the regional scale.

The peculiar geological history of the Mediterranean was at the origin of a distinct bioprovince, and justifies the necessity of a regional biostratigraphic zonation. During the

Neogene, the closure of marine connections with the Indian Ocean, to the East, and the reduction of comunications with the Atlantic, to the West, led to the progressive decrease of exchanges with oceanic waters. From the Oligocene and Miocene, the latitudinal northward shift was the main cause of the disappearance of tropical macroforaminifers, whereas the stronger climatic deterioration during the Upper Pliocene and the Pleistocene affected also the warm or warm-temperate water planktonic and benthic foraminifers. Aims of this paper are to document the application of foraminiferal biostratigraphy and ecobiostratigraphy to the Pliocene succession of Piedmont, that provide new data on the evolution of the Northwestern Mediterranean margin and its relation with the central basin.

2. Biostratigraphy and ecobiostratigraphy methods

From the second half of the last century, planktonic foraminifers provide fundamental tools for biostratigraphy, the part of stratigraphy using the fossil distribution to construct relative time scales and to correlate time-equivalent sediment layers. From the Jurassic, and mainly from their explosive diffusion in the Cretaceous, many planktonic foraminiferal species have had the quality for being successfully applied as index fossils: a world-wide diffusion in the marine realm, good preservation in sediments due to their calcareous test, common to abundant specimens in marine deposits, fastly evolving taxa with many short-lived species, and various morphological features, allowing taxonomic identification of different species and subspecies. In the last decades, the integration of micropaleontological, geomagnetic and radiometric data with orbital periodicities provided an accurate calibration of biostratigraphic units and events.

The biozone, formal unit of biostratigraphy, is defined as the stratigraphic interval characterized by a typical fossil content, different from those of the adjacent rock bodies (Salvador, 1994). Primary data for the definition of biozones and biostratigraphic scales are bioevents, such as:

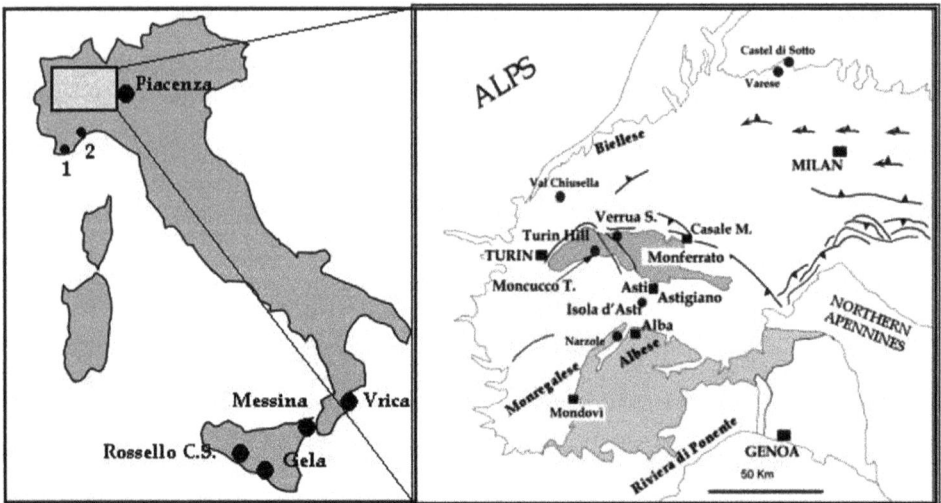

Fig. 1. Location of the Rossello Composite Section (Zanclean and Piacentian GSSP = Global Sections and Points), of cited sites and studied sections. 1: Ventimiglia, 2: Ceriale, Rio Torsero.

- the First Appearance Datum (FAD), defined as the first finding of a new taxon. The datum is an evolutive event, inferred geologically instantaneous in the speciation area.
- the Last Appearance Datum (LAD), defined as the last finding of the marker species. It corresponds to the end of the distribution range of the taxon.
- the First Occurrence (FO) and First Common Occurrence (FCO). The diffusion of a new species outside the origin area requires time and favourable environmental conditions. Therefore its occurrence is diachronous at different sites, and its FO and FCO will occurr at younger ages than the taxon FAD in the origin area. The taxon FO is often difficult to detect, for rareness of specimens, unfavourable paleoenvironment or sedimentary gaps. For these reasons, the FCO is preferred as a more reliable datum.
- the Last Occurrence (LO) and the Last Common Occurrence (FCO). Opposite to the FO and FCO, the LO and LCO register the local disappearance, or the last common recovery of a species.

All these bioevents are applied to define biozone boundaries or as datum planes or datum levels for the definition of biohorizons. Among the different types of biozones used by paleontologists, one of the more frequently used by micropaleontologists is the Interval Zone, defined as the interval, or body of strata, between two specified bioevents (FAD, LAD, FO, etc.). Other biozones of common application are the two type of Range Zones: the Taxon-Range Zone or Total Range Zone, defined as the total range of stratigraphic and geographic occurrence of the selected marker, and the Concurrent Range Zone, defined as the body of strata including the overlapping parts of the distribution of two marker taxa. Abundance Zones, frequently called Acme or Peak Zones, are defined as the sedimentary interval in which an unusual abundance of a peculiar taxon is registered. The species abundance is strongly influenced by local environmental factors and can occur at different times in different sites. Acme zones are more properly ecobiostratigraphical zones, strongly dependent on environmental factors (water depth and temperature, trophic resources, etc.), that influence the species diffusion and abundance (Iaccarino, 1985). Lineage Zones are defined as the sedimentary interval yielding specimens of a peculiar (total or partial) segment of an evolutionary lineage. Lineage Zones are considered the most reliable units for biostratigraphic correlations and their boundary can approximate those of chronostratigraphic units. They require confident philetic relations between the selected taxa and therefore their application is limited to few well known lineages. For the Pliocene, an example is provided by the evolution of *Globorotalia margaritae*. The taxon is an index species for the Mediterranean Pliocene foraminiferal biozonation and evolved in the equatorial Atlantic during the Messinian (Late Miocene), probably from the ancestral forms *Globorotalia scitula* and *G. juanai* (Kennett & Srinivasan, 1983). The species, originally described from sediments of Venezuela of dubious Miocene age (Bolli & Bermudez, 1965), was later entirely ascribed to the Pliocene (Bolli, 1970). Cita (1973) documented the morphological evolution of *G. margaritae* in Mediterranean deep-sea Pliocene assemblages and recognized two new subspecies, *Globorotalia margaritae primitiva* and *Globorotalia margaritae evoluta*, besides the typical *Globorotalia margaritae margaritae*. On the basis of their stratigraphical distribution, the author defined two lineage zones, comprised with the younger *Sphaeroidinellopsis* acme zone in a super zone, named *Globorotalia margaritae* Total Range Zone, that covered the entire Lower Pliocene (in the twofold division in Lower and Upper Pliocene).

Cita (1973) zonal scheme was substituted by the Mediterranean Pliocene zones, (MPl1 to 6), numbered from bottom to top (Cita, 1975a). Nevertheless, the *G. margaritae* lineage zones remain successfully applied to Mid-Atlantic (Bolli & Saunders, 1985) and Southern Atlantic successions (Coimbra et al., 2009). Cita (1973) also hypothesized that the *Globorotalia margaritae* Total Range Zone would cover an undefined interval of the late Miocene. In the upper Messinian, the closure of communications with the Atlantic prevented the diffusion of the taxon within the Mediterranean. The hypothesis of a Miocene origin of *G. margaritae* was confirmed by following researches, dating its FAD at 6.2 Ma (Chaisson & Pearson, 1997) in the central Atlantic Ocean. A taxon related to the *Globorotalia margaritae* plexus is *Globorotalia praemargaritae*, described in Early Messinian sections of Sicily (Catalano & Sprovieri, 1969; 1971), recognized also in coeval interval of land sections in Spain (Berggren et al., 1976). The presence of this unkeeled species testified influxes of the *G. margaritae* lineage ancestrals in the Mediterranean, before the onset of the Messinian Salinity Crisis.

Ecobiostratigraphy applies to stratigraphy paleobiological signals derived or influenced by the ecosystem, so increasing the biostratigraphic resolution and allowing detailed correlations. The FO, FCO, LO and LCO, as well as fluctuations in abundance of a taxon, are diachronous in different paleobioprovinces or in distant areas of the same geographical province and are influenced by paleoenvironmental factors, such as temperature, changes in water-mass circulation, basin depth, etc. Also variations of ecophenotypical characters such as the coiling pattern are successfully applied as bioevents and for local or world-wide correlations. This is the case of the coiling change of *Neogloboquadrina acostaensis* from sinistral to dextral, registered worl-wide in the Late Miocene (Messinian) (Bolli & Saunders, 1985). In the Mediterranean successions it marks the upper boundary of the *Globorotalia conomiozea* Zone (pre-evaporitic Messinian) (Iaccarino, 1985). Other coiling shifts of *N. acostaensis*, recognized in the lowermost Pliocene (Di Stefano et al., 1996), provide tools for correlations in the Mediterranean area (Iaccarino et al., 1999a; Pierre et al., 2006). A high resolution sampling and detailed quantitative studies are needed to obtain a reliable ecobiostratigraphic record. The most important feature of ecobiostratigraphic events is that they occurred in the same stratigraphic order in different sites. Integrated micropaleontological, cyclostratigraphic, isotopic and paleomagnetostratigraphic analyses precisely calibrated many ecobiostratigraphic events occurring in the Pliocene, that will be discussed in the following paragraphs.

3. The Pliocene chronostratigraphy

It is beyond the scope of this work to propose a detailed review of former biostratigraphic and chronostratigraphic researches on the Pliocene and its subdivisions. Wide discussion and references can be found in Berggren et al. (1985), Cita (1975b), Gradstein et al. (2004), Iaccarino et al. (1985), Rio et al. (1991) and Vai (1997). Here the present state-of-the-art will be presented, with a synthesis of selected previous studies.

The Pliocene series was introduced by Lyell (1833), accepting the "Sub Apennine Strata" in northern Italy (Brocchi, 1814) as its sedimentary documentation. During the XIX century, the Pliocene stages were also described in Italy, mainly on the basis of lithology and macrofossil assemblages preserved in the marine deposits. Following researches evidenced the poor definition of the original stages description and the necessity of formally defined chronostratigraphic subdivisions. The bloom of micropaleontological studies in the second half of the XX century provides tools for a detailed description of type sections as well as for the erection of biostratigraphic zonations. More recently, integrated biostratigraphy, magnetostratigraphy, cyclostratigraphy and climatostratigraphy greatly improved the age resolution of the Pliocene record. Meanwhile, these researches demonstrated that the original sections introduced as stratotypes for the historical Pliocene stages (Zanclean, Tabianian, Piacenzian, Astian) were inadequate, because now not accessible (Zanclean) or for gaps (Tabianian, Piacenzian). Other stages, as the Astian (Mayer-Eymar, 1868) and the Fossanian (Sacco 1887), originally proposed in Piedmont, later resulted to represent time-equivalent sedimentary facies, often heteropic to the Early Pliocene sediments. The Sicilian (Doderlein, 1870-1872), introduced as representing the older Pliocene, was soon referred as to the Pleistocene and now abandoned. Many chronostratigraphic subdivisions of the Pliocene series in two or three subseries were proposed, often with different chronological significance. A brief selection is summarized in Tab. 1. As a consequence, Pliocene stages and their boundaries were frequently revised and modified, increasing instability in stratigraphic nomenclature and confusion in correlations between different successions.

Biostratigraphic zones, based on datum planes such as FAD, LAD etc., are indipendent from fluctuating chronostratigraphic boundaries and provided more stable references. The continuous sediments cored in the Mediterranean Sea by the DSDP Leg 13 (Ryan et al., 1973), ODP Leg 107 (Kastens et al., 1990), Leg 160 (Robertson et al., 1998) and Leg 161 (Zahn et al., 1999) provided reference deep-sea sequences, with a micropaleontological record more complete and better preserved than in most Pliocene land-sections.

4. The Pliocene stages

4.1 Zanclean

The Zanclean stage was introduced by Seguenza (1868) on the Gravitelli (Sicily) outcrop of the Trubi Formation, whitish calcareous marls with abundant planktonic foraminifers. Gravitelli, at Seguenza's time a village near the city of Messina, in the following years was included in the city and now the described Trubi outcrop is not accessible.

PLIOCENE					REFERENCES
PIACENZIAN	ASTIAN		FOSSANIAN		Sacco, 1889-1890
PIACENZIAN	ASTIAN		SICILIAN		De Lapparent, 1906
EARLY		MIDDLE	UPPER		Ruggieri & Selli, 1949
TABIANIAN		PIACENZIAN			Barbieri, 1967
ZANCLEAN		PIACENZIAN			Cita & Gartner, 1973; Cita, 1975
EARLY		MIDDLE	UPPER	PLEISTOCENE	AGIP, 1982
ZANCLEAN		PIACENZIAN			Iaccarino, 1985
ZANCLEAN		PIACENZIAN	GELASIAN		Rio el al., 1994
ZANCLEAN		PIACENZIAN	GELASIAN		Gradstein et al., 2004
ZANCLEAN	PIACENZIAN		GELASIAN		Gibbard et al., 2010
PLIOCENE			PLEISTOCENE		

Table 1. A brief synthesis of chronostratigraphic divisions proposed for the Pliocene Series and Stages.

Cita & Gartner (1973) and Cita (1975b) proposed a neostratotype for the Zanclean stage in the Trubi section of Capo Rossello (Agrigento, Sicily), where calcareous microfossils were extensively studied (Sprovieri, 1978; Vismara Schilling & Stradner, 1977; Zachariasse et al., 1978 among others) and the Messinian/Zanclean boundary crops out. Following magnetostratigraphic, biostratigraphic and cyclostratigraphic researches in the Capo Rossello area resulted in the designation of the boundary-stratotype for the Lower Pliocene Zanclean stage and GSSP for the base of the Pliocene series (Van Couvering et al., 2000) at Eraclea Minoa, near Capo Rossello, on the basis of the exceptional continuity of paleomagnetic and astrochronologic signals and an accessible exposure (Hilgen, 1991; Hilgen & Langereis, 1988; Langereis & Hilgen, 1991; Lourens et al., 1996). The Rossello Composite Section, made from bottom to top by the Eraclea Minoa, Punta di Maiata, Punta Grande and Punta Piccola sections, represents the global reference section for the Zanclean and Piacenzian stages (Hilgen, 1991; Langereis & Hilgen, 1991; Lourens et al., 1996). The base of the Pliocene series, and Messinian/Zanclean boundary, is identified by the sharp contact between the underlying non-marine dark brown sands and marls of the Arenazzolo Formation and the whitish Trubi marls, deposited on open marine slope bottoms (Zachariasse et al., 1978), at inferred water depth of about 600-800 m (Sgarrella et al., 1997). The lithological discontinuity marking the Messinian/Zanclean boundary is widely recognized throughout the Mediterranean area and is assumed to derive by the sudden infilling of the totally or partyally desiccated basin after the Messinian Salinity Crisis (Hsü et al., 1973). The return to deep marine conditions appears to have been synchronous at the geological scale in different and very far areas (Iaccarino et al., 1999a; Pierre et al., 2006; Spezzaferri et al., 1998; Sturani, 1978; Trenkwalder et al., 2008). Outside the Mediterranean marine sedimentation occurred during the Messinian Salinity Crisis and the boundary does not show lithological evidences. An international debate on global correlations of the Messinian/Zanclean boundary produced a great amount of publications and more detailed and high resolution studies on the Miocene/Pliocene succession of the Atlantic coast of Morocco (Benson & Rakic El-Bied, 1996; Suc et al., 1997) as well as of Sicily and Crete (Greece) (Di Stefano et al., 1996; Hilgen & Langereis, 1988; Lourens et al., 1996; Zachariasse, 1975 among others). These researches supported the definition of the Eraclea Minoa as reference section for the Early Pliocene and the validity to mantain Pliocene stratotypes in Italy, whert the Pliocene Series was originally designated. The base of the Zanclean stage is defined at the base of the carbonate bed of the small-scale lithostratigraphic cycle 1 (Hilgen & Langereis, 1988; Langereis & Hilgen, 1991), coincident with the insolation cycle 510, dated at 5.33 Ma (Lourens et al., 1996). The base of the Thvera magnetic chron (C3n.4n of Cande & Kent, 1992; 1995), dated at 5.236 Ma (Lourens et al., 1996), approximates the boundary. Biostratigraphic events for global correlations are mainly provided by calcareous nannofossils: the *Discoaster quinqueramus* LO, dated at 5.537 Ma, and the first occurrence of *Ceratolithus acutus* dated at 5.37 Ma in the equatorial Atlantic slightly predating the Messinian/Zanclean boundary (Backman & Raffi, 1997). The disappearance of *Triquetrorhabdulus rugosus* was recorded in Mediterranean deep-sea cores (Castradori, 1998; Di Stefano et al., 1996) and in the equatorial Atlantic (Backman & Raffi, 1997) at 5.23 Ma. Planktonic foraminiferal events, such as the *Sphaeroidinellopsis* spp. acme, and the *Globorotalia margaritae* FCO can be recognized only in the Mediterranean area.

4.2 The Piacenzian stage

The Piacenzian, which now is the second Pliocene stage, was originally established by Mayer-Eymar (1858) as representative of the older interval of the Pliocene. The same author later proposed the stage Astian for the upper and younger Pliocene portion (Mayer-Eymar, 1868). The author indicated the gray-blue clays of the Argille Azzurre (Blue Clays) Formation cropping out in Castell'Arquato and Lugagnano, near Piacenza, as typical of the Piacenzian stage. The Argille Azzurre Formation is well known in literature for its abundant and diversified macrofossils (mainly molluscs). Following studies demonstrated an older than earliest Pliocene age for the Argille Azzurre sediments, confirmed the importance of the fossiliferous record and extended the use of the term Piacenzian to stratigraphical correlations in Italy. The necessity of formal definition for stratigraphical units was object of increasing scientific debate in the second half of the last century. In the meanwhile, micropaleontology became widely applied to stratigraphical studies, foraminiferal and calcareous nannofossil biozonations were proposed and amended, providing higher resolution biostratigraphic tools than bivalvs and gastropods. Barbieri (1967) designated a unit-stratotype in the Castell'Arquato section and proposed the disappearance of the planktonic foraminifers *Globorotalia margaritae* as indicator of the Piacenzian base. Further micropaleontological studies (Raffi et al., 1989; Rio et al., 1988) demonstrated a hiatus at the very base of the Castell'Arquato section, corresponding to a large part of the lower Piacenzian, and the lack of the Zanclean/Piacenzian boundary. Therefore the proposed stratotype section resulted inadequate to represent the Piacenzian stage and the foraminiferal event (*G. margaritae* LA) was devoid of biostratigraphic value. A complete sedimentary documentation of the Piacenzian stage was recognized in the Punta Piccola section (Lourens et al., 1996; Spaak, 1983; Sprovieri, 1992, 1993; Sprovieri et al., 2006), the upper part of the Rossello Composite section, located in Southern Sicily (Cita et al., 1996). Here the GSSP of the Zanclean/Piacenzian boundary and the Piacenzian stratotype section have been ratified (Castradori et al., 1998). In the well exposed succession at Punta Piccola, the Zanclean Trubi calcareous marls crop out at the bottom and gradually pass upward to the more laminated marls of the Monte Narbone Formation. The Piacenzian/Gelasian boundary was detected in the upper part of the section. Micropaleontological data testify an upper epibathyal setting and a water depth of about 800-1000 m. The Piacenzian GSSP, and the coincident Zanclean/Piacenzian boudary, was defined at the base of the small-scale carbonate cycle 77, correlated to the astrochronological insolation cycle 347, dated at 3.6 Ma. The Gilbert-Gauss magnetic reversal, dated at 3.596 Ma, approximates the boundary (Lourens et al., 1996). The foraminiferal event applied to recognize the lower Piacenzian boundary is the disappearance of *G. puncticulata*, dated at 3.57 Ma (Channel et al., 2009; Lourens et al., 1996). Among calcareous nannofossils, the end of the paracme interval of *Discoaster pentaradiatus*, dated at 3.61 Ma, approximates the boundary, the *Sphenolithus* spp. LO, dated at 3.70 Ma in the Mediterranean (Lourens et al., 1996), was recognized at 3.66 in the Atlantic Ocean (Shackleton et al., 1995). A bioevent previously proposed as useful for global correlations, such as the LO of *Globorotalia margaritae*, was recently recalibrated, from 3.58 (Berggren et al. 1995; Castradori et al., 1998) to 3.81 Ma (Channel et al., 2009).

4.3 The Gelasian stage

The Gelasian stage was proposed by Rio et al. (1994; 1998) to represent the uppermost and third interval of the Pliocene Series, corresponding to the strong climatic deterioration due to orbitally controlled glacial cycles in the Northern Hemisphere. The stage was named from the Greek name of the town of Gela, near to the Monte S. Nicola section where the Gelasian GSSP was established. This location was preferred to the also detailed studied Singa section

(Calabria, Southern Italy) (Negri et al., 2003; Zijderveld et al., 1991) for its better paleomagnetic signal, exposition and accessibility. In the Monte S. Nicola section an undisturbed and complete Pliocene-Pleistocene succession crops out. From the bottom, marls and limestones of the Trubi Formation, documenting the upper Zanclean and the Piacenzian, grade upward to the silty marls of the Monte Narbone Formation, deposited from the Piacenzian to the Pleistocene (Rio et al. 1994; Spaak, 1983; Sprovieri, 1992, 1993). In the Gelasian time interval of the Monte S. Nicola section, micropaleontological data suggested a deposition in a slope-basin setting, at depths ranging from 500 to 1000 m. The Piacenzian/Gelasian boundary ("golden spike") was defined at the base of the marly layer overlying sapropel MPRS 250 (Mediterranean Precession Related Sapropel), and corresponds to the insolation cycle 250, astronomically dated at 2.588 Ma (Lourens et al., 1996). The base of the Gelasian falls in isotopic stage 103 (Raymo et al., 1989) and predates by about 60 ky the cold isotopic stage 100, marking the major climatic deterioration in the Northern Hemisphere. The Gauss/Matuyama reversal boundary predates of about 20 ky the Gelasian base. Calcareous microfossil events approximating the lower Gelasian boundary are the LO of *Globorotalia bononiensis* and *Neogloboquadrina atlantica*, dated at 2.41 Ma and nearly isochronous in Mediterranean and central Atlantic (Gradstein et al. 2004; Sierro et al. 2009), the LO of *Discoaster surculus*, dated at 2.53 Ma and the LO of *Discoaster pentaradiatus*, dated at 2.53 Ma (Sprovieri et al., 1998; Rio et al. 1998). Other biostratigraphic tools for global correlations are provided by siliceous microfossils. The LO of the radiolarian *Stichocorys peregrina*, the FO of the diatom *Nitzschia jouseae* and the LO of *Denticulopsis kamtschatica* approximate the Gauss/Matuyama boundary. In open-marine successions, other biovents occurring during the Gelasian are the FO of *Globorotalia truncatulinoides*, dated at 2.03 Ma in the Atlantic Ocean (Sierro et al. 2009), the LO of *Globigerinoides extremus*, dated at 1.98 Ma, and the FCO of *Neogloboquadrina pachyderma* sinistral, registered at 1.79 Ma (Gradstein et al., 2004; Rio et , 1998).

The upper boundary of the Gelasian was originally defined as coincident with the base of the Pleistocene Series and Calabrian stage, dated to 1.806 Ma (Lourens et al., 2004) in the thoroughly studied Vrica section (Calabria, Southern Italy), where the Pliocene/Pleistocene GSSP was proposed (Aguirre & Pasini, 1985) and formally accepted by the IUGS in 1984 (Bassett, 1985; Cita, 2008). In 2009, the IUGS formally ratified the proposal to lower the base of the Quaternary System and Pleistocene Series to the GSSP of the Gelasian stage (Gibbard et al., 2010). As a consequence, the Gelasian represents the first stage of the revised Pleistocene, the GSSP at Vrica section remains valid as the base of the Calabrian, now the second stage of the Pleistocene.

5. The Mediterranean Pliocene foraminiferal biozonation

During the 1960's and 1970's many foraminiferal biozonation for the Mediterranean Pliocene have been proposed (Bizon & Bizon, 1972; Cita, 1975a; Colalongo et al., 1982, among others), most of them were summarized by Iaccarino, 1985. The most successful was the zonal scheme of Cita (1975a), based on six foraminiferal zones, designated with the initials MPl (Mediterranean Pliocene) and numbered from 1 (the oldest) to 6 (the youngest). Rio et al (1984) integrated the foraminiferal and calcareous nannofossil Mediterranean Pliocene biozonations. Sprovieri (1992) amended the Cita (1975) scheme, introducing four new foraminiferal Subzones and so increasing the biostratigraphical resolution in the MPl4 and

MPl5 Zones (Tab. 2). The definition of each foraminiferal Zone and Subzone will be briefly described. Datum planes absolute ages are based on Gradstein et al. (2004).

Ma	STRATIGRAPHY		FORAMINIFERAL ZONES		FORAMINIFERAL BIOEVENTS	CALCAREOUS NANNOFOSSIL ZONES	
	Gradstein et al., 2004	Walker et al., 2009	Cita, 1975 Sprovieri,1992	Iaccarino, 1985		Rio et al., 1990	
	PLEISTOC.	CALABR.	G. cariacoensis	G. cariacoensis	FAD G. cariacoensis 1.75	MNN19b	C. macintyrei
2.0			MPL6	Globorotalia inflata	FCO N. pachyder. s. 1.79	MNN19a	D. productus
		GELASIAN			FAD G. inflata 2.09		
			MPL5b			MNN18	Discoaster brouweri
2.5					LAD G.bononiensis 2.41 LO N. atlantica 2.41		
						MNN16b/17	Discoaster pentaradiatus
			MPL5a	Globorotalia aemiliana	FO N. atlantica 2.72		
3.0		PIACENZIAN					
			MPL4b		LAD Sphaer. spp. 3.19 FAD G.bononiensis 3.31	MNN16a	Discoaster tamalis
3.5				Globorotalia puncticulata	LAD G. puncticulata 3.57		
	PLIOCENE		MPL4a				
4.0		PLIOCENE			LCO G. margaritae 3.98	MNN14/15	Reticulofenestra pseudoumbilicus
			MPL3	Globorotalia puncticulata & Globorotalia margaritae		MNN13	Ceratolithus rugosus
4.5		ZANCLEAN			FO G. puncticulata 4.52		
			MPL2	Globorotalia margaritae		MNN12	Amaurolithus tricorniculatus
5.0					FCO G. margaritae 5.08		
05:33			MPL1	Sphaeroidinell. acme			

Table 2. Planktonic foraminifer and calcareous nannofossil integrated biostratigraphic scheme for the Mediterranean Pliocene.

5.1 MPI1 Zone (*Sphaeroidinellopsis* spp. Acme-Zone, Cita, 1975a)

Lower boundary - the return to open marine conditions, related to the Mediterranean infilling after the Messinian Salinity Crisis. It coincides with the Pliocene base, dated at 5.33 Ma.

Upper boundary - *Globorotalia margaritae* FCO, dated at 5.08 Ma.

The biozonal marker *Sphaeroidinellopsis*, a deep mesopelagic taxon, is absent at the very base of the MPl1 Zone and is often rare in this interval both in deep-sea and land successions (Cita, 1973; Cita. 1975b; Di Stefano et al., 1996; Iaccarino et al., 1999b; Pierre et al., 2006). The *Sphaeroidinellopsis* increase in relative abundance, corresponding to its acme, is dated at the interval from 5.29 to 5.20 Ma (Iaccarino et al., 1999b) and extends from lithological cycle 2 to cycle 6 (Di Stefano et al., 1996). Acme zones are more properly ecobiostratigraphical zones, strongly dependent from environmental factors (water depth and temperature, trophic resources, etc.), that influence the species diffusion and abundance. In particular, *Sphaeroidinellopsis* specimens did not occur in shelf deposits, they were very rare and sporadic in upper epibathyal assemblages too.

Many ecobiostratigraphic events have been recognized in MPl1 Zone, and provide tools for correlations in the Mediterranean area, from bottom to top:

- an abundance peak of *Globigerina nepenthes* was recognized at the very base of the Pliocene succession, in Southern Italy (Zachariasse & Spaak, 1983) and in the Western Mediterranean (Iaccarino et al., 1999b) as well as in Northwestern Italy (Trenkwalder et al., 2008).

- two sinistral shifts of *N. acostaensis* were identified below the *Sphaeroidinellopsis* spp. acme in the Roccella Ionica-Capo Spartivento (Southern Italy) composite section (Di Stefano et al., 1996), where the basal Zanclean succession is completely preserved: the first and older between the lithological cycles 1-2, the second and younger between the cycles 2-3, near the base of the *Sphaeroidinellopsis* spp. acme, which encompasses cycles 2 to 6.

- the Common Occurrence (CO) of dextral coiling *Globorotalia scitula*, was registered in cycle 6 of the basal Zanclean in the Western Mediterranean and interpreted as a "delayed invasion event" (Iaccarino et al., 1999b).

- the re-immigration of the epibathyal to mesobathyal benthic foraminifer *Siphonina reticulata*, a nearly synchronous event, recognized at considerable geographic distance (Iaccarino et al., 1999b; Pierre et al., 2006; Spezzaferri et al., 1998; Violanti et al., 2009; 2011). *S. reticulata* was proposed as a Mediterranean quasi-endemic form, indicative of Early Pliocene Mediterranean Intermediate Water (EPMIW) (Sgarrella et al., 1997, 1999). Its re-diffusion after the Messinian Dessication Event was correlated to lithological cycle 6 (Di Stefano et al., 1996).

- the subsequent re-entry of the deep, oxyphilic *Cibicidoides robertsonianus*. The benthic taxon is a typical NADW (Nord Atlantic Deep Water) species. Its delayed diffusion was related to the basin deepening and establishment of deep oceanic-type conditions during the early Zanclean (Hasegawa et al., 1990; Spezzaferri et al., 1998).

5.2 MPl2 Zone (*Globorotalia margaritae margaritae* Interval-Zone, Cita, 1975a)

Lower boundary - *Globorotalia margaritae* FCO, dated at 5.08 Ma. The zonal marker is generally rare in the lower part of the zone and became common in bathyal assemblages upwards.

Upper boundary - *Globorotalia puncticulata* FO, dated at 4.52 Ma.

No ecobiostratigraphic events are reported in this interval zone.

5.3 MPl3 Zone (*Globorotalia margaritae* - *Globorotalia puncticulata* Concurrent range-Zone, Cita, 1975a)

Lower boundary - *Globorotalia puncticulata* FO, dated at 4.52 Ma.

Upper boundary - *Globorotalia margaritae* LCO, dated at 3.98 Ma.

This concurrent range zone is well recognizable in bathyal assemblages, less detectable in
outer shelf assemblages, for the local absence of one or both of the two zonal markers.

Cita (1973), on the basis of morphological statistical studies, distinguished the new
subspecies G. *margaritae primitiva*, a never keeled form, characterized by small, almost non-
lobate tests, and G. *margaritae evoluta*, a large, strongly lobate form with an imperforate keel
evident almost on the two last chambers, from the typical G. *margaritae margaritae*, with
intermediate characteristics. Cita (1973) proposed a Lower Pliocene biozonation on the basis
of the G. *margaritae* lineage. The author pointed out that G. *margaritae margaritae* and G.
margaritae primitiva commonly co-occur in all the interval of the G. *margaritae margaritae*
Lineage Zone (corresponding to the MPl2 Zone, Cita, 1975a). G. *margaritae evoluta* is
common only in the upper part of the species range, in the G. *margaritae evoluta* Lineage
Zone (corresponding to the MPl3 Zone, Cita, 1975a).

5.4 MPl4 Zone (*Sphaeroidinellopsis subdehiscens* Interval-Zone, Cita, 1975a)

Lower boundary - *Globorotalia margaritae* LCO, dated at 3.98 Ma.

Upper boundary - *Sphaeroidinellopsis* spp. LAD, dated at 3.19 Ma.

Cita (1973; 1975a) noted that the zonal marker was not abundant in the deep-sea reference
cores (Site 125 and 132, DSDP Leg 13). In continental successions, *Sphaeroidinellopsis* spp. are
absent or occur very randomly. G. *puncticulata* is common in the lower portion of the
interval. *Globigerina (Globoturborotalita) apertura*, *Globigerina (Globoturborotalita) decoraperta*
and *Globigerinoides extremus* are still common. Sprovieri (1992) formalized previous
suggestions (Spaak, 1983) and proposed the division of the MPl4 Zone in two subzones:

5.4.1 MPl4a Subzone (*Globorotalia puncticulata* Interval-Subzone, Sprovieri, 1992)

The lower MPl4a Subzone extends from the *Globorotalia margaritae* LCO up to the G.
puncticulata LAD, now dated at 3.57 Ma and therefore nearly approximates the
Zanclean/Piacenzian boundary, dated at 3.60 Ma.

5.4.2 MPl4b Subzone (*Globorotalia planispira* Interval-Subzone, Sprovieri, 1992)

The upper MPl4b Subzone is comprised between the *Globorotalia puncticulata* LAD and the
Sphaeroidinellopsis spp. LAD. The *Globorotalia bononiensis* FAD, dated at 3.31 Ma, an
isochronous event in the Mediterranean, occurs in the upper part of this subzone. Also
Globorotalia aemiliana, a taxon of the *Globorotalia crassaformis* plexus, firstly occurs nearly in
the same horizon of *Globorotalia bononiensis* FAD. Iaccarino (1985) proposed G. *aemiliana* as
index form of her homonimous zone, extending from the appearance of the zonal marker up
to the appearance of *Globorotalia inflata*, encompassing part of the MPl4b Subzone and all the
MPl5 Zone of Cita (1975a), amended Sprovieri (1992). Following studies evidenced a G.
aemiliana diachronous diffusion in the Mediterranean and as a consequence the taxon
inadequacy as zonal marker.

5.5 MPl5 Zone (*Globigerinoides elongatus* Interval-Zone, Cita, 1975a)

Lower boundary - *Sphaeroidinellopsis* spp. LAD, dated at 3.19 Ma.

Upper boundary - *Globorotalia inflata* FAD, dated at 2.09 Ma.

In the same time interval Cita (1973) proposed the *Globigerinoides obliquus extremus* Zone, defined as the interval between the *Sphaeroidinellopsis* spp. LAD and the extinction of the zonal marker. Cita (1975a) named the MPl5 Zone as *Globigerinoides elongatus* Interval Zone, for the rareness of the previous zonal marker in the upper part of the interval, in which it is substituted in abundance by *G. elongatus*. Sprovieri (1992) introduced two subzones:

5.5.1 MPl5a Subzone (*Globorotalia bononiensis* Interval-Subzone, Sprovieri, 1992)

The lower MPl5a Subzone is defined as the interval comprised between the *Sphaeroidinellopsis* spp. LAD and the *Globorotalia bononiensis* LAD, dated at 2.41 Ma. The *Neogloboquadrina atlantica* FO in the Mediterranean Pliocene is dated at 2.72 Ma, its LO is dated at 2.41, isochronous to the *Globorotalia bononiensis* LAD. The *G. crassaformis* group (*G. aemiliana*, *G. crassaformis*, *G. crassula*) is often common. In particular, *G. aemiliana*, an easily recognizable species for its typical morphology, is often more frequent than *G. bononiensis* and can help the biostratigraphic assessment of poorly diagnostical assemblages.

5.5.2 MPl5b Subzone (*Globorotalia incisa* Interval-Subzone, Sprovieri, 1992)

The upper MPl5b Subzone is defined as the interval comprised between the *Globorotalia bononiensis* LAD and the *Globorotalia inflata* FAD. Assemblages are poor of useful biostratigraphic taxa. *G. (Gt.) apertura*, *G. (Gt.) decoraperta*, *G. obliquus*, and *G. extremus* are rare and become extinct at the top of this subzone or soon after in the Gelasian.

5.6 MPl6 Zone (*Globorotalia inflata* Interval-Zone, Cita, 1975a)

Lower boundary - *Globorotalia inflata* FAD, dated at 2.09 Ma

Upper boundary - *Neogloboquadrina pachyderma* FCO, dated at 1.79 Ma

Lourens et al. (2004) introduced a subdivision of the MPl6 Zone in two subzones, MPl6a and MPl6b respectively, separated on the basis of the appearance of *Globorotalia truncatulinoides*, dated at 2.0 Ma. In many on-land successions *G. truncatulinoides* very seldom occurs and the subzones can hardly be recognized.

The upper boundary of the MPl6 Zone was coincident with the top of the Gelasian stage, correlated with the Pliocene/Pleistocene boundary (Gradstein et al., 2004 and literature until 2009). On the contrary, if following the recent controversial revision of this boundary ratified by the International Union of Geological Sciences (IUGS) (Gibbard et al., 2010), the entire MPl6 Zone pertains to the Pleistocene Series (see Tab. 2).

6. Previous foraminiferal studies on the Northwestern Italy Pliocene

Pliocene marine sediments of Northwestern Italy widely crop out in the central part of Piedmont (Astigiano and Monferrato) and are well know in literature for their rich macrofossil assemblages. Foraminiferal assemblages were also analyzed in progressively increasing detail, from the first taxonomical descriptions carried out in the region by Dervieux (1892). Small and discontinuous Pliocene marine deposits crop out in the northern part of the sector, along the Alps margins, and to the south, along the western Ligurian coast (Fig. 1). The northernmost Pliocene foraminiferal assemblages along the southern side of the Alps were described at Castel di Sotto (Canton Ticino, Switzerland) few kilometers outside the Italian border. Sediments were referred to the upper part of the Early Pliocene on the occurrence of *Globigerinoides extremus*, *G. obliquus*, and *Bolivina placentina* (Premoli Silva, 1964; Violanti, 1994). A more precise biostratigraphic correlation was hampered by the absence of biozonal planktonic markers. Statistical analysis of the abundant benthic taxa documented deep outer neritic, disoxic muddy bottoms, dominated by the stress-tolerant *Globobulimina affinis*, and affected by high sedimentation rates of terrigenous and vegetal debris. Similar assemblages, with scarce and long-ranging planktonic forms, dominant benthic taxa indicative of outer neritic to inner neritic paleoenvironment, characterize other nearby Pliocene subsurface deposits (Morbio (Canton Ticino) core, Violanti, unpublished data) as well as cropping out in the Varese area (Cremaschi et al., 1985; Lualdi 1981; Martinis, 1950).

In the Biellese, a Piedmont region at the foothills of the Western Alps, Barbieri et al. (1974) and Aimone & Ferrero Mortara (1983) described foraminiferal assemblages typical of inner neritic to outer neritic bottoms, referred by the authors to the Middle-Upper Pliocene. Basilici et al. (1997), in their interdisciplinary study of Val Chiusella outcrops, dated the lower deposits of the marine to continental succession to the uppermost Zanclean, on the occurrence of the benthic foraminifer *Buccella granulata* and of the gastropod *Bufonaria marginata*, which disappeared in correspondence with the LAD of *Globorotalia puncticulata*, marking the upper boundary of the MPl4a Zone. Foraminiferal and mollusc assemblages were indicative of an inner neritic paleoenvironment.

In the central part of Piedmont, Martinis (1954) analyzed the foraminiferal assemblages of the Pliocene succession in a number of localities, from Verrua Savoia to the North, to Villalvernia to the Southeast, to the Astigiano and the Monregalese to the Southwest, collected both from the Argille di Lugagnano (now reported as Argille Azzurre, Blue Clays) and from the Sabbie di Asti (Asti Sands). The author dated the Argille Azzurre, with generally abundant planktonic foraminifers, indicative of neritic paleoenvironment, to the Early Pliocene and referred the outer neritic to inner neritic assemblages of the Sabbie di Asti to the Middle and Upper Pliocene.

Casnedi (1971) presented another detailed review of Lower Pliocene assemblages from the Southern and Western Piedmont Argille Azzurre, characterized by sometimes common *Globorotalia margaritae* and/or *G. puncticulata* and by deep outer neritic to bathyal benthic species such as *Anomalinoides helicinus*, and *Uvigerina rutila*. Poorly diversified assemblages, dominated by shallow water foraminifera (*Ammonia beccarii*, *Elphidium* spp. and *Cibicides* spp.) were described by the same author from the sandy deposits cropping out in the surroundings of Asti. In the Albese area, Montefameglio et al. (1979) dated the clayey marls and sandy clays to the Lower Pliocene MPl3 Zone on the occurrence of locally common *Globorotalia puncticulata, Anomalinoides helicinus, Bulimina minima* and *Uvigerina rutila*. In the South-Western Piedmont Boni et al. (1987) documented deep water Lower Pliocene assemblages, yielding *Globorotalia margaritae*, overlaid by shallow water deposits, referred by the authors to the Upper Pliocene-Pleistocene, along the Pesio Stream (Monregalese). In other nearby outcrops Pavia et al. (1989) pointed out the absence of globorotalias and the presence of typical Lower Pliocene taxa such as *Bolivina leonardii* (Breolungi, Mondovì). The absence of planktonic biostratigraphic markers was also evidenced in the Monregalese assemblages (Violanti & Giraud, 1992), dated at the Lower-Middle Pliocene mainly on the occurence of benthic taxa.

Violanti (2005) provided data on foraminiferal assemblages from the North-Eastern Monferrato, Astigiano, Langhe and Monregalese. The lowermost MPl1 Zone was unequivocally documented only in the Moncucco Torinese quarry (Turin Hill). Rich and diversified assemblages referable to the MPl2 Zone, with *Globorotalia margaritae*, and to the MPl3 Zone, with *G. margaritae* and *G. puncticulata*, indicative of upper epibathyal depths, suggested paleoenvironmental conditions similar to those of coeval pelagic successions of Sicily and of the Tyrrhenian Sea. Basin shallowing and increasing transport from emerged areas and shallow water bottoms were documented during the MPl3 Zone, and chiefly MPl4 Zone. Following researches excluded a marine sedimentation during the following MPl5 Zone in the central Piedmont basin, tentatively proposed by Violanti (2005). The youngest marine succession of the region, ranging from the MPl4a to the MPl5a Subzones, was described from sediments cored at Casale Monferrato (AL), at the margin of Monferrato (Violanti & Sassone, 2008). Here, *Globorotalia puncticulata* and *G. puncticulata padana* occurred in the lower layers, whereas *Globorotalia aemiliana* and *G. bononiens* specimens were detected upwards. Recent researches in the Piedmont region, focusing on the Messinian paleoenvironmental evolution (Clari et al., 2008; Dela Pierre et al., 2011; Lozar et al. 2010), revealed an unexpected nearly complete documentation of the Early Pliocene in the Turin Hill (Moncucco Torinese quarry: Trenkwalder et al., 2008; Violanti et al., 2011) and in the Albese area (Narzole core: Violanti et al., 2009). The Moncucco T. section, spanning from the uppermost non-marine Messinian (Lago-Mare deposits) to the upper Zanclean MPl4a Zone, will be discussed and integrated by new data in the following pages. The Narzole core also encompasses the Messinian/Zanclean boundary, marked by a 0.50 cm thick barren arenitic layer, dark brown to black, similar to that recovered in the same stratigraphic position at Moncucco T. Microfossils (foraminifers, calcareous nannofossils and ostracods) documented the MPl1 and MPl2 foraminiferal Zones and the MNN12 calcareous nannofossils Zone of the Early Pliocene. Many bioevents, which occurred during the MPl1 Zone and previously discussed, such as one sinistral coiling shift of *Neogloboquadrina acostaensis*, the *Globorotalia scitula* sinistral CO, the re-immigration of *Siphonina reticulata*, the first influx of the Nord Atlantic Deep Water (NADW) taxon *Cibicidoides robertsonianus*, were recognized in the Narzole core and allowed correlations with the Moncucco T. section and the Mediterranean deep-sea succession. Integrated studies are in progress on subsurface and surface deposits of the region, mainly focusing on the Messinian/Zanclean boundary and the lower Pliocene paleonvironmental evolution.

On the southern margin of the Alps, Pliocene sediments discontinuously crop out along the coast of the western Ligury. The thickest and fossiliferous deposits were described between Ventimiglia, to the West, and Ceriale (Rio Torsero), to the East. Rich foraminiferal assemblages, yielding common specimens of *Sphaeroidinellopsis* spp., were described in the older Pliocene mudstones and marls. Less diversified assemblages, poor of biostratigraphic indices, characterized the upper deposits (Boni et al., 1985; Giammarino & Tedeschi, 1980; Giammarino et al., 1984). Recently, Breda et al. (2009) analyzed the sedimentological and tectonic evolution of Ventimiglia incised valley and described a Pliocene succession spanning from the Zanclean MPl1 and MNN12 Zones to the Piacenzian MPl5 and

MNN16b/17 Zones. A successions dated from the MPl3 Zone, by the co-occurrence of *Globorotalia margaritae* and *Globorotalia puncticulata*, to the MPl4 Zone, with *G. puncticulata*, was described along the Rio Torsero (Violanti, 1987), near Ceriale.

7. Recent and new data on the Piedmont Pliocene

Three sections, from Moncucco Torinese, Isola d'Asti and Verrua Savoia, will be analyzed, as representative of the Pliocene biostratigraphic succession and paleoenvironmental evolution in the central Piedmont, from the very Pliocene base to the upper Zanclean MPl4a Subzone.

7.1 Material and methods

Foraminiferal analyses of Pliocene Piedmont assemblages were carried out on sieving residues prepared following standard procedures (Haynes, 1981). 500-100 g of dry sediment (clay, silt and sand) were disaggregated with water and a small amount of hydrogen peroxide (H_2O_2) or by gently boiling with water for 1-2 hours, rather consolidated sediments as some calcarenites were partially disaggregated by boiling with water and a small quantity of soda (Na_2CO_3). When disaggregated, the material was gently washed, sieved into grain size fractions greater than 250 µm, 125-250 µm and 125-63 µm, dried at 50° C and weighed. Percentages of the separate grain size fractions were calculated, the total >63 µm residue percentages will be discussed here. Assemblages of well consolidated calcarenites were studied on thin sections.

Qualitative foraminiferal analyses were made on the three grain size fractions in order to identify the often rare biostratigraphic markers. Quantitative analyses were carried out on the whole of the >125 µm fraction, split into aliquots containing at least 300-400 well preserved foraminiferal tests. Semiquantitative studies were performed on thin sections and when the tests poor preservation or abundant reworked forms prevented the quantitative analysis.

Taxonomy is according to Kennett & Srinivasan (1983) and Stainforth et al. (1975), for planktonic species, to AGIP (1982), Loeblich and Tappan (1988) and Van Morkhoven et al. (1986) for benthic species. The biostratigraphic scheme here adopted is that of Cita (1975a), amended by Sprovieri (1992). Selected taxa are shown in Plate 1.

7.2 The Moncucco Torinese section

The Pliocene succession exposed in the Moncucco Torinese (Torino Hill, Tertiary Piedmont Basin) quarry, represents, at the present state of study, the more complete documentation of Early Pliocene (Zanclean) upper epibathyal deposits in the Piedmont area as well as in the Northwestern Italy (Trenkwalder et al., 2008; Violanti et al., 2011). In the Moncucco T. quarry a thick Messinian evaporitic and post-evaporitic succession is overlaid by a 10-50 cm thick barren black arenitic layer, marking the Messinian/Zanclean boundary, followed by about 26 meters of Zanclean gray to whitish marly clays, pertaining to the Argille Azzurre (Blue Clays) Formation. The Zanclean unit is rather uniform, without evident bedding, a thin silty layer was detected in the middle/upper section, two biocalcarenitic layers, 0.50 cm thick each ones, are interbedded in its upper part. Pleistocene continental deposits cover with an erosional surface the Zanclean marly clays.

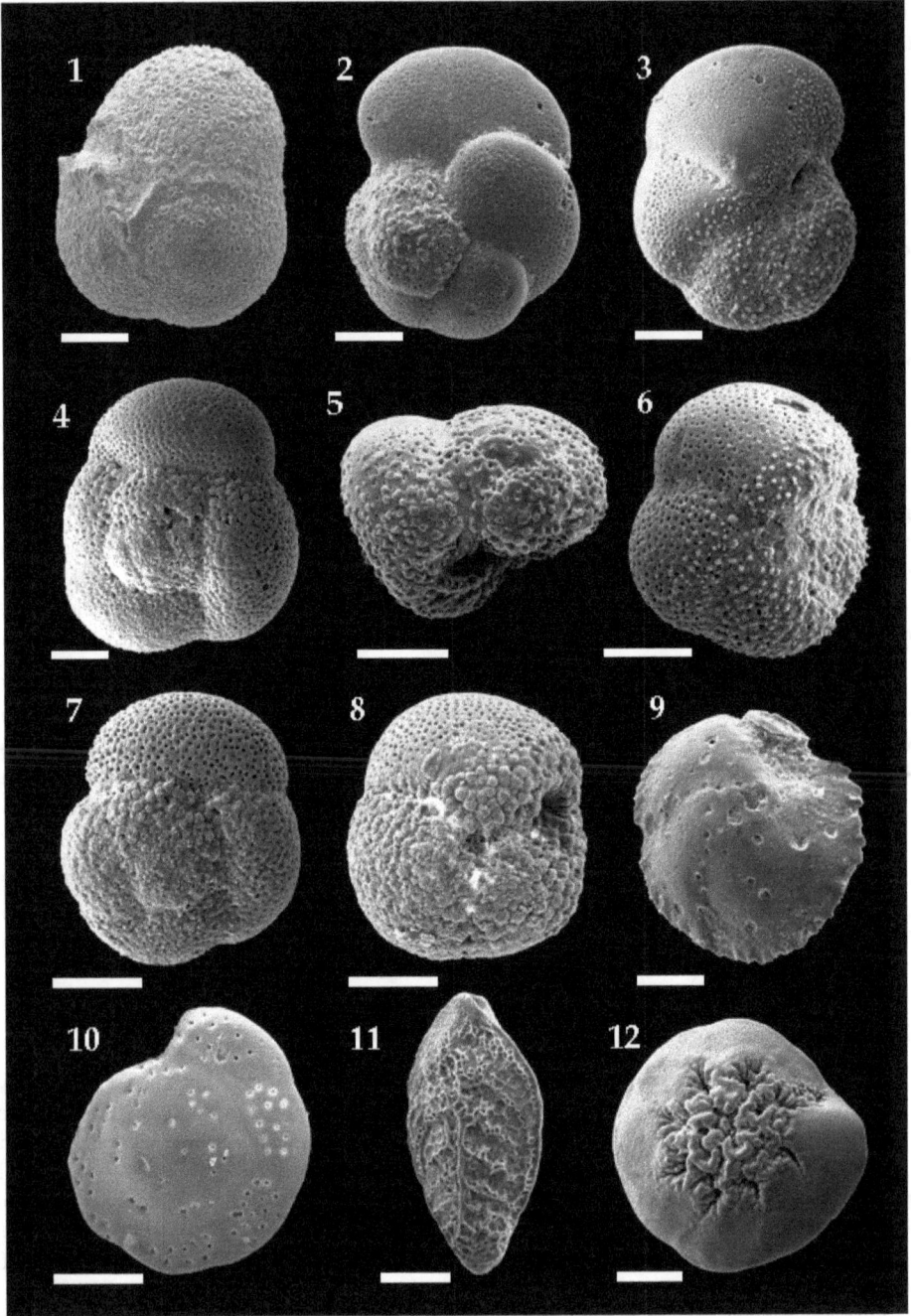

Plate 1. Fig. 1: *Sphaeroidinellopsis subdehiscens* Blow, sample M7, umbilical view. Figs. 2- 3: *Globorotalia margaritae* Bolli & Bermudez, sample Is12: Fig. 2, spiral view, Fig. 3, umbilical

Pliocene Mediterranean Foraminiferal Biostratigraphy: A Synthesis and Application to
the Paleoenvironmental Evolution of Northwestern Italy

173

view,. Figs. 4-6: *Globorotalia puncticulata* (Deshayes), sample Is21: Fig. 4, spiral view, Fig. 5,
lateral view, Fig. 6, umbilical view. Figs. 7-8: *Globorotalia puncticulata padana* Dondi &
Papetti, sample Is35: Fig. 7, spiral view, Fig. 8, umbilical view. Fig. 9: *Siphonina reticulata*
(Czjzek), sample Is10, umbilical view. Fig. 10: *Cibicidoides robertsonianus* (Brady), sample
M26, spiral view. Fig. 11: *Bolivina leonardii* Accordi & Selmi, sample Is7, lateral view. Fig. 12:
Buccella granulata (Di Napoli), sample Vc39, umbilical view. Scale bar = 100 µm

Primary biostratigraphic data are given by occurrence and frequency of the zonal markers.
The basal MPl1 Zone (*Sphaeroidinellopsis* spp. acme Zone, Cita, 1975a) is documented by the
occurrence of *Sphaeroidinellopsis* spp. from about 1 m above the marly clays bottom.
Sphaeroidinellopsis spp. reach their abundance peak (acme) in a short interval, encompassing
sample 6 to 11 and extending from 5.29 Ma, base of the *Sphaeroidinellopsis* acme (Di Stefano
et al., 1996; Iaccarino et al., 1999b), to 5.17 Ma (Di Stefano et al., 1996) or 5.20 Ma (Iaccarino et
al., 1999a; 1999b), top *Sphaeroidinellopsis* acme (Fig. 2). The absence of *Sphaeroidinellopsis* spp.
characterizes the very base of the Pliocene succession and was described both in
Mediterranean deep sea sediments (Cita, 1973; Iaccarino et al., 1999a, 1999b; Spezzaferri et
al., 1998), as well in outcrops (Di Stefano et al., 1996; Pierre et al., 2006; Spaak, 1983;
Sprovieri, 1992; Sprovieri, 1993; Zachariasse, 1975). At Moncucco T., as well as in the rather
nearby Narzole core (Alba, Piedmont) (Violanti et al., 2009), specimens of the zonal marker
are smaller and rarer in comparison with those occurring in the Sicily Trubi Formation
(Sprovieri, 1992; Vismara Schilling & Stradner, 1977; Violanti, 1989) and suggest local
paleoenvironmental conditions non optimal for the taxon. Percentages of the zonal marker
are very low, always less then 2%, but are similar to those registered in other Mediterranean
sites (Iaccarino et al., 1999a; Spaak, 1983). The upper boundary of the MPl1 Zone was
indicated by the *Globorotalia margaritae* FCO (5.08 Ma, Gradstein et al., 2004), recorded in
sample 26, about 11.50 m above the marly clays bottom (Fig. 2). The taxon showed two
abundance peaks (5-6%) at about 15 m and 20.5 m from the bottom and was detected up to
sample 48. Rare tests of *Globorotalia puncticulata* firstly occurred in sample 40, about 18 m
above the bottom and became common in most of the upper samples (Fig. 2). Nevertheless,
the upper boundary of the MPl2 Zone (*G. margaritae* zone), or lower boundary of the MPl3
Zone (*G. margaritae/G. puncticulata* Concurrent Zone) was proposed in correspondence to the
lower sample 39, where the *Helicosphaera sellii* FO was recorded (Trenkwalder et al., 2008).
This calcareous nannofossil bioevent indicates the MNN12/MNN13 boundary (Rio et al.,
1990) and is coeval with the Mediterranean FO of *Globorotalia puncticulata* (4.52 My, Hilgen,
1991). Finally, the upper boundary of the MPl3 Zone, corresponding to the lower boundary
of the MPl4a Subzone (Cita, 1975a, amended Sprovieri, 1992) was recorded in sample 48,
where the *G. margaritae* LCO (3.98 My, Gradstein et al., 2004) was registered (Fig. 2).

Other secondary data are given by ecobiostratigraphical events, recognized at Mediterranean
scale and also in the Narzole (Albese area) core (Violanti et al., 2009). They characterize the
lowermost Pliocene succession, referable to the MPl1 Zone, and are related to the re-
establishment of deep marine conditions in the Mediterranean basin. In stratigraphic order
they are (Fig. 2): 1) an abundance peak of *Globigerina nepenthes*, recorded at the very base of the
Pliocene succession, reported also in the Western Mediterranean (Iaccarino et al., 1999b) as
well as in Southern Italy (Zachariasse & Spaak, 1983); 2) one sinistral shift of *Neogloboquadrina
acostaensis*, observed in the same basal layer (sample 4). Two sinistral shifts of *N. acostaensis*
have been described below the *Sphaeroidinellopsis* spp. acme within complete successions (Di

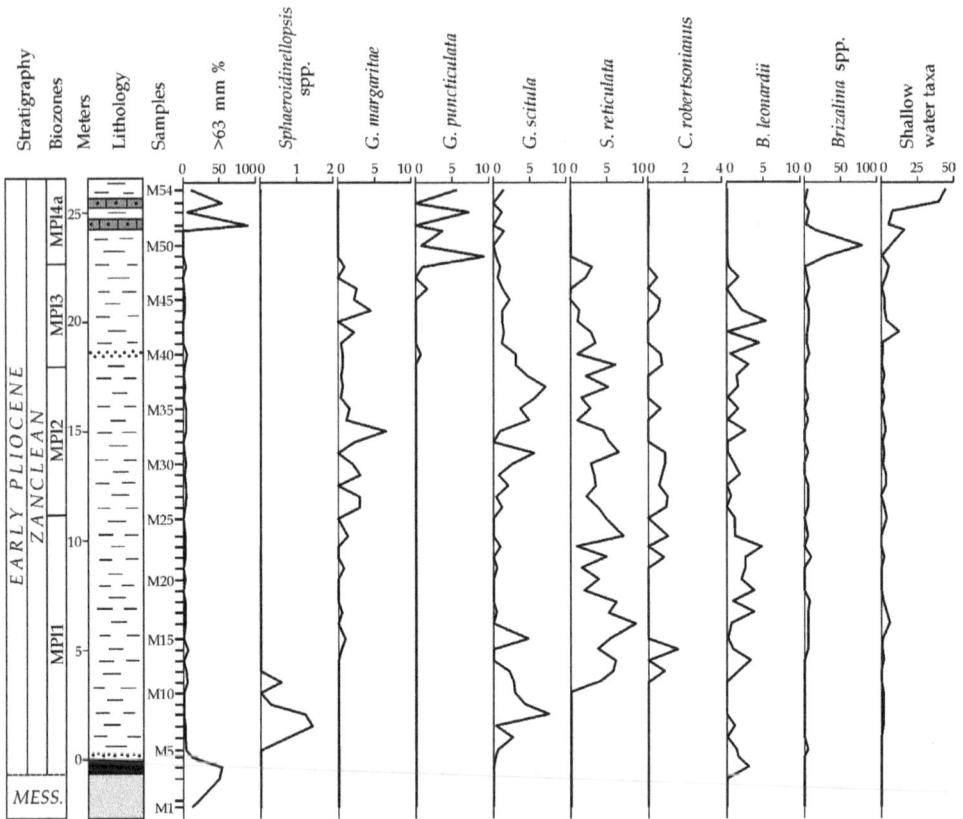

Fig. 2. Moncucco T. section. Chronostratigraphic synthesis, lithological column, samples position, percentage variations of the grain sized > 63 µm fraction, of planktonic (*Sphaeroidinellopsis* spp., *Globorotalia margaritae*, *Globorotalia puncticulata* and *Globorotalia scitula*) and benthic (*Siphonina reticulata*, *Cibicidoides robertsonianus*, *Bolivina leonardii*, *Brizalina* spp. and Shallow water taxa) foraminiferal taxa.

Stefano et al., 1996; Iaccarino et al., 1999b). A short hiatus in the basal Pliocene of Moncucco T. was therefore inferred, also on the basis of *Triquetrorhabdulus rugosus* absence, a calcareous nannofossil taxon usually recorded in the lowermost Zanclean (Castradori, 1998); 3) the *Globorotalia scitula* dextral Common Occurrence (CO), a "delayed invasion event", was recorded in sample 8 of the Moncucco T. section and in the basal Early Pliocene of the Western Mediterranean (Iaccarino et al., 1999b); 4) the re-immigration of the epibathyal to mesobathyal benthic foraminifer *Siphonina reticulata*, and the subsequent re-entry of the deep, oxyphilic benthic *Cibicidoides robertsonianus*, nearly synchronous events, recognized at considerable geographic distance (Iaccarino et al., 1999b; Pierre et al., 2006; Spezzaferri et al., 1998) in the Mediterranean basin. The re-immigration of bathyal benthic species, disappeared from the Mediterranean before the Messinian Salinity Crisis (Kouwenhoven et al., 1999), was related to the establishment of deep oceanic-type conditions during the early Zanclean (Hasegawa et al., 1990; Spezzaferri et al., 1998).

In the upper MPl2 Zone and in the following MPl3 Zone the decrease to disappearance of bathyal benthic taxa (*C. robertsonianus*, *S. reticulata* etc.) is opposite to the frequency increase of infaunal, low-oxygen tolerant taxa such as *Brizalina* spp. (Fig. 2), suggesting more disaerobic and probably rather shallower bottoms. High percentages of winnoved shallow water forms (*Ammonia beccarii*, *Cibicides lobatulus*, *Elphidium* spp.) characterized the uppermost calcarenitic levels and document enhancing trasport from the inner shelf during the MPl4a Zone. In many samples the distribution of *Bolivina leonardii*, a typical Pliocene taxon, shows an opposite pattern to that of *Brizalina* spp. (Fig. 2) and a positive correlation with the deep, oxyphilic taxa.

Trenkwalder et al. (2008) inferred an only partial documentation of the MPl2 and MPl3 Zones in the Moncucco T. section, due to the reduced thickness of sediments representing these intervals in comparisong with those preserved in on-land sections (Rossello Composite Section, Sgarrella et al., 1997; Sprovieri, 1992; Zachariasse et al., 1978) and deep-sea cores (Cita, 1973, 1975a; Cita et al., 1999; Pierre et al., 2006). Planktonic and benthic assemblages suggested paleodepths similar to those proposed in the correspondent interval of the previously cited Southern Mediterranean successions: upper epibathyal bottoms, probably not deeper than 500 m, at the very base of the MPl1 Zone, a basin deepening, probably to about 800-1000 meter depth during the same MPl1 Zone and stable deep bottoms during most of the MPl2 Zone. Upwards, a shallowing to deep outer neritic or uppermost epibathyal zone, and a correspondent increase of transport from the inner shelf, was suggested at Moncucco T. by the progressive disappearance of deep bathyal taxa and by the higher frequency of shallow water taxa (Fig. 2).

7.3 The Isola d'Asti section

The Isola d'Asti section was studied for its good documentation of upper epibathyal sediments in the Astigiano area, where silty or sandy deposits indicative of outer neritic to inner neritic bottoms, very poor of biostratigraphic indices, dominantly occur.

The Isola d'Asti section was sampled behind the Merlino brickyard (now shutted down), at about 1,5 km NE of Isola d'Asti and 8 km SW of Asti (Fig. 1) The succession pertains to the Argille Azzurre (Blue Clays) Formation and is represented by dominant bioturbated and massive marly-silty clays. Four layers of finely laminated pelites, few centimeters or decimeters thick, are interbedded to the clays. Two coarse sandy levels, rich in macrofossils (mainly bivalvs, gastropods), interpreted as turbiditic deposits, occurred from 6.5-7.2 m above the bottom of the outcrop and are separated by a 0.2 m thick layer of silty-sandy clays. Samples for foraminiferal analyses were collected at distance of 5-10 cm each others in the laminated layers, of 0.5-1 m in the bioturbated silty clays.

Foraminiferal preservation is very good, dominant planktonic species in the >125 μm fraction are *Globigerina* (*Globoturborotalita*) *apertura*, *Globigerina bulloides*, *G.* (*Gt.*) *decoraperta*, *Globigerinella obesa*, *Globigerinoides obliquus*, *G. extremus*. *Globorotalia scitula* and *Orbulina universa* are often rare. The mesopelagic globorotalias are represented by rare but large specimens of *G. margaritae* (*G. margaritae margaritae* and *G. margaritae evoluta*) occurring in the lower samples up to sample 17 and by *G. puncticulata* and *G. puncticulata padana*, randomly occurring from sample 5 upwards and frequent in the interval between the samples 16 and 23 (Fig. 3). On the basis of *G. margaritae* and *G. puncticulata* distribution, the

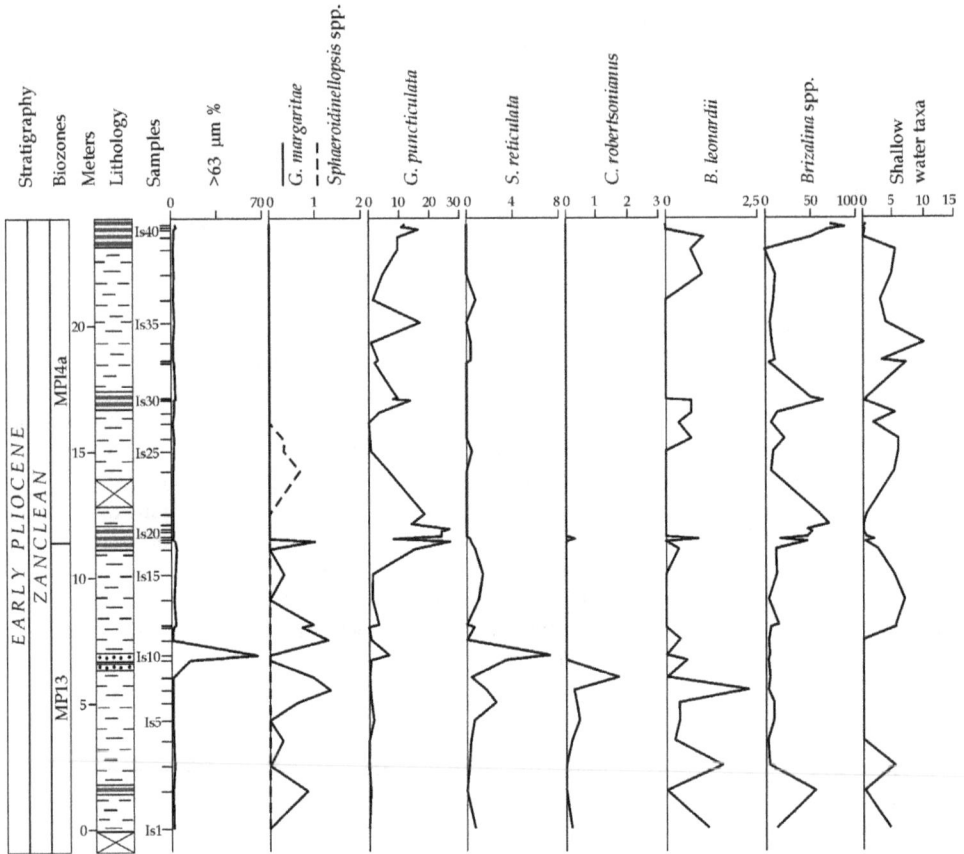

Fig. 3. Isola d'Asti section. Chronostratigraphic synthesis, lithological column, samples position, percentage variations of the grain sized > 63 μm fraction, of planktonic (*Sphaeroidinellopsis* spp., *Globorotalia margaritae* and *Globorotalia puncticulata*) and benthic (*Siphonina reticulata*, *Cibicidoides robertsonianus*, *Bolivina leonardii*, *Brizalina* spp. and Shallow water taxa) foraminiferal taxa.

MPl3 and MPl4a Zones are documented, the first in the lower part of the section, from the bottom to 11.4 m, the latter in the upper part of the section. Calcareous nannofossil assemblages (Lozar, pers. comm.) allow the correlation to the MNN14/15 Zone (Rio et al., 1990), coeval of the upper MPl3 and lower MPl4a foraminiferal Zones. Very rare specimens of *Sphaeroidinellopsis subdehiscens*, a deep mesopelagic taxon, were collected in few samples (Fig. 3) and suggest a deep water colomn. Benthic foraminiferal assemblages are rich and well diversified in the massive clays. Deep outer neritic to bathyal taxa, such as *Cibicidoides pseudoungerianus*, *Hoeglundina elegans*, *Siphonina reticulata*, *Uvigerina peregrina*, and Nodosariidae (Van Morkhoven et al., 1986; Wright 1978) are common, more frequent in the MPl3 interval. Many taxa that go extinct during the MPl5 Zone (*Anomalinoides helicinus*, *Bolivina leonardii*, *B. placentina*, *Bulimina minima*, *Uvigerina rutila*) are present or common (AGIP, 1982; Sprovieri 1986). In particular, the epifaunal *S. reticulata*, indicative of Early

Pliocene Mediterranean Intermediate Water (EPMIW) (Sgarrella et al., 1997) reach its frequency peaks in the central samples dated to the MPl3 Zone (Fig. 3) and occurs also, even if very rare, in some upper levels referred to the MPL4a Subzone. Another bathyal species, just occurring in the Moncucco MPl1 to MPl3 interval, is *C. robertsonianus*, indicative of well-oxygenated deep paleoenvironment (Kouwenhoven et al., 1999). At Isola d'Asti, its last recovery immediately follows the *G. margaritae* LO (sample 18 and 17, respectively). Specimens of *B. leonardii* are more common in the lower section and occur mainly below the laminated sapropelitic layers or in the basal laminite samples.

In contrast with the high diversity of benthic foraminifers in the massive clays, nearly oligotypic benthic assemblages characterize the laminated layers. *Brizalina* spp. (*B. aenaeriensis*, *B. dilatata* and *B. spathulata*) is strongly dominant, followed by other infaunal, stress tolerant taxa as Buliminids, *Fursenkoina schreibersiana*, and *Stainforthia complanata*, widespread in organic matter rich sediments. Disaerobic bottom conditions are responsible for the origin of the laminated, sapropelitic intervals recognized in many Pliocene outcrops and correlated to the deep sea Mediterranean succession (Capozzi & Picotti, 2003; Negri et al., 2003; Rio et al., 1997; Thunell et al., 1984).

At Isola d'Asti, the second laminated layer from the bottom occurs just above the *G. margaritae* LO at the base of the MPl4a Zone. A generally positive correlation of the *Brizalina* spp. and *G. puncticulata* curves is also registered. The displaced shallow water forms (mainly epiphytic species, such as *Cibicides lobatulus* and *Rosalina globularis*) randomly occur in the lower samples and become common to frequent upwards. They are absent or extremely rare in the laminated layers, and suggest a reduced transport from the inner shelf during the sapropelitic levels deposition.

Planktonic and benthic assemblages document an open-sea setting for all the succession, on epibathyal bottoms probably at depth of at least 600-800 m up to its middle part, in which *Sphaeroidinellopsis* occurred. A slight shallowing is inferred in the upper MPl4a zone, by the decrease or disappearance of the deep benthics *S. reticulata* and *C. robertsonianus*.

7.4 The Verrua Savoia composite section

The studied site is located in central Piedmont, in the area of transpressive faults that separates the NE Monferrato, of North Apennine affinity, from the Turin Hill, of Alpine affinity (Piana & Polino, 1995) (Fig. 1). The Verrua Savoia succession was known in literature as being the thicker Pliocene outcrop of the central Piedmont (Bonsignore et al., 1969). Foraminiferal assemblages of few samples from the Verrua S. area were described by Martinis (1954) and Zappi (1961). A detailed sampling in the Cementi Vittoria quarry, opened at the foot of Verrua S. hill, was carried out during the Sheet 157 Trino, 1:50.000 of the Geological Map of Italy (Clari and Polino, 2003; Dela Pierre et al., 2003) survey. Preliminary data on planktonic and benthic foraminifers were presented by Bove Forgiot et al. (2005). The composite section here analyzed is represented by two transects: a lower transect, cropping out in the northern side of the quarry and a second upper transect, sampled at the top of the hill, below the Verrua S. castle (Fig. 4). At the bottom, Pliocene sediments overlay with an erosional surface Upper Eocene gray calcareous marls (sample Va1 and Va2), the boundary is evidenced by bioturbations infilled by the overlying Pliocene grayish marly silts. The Pliocene succession is represented by about 80 meters of well stratified marly clays, silts and sandy silts, affected by faults, with interbedded sand and calcarenite banks, progressively more frequent and thicker upwards. Only sandy silts, sands and calcarenites are represented in the upper transect. Foraminiferal assemblages were analyzed on washing residues, prepared from silty or sandy samples and from the less consolidated calcarenites, and on thin sections. Only semiquantitative micropaleontological analyses were carried out, because the abundant reworked tests, of Eocene forms in the lower section, and of Miocene to lower Pliocene taxa in the middle and upper layers, prevented the quantitative study. Foraminiferal preservation was generally poor, tests were encrusted or partially included in calcareous aggregates. The planktonic foraminiferal assemblages are nearly similar to those described in the Moncucco T. and Isola d'Asti section: they are mainly given by Miocene to Pliocene taxa such as *Globigerina (Globoturborotalita) apertura, G. (Gt.) decoraperta, Globigerinoides obliquus, G. extremus* and less common *Neogloboquadrina acostaensis*. Well preserved specimens of *G. margaritae* occur from sample Va3, base of the Pliocene succession, to sample Va31. *G. puncticulata* and *G. puncticulata padana* were recognized in all the samples, more frequent in the lower, thinner layers. Empty well preserved tests occur also in some uppermost samples, together with infilled, partially diagenized tests of the same species, interpreted as slightly older and reworked. Therefore, the lower part of the marly clays and silts was correlated to the MPl3 Zone and the following sandy silts and calcarenites to the MPl4a Zone. The reasonably *in situ* (not reworked) benthic foraminiferal assemblages of the Verrua S. silts are very diversified and yield common deep outer neritic to bathyal taxa (*Anomalinoides helicinus, Cassidulina carinata, Cibicidoides pseudoungerianus, Hoeglundina elegans, Planulina ariminensis, Siphonina reticulata, Uvigerina peregrina, U. rutila*) (Van Morkhoven et al., 1986; Wright 1978). In particular, *S. reticulata* occurs in the silty layers of the first transect, whereas it is absent in the upper calcarenites (Fig. 4). Very rare, small empty specimens of the deep water *C. robertsonianus* are present only in few lower samples, in which *S. reticulata* is also common. The infaunal, stress tolerant *Bolivina* (*B. placentina* and mainly *B. leonardii*), *Brizalina* (*B. aenaeriensis, B. spathulata*) and *Bulimina* spp., preferential of muddy sediments, are scarce.

In the uppermost sands and calcarenites rare test of *Buccella granulata* were recognized. The species, originally described as *Eponides frigidus granulatus* (Di Napoli Alliata, 1952), is living in inner neritic sandy bottoms and cool waters (Serandrei Barbero et al., 1999) and was firstly reported in the Early Pliocene MPl4 Zone (Basilici et al., 1997; Rio et al., 1988). The biostratigraphic application of this taxon is limited by its ecological preference for sandy sediments, devoid of biostratigraphic markers, and the precise calibration of its FO is dubious. The recovery of *B. granulata* only in the uppermost deposits and its absence in the lower sandy and calcarenitic layers of the Verrua S. section, suggest an occurrence during the MPl4a Subzone, but excluding its lowermost interval. Other peculiar characteristics of the Verrua S. assemblages are the very low recovery of *Brizalina* spp., absent or rare in most samples and the high frequency of shallow water forms, showing increasing abundances upwards. Epiphitic taxa (*Elphidium crispum, Neoconorbina terquemi* and *Rosalina* spp.) are dominant in the calcarenitic layers. The progressively coarser sediments and higher amounts of winnowed shallow water forms testify nearby inner shelf vegetated bottoms as well as the increasing frequency of turbiditic episodes. A basin shallowing to upper epibathyal and to outer neritic depths is inferred.

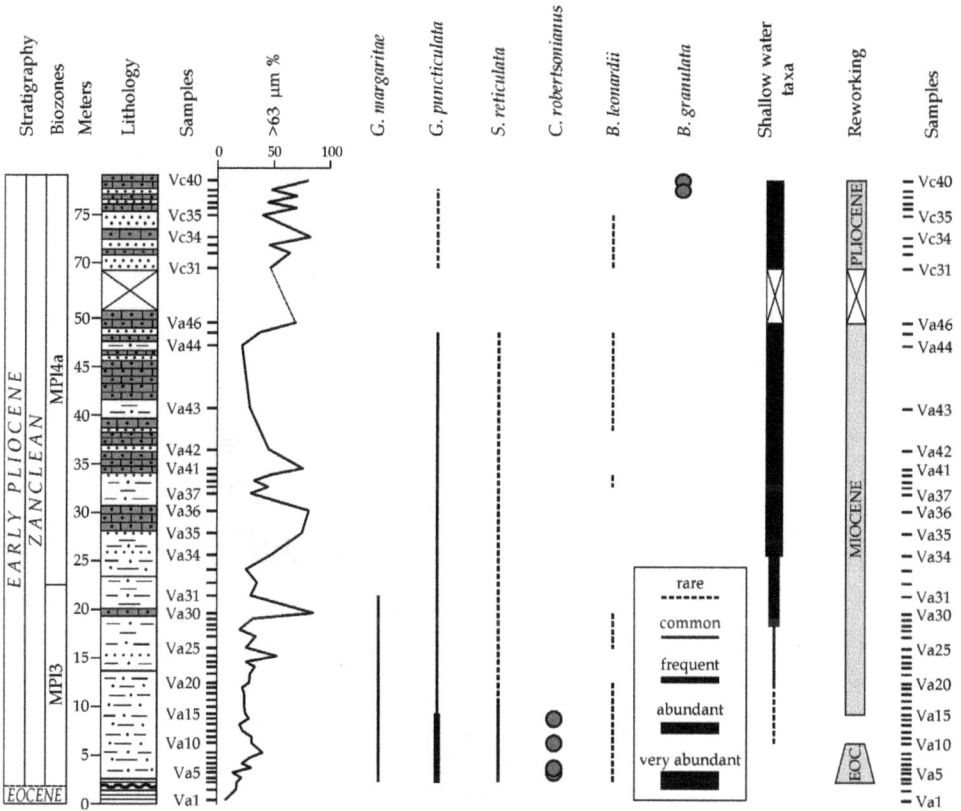

Fig. 4. Chronostratigraphic synthesis, lithological column, samples position, percentage variations of the grain sized > 63 μm fraction, qualitatitive distribution and frequency of planktonic (*Globorotalia margaritae* and *Globorotalia puncticulata*) and benthic (*Siphonina reticulata, Cibicidoides robertsonianus, Bolivina leonardii, Buccella granulata* and Shallow water taxa) foraminiferal taxa, age and abundance of reworked specimens.

7.5 Comparison and correlations

Biostratigraphic and paleoenvironmental data provide tools for comparisons and correlations between the three previously discussed sections, useful for the recontruction of the regional geological history. Moreover, biostratigraphic and ecobiostratigraphic events allow detailed correlations between the Piemont area and Mediterranean area.

The Messinian/Zanclean boundary and the early Zanclean are well preserved at the Moncucco T. section. During the lowermost MPl1 Zone, the recognized biovents (the abundance peak of *Globigerina nepenthes*, the sinistral shift of *Neogloboquadrina acostaensis*, the *Sphaeroidinellopsis* spp. acme, the *Globorotalia scitula* dextral CO, and the re-immigration of the benthic species *Siphonina reticulata* and *Cibicidoides robertsonianus*) occurred exactly in the same order and in similar stratigraphic position as in the Mediterranean reference successions (Di Stefano et al., 1996; Iaccarino et al., 1999a; 1999b; Spezzaferri et al., 1998).

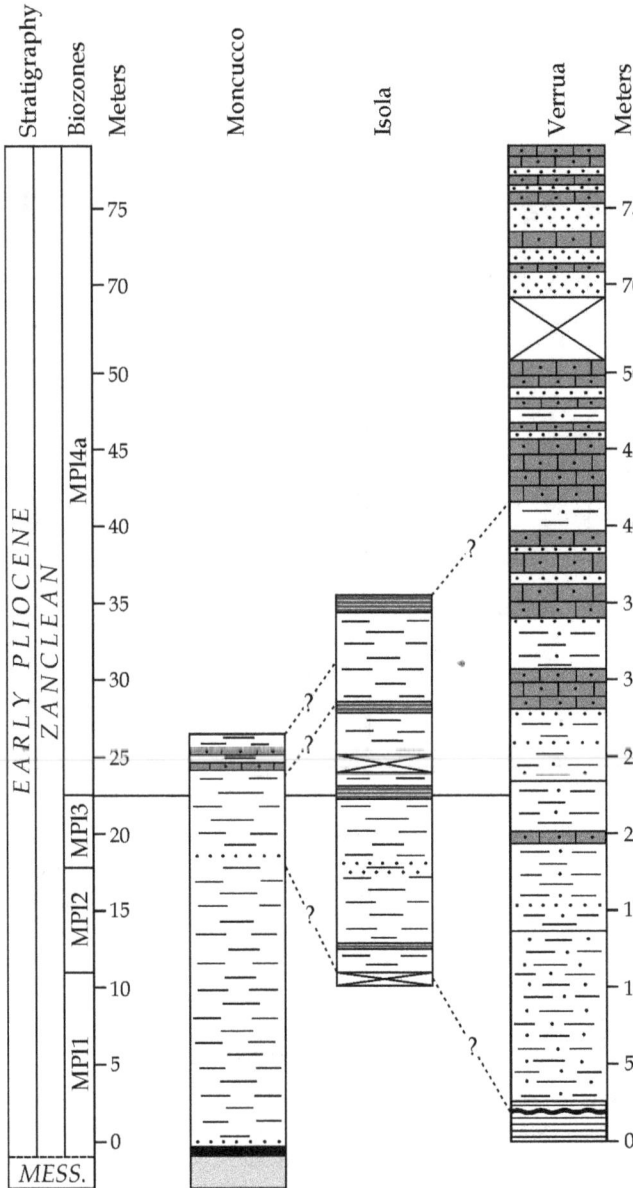

Fig. 5. Tentative correlation between the Moncucco T., Isola d'Asti and Verrua S. sections.

The paleodepths in the upper epibathyal zone, initially to about 500 m, then fastly deepening to about 800-1000 m, suggested in the Verrua S. section by the benthic foraminiferal assemblages during the MPl1 and MPl2 Zones, are similar to those proposed for the coeval interval of the Rossello Composite Section (Van Couvering et al., 2000) and of the Capo Rossello area (Barra et al., 1998; Sgarrella et al., 1997; 1999). All these data

Pliocene Mediterranean Foraminiferal Biostratigraphy: A Synthesis and Application to
the Paleoenvironmental Evolution of Northwestern Italy

181

document deep open sea conditions in the Northwestern Italy in the early Zanclean, and very wide connections with the Mediterranean basin, that can explain the inferred synchronous record of the flooding event at the base of the Zanclean, as well as the following planktonic and benthic foraminifer bioevents.

Whereas the lower Zanclean is preserved only in the Moncucco T section, the MPl3 and MPl4a Zones are well documented in all the studied sections. Biostratigraphic data, provided by the occurrence of the zonal markers (*G. margaritae* and *G. puncticulata*), and in particular by the *G. puncticulata* FO, marking the MPl3/MPl4a boundary, are the main tools for correlation of the studied sections. A tentative correlation between the interval encompassing the MPl3 and MPl4a Zones (upper Moncucco T. section, Isola d'Asti and Verrua S. sections) is here proposed (Fig. 5).

Other elements for ecobiostratigraphic correlations and for the paleonvironmental recontruction are provided by benthic foraminifer distribution and frequencies. Marly clays pertaining to the MPl3 and MPl4a Zones in the Moncucco T. and Isola d'Asti sections, and to the MPl3 Zone at Verrua S., are inferred to be deposited at very similar epibathyal depths, on the basis of their similar planktonic (rare to frequent globorotalias, mainly *G. puncticulata*, most frequent at Isola d'Asti) and benthic assemblages (common epibathyal taxa, such as *P. ariminensis, S. reticulata, U. peregrina, U. rutila*). At Moncucco T., laminated layers were not evident, but in the MPl4a Zone the frequency peak of the infaunal, low-oxygen tolerant *Brizalina* spp. suggests a brief interval of disaerobic bottom condition, immediately below the increase in abundance of shallow water taxa, and the deposition of the first calcarenitic layer, both indicative of active transport from the inner shelf. At Isola d'Asti, four laminated layers are interbedded to the massive marly clays and testify repeated episodes of dysoxia at the sea bottom. All layers are characterized by nearly oligotypic benthic assemblages, dominated by the deep infaunal *Brizalina* spp. and by the rareness or absence of shallow water displaced forms. The first sapropelitic level occurs in the MPl3 Zone, the others are in coincidence with the *G. margaritae* LCO, marking the MPl3/MPl4a boundary, and in the MPL4a Zone. The third sapropelitic layer of Isola d'Asti could be correlated to the *Brizalina* spp. peak at Moncucco T. (Fig. 5), on the basis of their stratigraphic position and foraminiferal composition. In the same time interval, bottoms influenced by stagnant episodes are better documented at Isola d'Asti than at Moncucco T. In this latter succession, turbidic accumulation from shelfal areas appears to have been subsequent to the dysoxic episode (*Brizalina* spp. peak). Roveri & Taviani (2003) suggested a close relationship between the shallow water calcarenitic bodies, and the deep water sapropels, widespread in Northern Apennine (Castell'Arquato Basin) and Sicily (Belice and Caltanissetta Basins) upper Pliocene. Calcarenite and sapropel clusters appeared together and became more developed starting from 3.1 Ma. Roveri & Taviani (2003) proposed that the calcarenite and sapropel clusters formation were driven by periodical changes in orbital parameters and directly linked to the cooling trend culminating in the onset of the Northern Hemisphere glaciation. The authors left open the question if calcarenites and sapropels developed in opposite phases of a same climatic cycle or were synchronous. Data from the Moncucco T. and Isola d'Asti sections document very low or nearly absent downslope trasport of shallow water material during the sapropels deposition. Therefore, an opposite, diachronous relation between the two lithologies is here tentatively proposed.

Coarsening upward sediments and increasing frequence of turbiditic episodes characterize the Verrua S. section, nearly coeval to the Isola d'Asti section, also encompassing the MPl3/MPl4a Zones. At Verrua S., a deep epibathyal setting, similar to that inferred at Moncucco T. and Isola d'Asti, can be suggested only for the lower part of the section, dated to the MPl3 Zone, on the basis of the total benthic assemblage, of the occurrence of common to frequent G. *margaritae* and G. *puncticulata* and of rare deep benthic taxa (*S. reticulata* and *C. robertsonianus*). During the MPl4a zone, the basin shallowing is more evident at Verrua S. than in the other sites and the upper silty layers are inferred to have been deposited in the outer neritic zone. Peculiar characteristics of the Verrua S. section are the common to abundant terrigenous debris, the high amount of reworked foraminiferal tests, of Upper Eocene age in the lowermost part of the section, of Miocene to Pliocene age upwards and the progressively dominant calcarenitic layers. All these features document a depositional setting on the upper slope to the shelf margin, rather near to the inner shelf or to emerged areas, sources of the reworked specimens, just during the MPl3 Zone. Turbiditic episodes became more frequent during the MPl4a Zone and high amounts of terrigenous and biogenic debris of inner shelf source were accumulated on the seafloor. As a consequence, the Verrua S. succession could represent a more marginal counterpart of the deeper, open sea Moncucco T. and Isola d'Asti deposits. The tectonic activity of the area, where a succession of Apennine affinity (Monferrato) was translated and is separated by transpressive faults by the Alpine affinity Turin Hill (Piana & Polino, 1995), also favoured the turbidites accumulation.

8. Conclusions

Recent biostratigraphical and ecobiostratigraphical studies on Pliocene foraminiferal assemblages of Piedmont (Northwestern Italy) provide results concerning both correlations at the Mediterranean scale and applications to the regional tectonic and paleoenvironmental evolution.

In the Northwestern sector of the Pliocene Mediterranean basin, data from the Moncucco T. and Isola d'Asti sections document open-sea bottoms at similar depths than in the Southern Mediterranean areas. Foraminiferal biovents registered in the Piedmont sections and in the reference on-land or deep sea successions are inferred to be isochronous in the MPl1 Zone (the sinistral shift of *Neogloboquadrina acostaensis*, the *Sphaeroidinellopsis* spp. acme, etc.) up to the MPl3 Zone (progressive disappearance of deep benthic taxa) (Iaccarino et al., 1999a; Pierre et., 2006; Sgarrella et al., 1999; Spezzaferri et al., 1998, among others).

At a regional scale, the described sections can be correlated to the MPl3/MPl4a interval and represent different but related depositional settings: epibathyal well oxygenated bottoms in most of the Moncucco T. succession, similar epibathyal bottoms, but registering dysaerobic conditions and sapropelitic deposition at Isola d'Asti, and a more marginal setting, affected by turbidic transport from the inner shelf at Verrua S.

Moreover, biostratigraphic and paleoenvironmental data on epibathyal deposition up to the MPl4a Zone in the Turin Hill area, where the Moncucco T. section is located, provide data on a younger than Zanclean uplift of the sector and enlight the tectonical evolution of this sector, strongly related to the Alps and Northern Apennines uplift.

Data from the Moncucco T. and Isola d'Asti succession suggest an opposite correlation between laminated, sapropelitic layers, in which material from shallow waters is nearly absent, and calcarenitic layers, composed by terrigenous and biogenic debris of shelfal origin.

Nevertheless, this hypothesis needs to be verified by further researches, that should be extended to other Piedmont and Northern Apennines successions.

High resolution, integrated studies are also in progress in the central part of Piedmont to improve the biostratigraphic data and to verify the hypothesis of a stratigraphical gap between the MPl3/MPl4a Zones and its possible relationship with erosional surfaces.

9. Taxonomic list of cited taxa

9.1 Planktonic foraminifers

Globigerina bulloides d'Orbigny, 1826
Globigerina (Globoturborotalita) apertura Cushman, 1918
Globigerina (Globoturborotalita) decoraperta Takayanagi & Saito, 1962
Globigerina nepenthes Todd, 1957
Globigerinella obesa (Bolli, 1957) (= *Globorotalia obesa* Bolli)
Globigerinoides elongatus (d'Orbigny, 1826) (= *Globigerina elongata* d'Orbigny)
Globigerinoides extremus Bolli & Bermudez, 1965 (= *Globigerinoides obliquus extremus* Bolli & Bermudez)
Globigerinoides obliquus Bolli, 1957 (= *Globigerinoides obliquus obliquus* Bolli)
Globorotalia aemiliana Colalongo & Sartoni, 1967
Globorotalia bononiensis Dondi, 1962
Globorotalia crassaformis (Galloway & Wissler, 1927) (= *Globigerina crassaformis* Galloway & Wissler)
Globorotalia crassula Cushman, Stewart & Stewart, 1930
Globorotalia margaritae Bolli & Bermudez, 1965
Globorotalia margaritae evoluta Cita, 1973
Globorotalia margaritae primitiva Cita, 1973
Globorotalia praemargaritae Catalano & Sprovieri, 1969
Globorotalia puncticulata (Deshayes, 1832) (= *Globigerina puncticulata* Deshayes)
Globorotalia puncticulata padana Dondi & Papetti, 1968
Globorotalia scitula (Brady, 1882) (= *Pulvinulina scitula* Brady)
Neogloboquadrina acostaensis (Blow, 1959) (= *Globorotalia acostaensis* Blow)
Neogloboquadrina atlantica (Berggren, 1972) (= *Globigerina atlantica* Berggren)
Orbulina universa d'Orbigny, 1839
Sphaeroidinellopsis subdehiscens Blow, 1969

9.2 Benthic foraminifers

Ammonia beccarii (Linnaeus, 1758) (= *Nautilus beccarii* Linnaeus)
Anomalinoides helicinus (Costa, 1857) (= *Nonionina helicina* Costa)
Bolivina leonardii Accordi & Selmi, 1952
Bolivina placentina Zanmatti, 1957
Brizalina aenaeriensis Costa, 1856

Brizalina dilatata (Reuss, 1850) (*Bolivina dilatata* Reuss)
Brizalina spathulata (Williamson, 1858) (= *Textularia variabilis spathulata* Williamson)
Buccella granulata (Di Napoli, 1952) (= *Eponides frigidus granulatus* Di Napoli)
Bulimina minima Tedeschi & Zanmatti, 1957 = *Bulimina aculeata minima* Tedeschi & Zanmatti
Cassidulina carinata (Silvestri, 1896) (= *Cassidulina laevigata carinata* Silvestri)
Cibicides lobatulus (Walker & Jacob, 1798) (= *Nautilus lobatulus* Walker & Jacob
Cibicidoides pseudoungerianus (Cushman, 1922) (= *Truncatulina pseudoungeriana* Cushman)
Cibicidoides robertsonianus (Brady, 1884) (= *Truncatulina robertsoniana* Brady)
Elphidium crispum (Linnaeus, 1758) (= *Nautilus crispus* Linnaeus)
Fursenkoina schreibersiana (Czjzek, 1848) (= *Virgulina schreibersiana* Czjzek)
Globobulimina affinis (d'Orbigny, 1839) (= *Bulimina affinis* d'Orbigny)
Hoeglundina elegans (d'Orbigny, 1826) (= *Rotalia elegans* d'Orbigny)
Neoconorbina terquemi (Rzehak, 1888) (= *Discorbina terquemi* Rzehak)
Planulina ariminensis d'Orbigny, 1826
Rosalina globularis d'Orbigny, 1826
Siphonina reticulata (Czjzek, 1848) (= *Rotalina reticulata* Czjzek)
Stainforthia complanata (Egger, 1893) (= *Virgulina schreibersiana complanata* Egger)
Uvigerina peregrina Cushman, 1923
Uvigerina rutila Cushman & Todd, 1941

10. Acknowledgments

The author sincerely thanks Elena Vassio and Magda Minoli for technical help, Simona Cavagna for help in SEM analyses, Francesco Dela Pierre, Giulio Pavia and Stefania Trenkwalder for scientific collaboration. The authors is greatly indebted with Francesca Lozar, for useful suggestions and for the revision of the english text.

11. References

AGIP (1982). *Foraminiferi padani (Terziario e Quaternario). Atlante iconografico e distribuzioni stratigrafiche*, Second Edition, 52 pls., AGIP Mineraria, S. Donato Milanese, Italy.

Aguirre, E. & Pasini, G. (1985). The Pliocene-Pleistocene boundary. *Episodes*, Vol.8, pp. 116-120.

Aimone, S. & Ferrero Mortara, E. (1983). Malacofaune plioceniche di Cossato e Candelo (Biellese, Italia NW). *Bollettino del Museo Regionale di Scienze Naturali di Torino*, Vol.1, No.2, pp. 279-328.

Backman, J. & Raffi, I. (1997). Calibration of Miocene nannofossil events to orbitally tuned cyclostratigraphies from Ceara Rise, In: *Proceedings of the Ocean Drilling Program, Scientific Results*, N.J. Shackleton , W.B. Curry, C. Richter, & T.J. Bralower,. (Eds.), Vol.154, pp. 83-99, College Station, Texas, U.S.A.

Barbieri, F. (1967). The foraminifera in the Pliocene section Vernasca-Castell'Arquato including the "Piacenzia Stratotype". *Memorie della Società Italiana di Scienze Naturali, Museo Civico di Scienze Naturali di Milano*, Vol.15, pp. 145-163.

Barbieri, F.; Carraro F. & Petrucci, F. (1974). Osservazioni micropaleontologiche e stratigrafiche sulla serie marina e glaciomarina della Valchiusella (Ivrea, Prov. di Torino). *L'Ateneo Parmense - Acta Naturalia*, Vol.10, No.1, pp. 5-14.

Barra, D.; Bonaduce, G. & Sgarrella, F. (1998). Palaeoenvironmental bottom water conditions
    in the early Zanclean of the Capo Rossello area (Agrigento, Sicily). *Bollettino della
    Società Paleontologica Italiana*, Vol.37, No.1, pp. 61-98.

Basilici, G.; Martinetto, E.; Pavia G. & Violanti D. (1997). Paleoenvironmental evolution in
    the Pliocene marine-coastal succession of Val Chiusella (Ivrea, NW Italy). *Bollettino
    della Società Paleontologica Italiana*, Vol.36, No.1-2, pp. 23-52.

Bassett, M.G. (1985). Towards a "common language" in stratigraphy. *Episodes*, Vol.8, pp. 87-
    92.

Benson, R.H. & Rakic El-Bied, K. (1996). The Bou Regreg Section, Morocco: proposed global
    boundary stratotyper section and point of the Pliocene. *Service Géologique de Maroc,
    Notes and Mémoires*, Vol. 383, pp. 51-150.

Berggren, W.A. (1973). The Pliocene time scale: calibration of planktonic foraminifera and
    calcareous nannoplankton zones. *Nature*, Vol. 241, No. 5407, pp. 391-397.

Berggren, W.A. & Haq, B.U. (1976). The Andalusian Stage (Late Miocene): Biostratigraphy,
    biochronology and paleoecology. *Palaeogeography, Palaeoclimatology, Palaeoecology*,
    Vol.20, pp. 67-129.

Berggren, W.A.; Kent, D.V. & Van Couvering, J.A. (1985). Neogene geochronology and
    chronostratigraphy. In: *The Chronology of the Geological Record*, N.J. Snelling (Ed.),
    Geological Society of London Memoir, Vol.10, pp. 211-260.

Berggren, W.A.; Hilgen, F.J.; Langereis, C.G.; Kent, D.V.; Obradovich, J.D.; Raffi, I.; Raymo,
    M.E. & Shackleton, N.J. (1995). Late Neogene (Pliocene-Pleistocene) Chronology:
    New perspectives in High Resolution Stratigraphy. *Geological Society of America
    Bulletin*, Vol.107, pp. 1272-1287.

Bizon, G. & Bizon, J.J. (1972). *Atlas des principaux foraminifères planctoniques du bassin
    méditerranéen. Oligocene à Quaternaire*. pp. 1-316, Edtions Technip, Paris.

Bolli, H.M. (1970). The foraminifera of Sites 23-31, Leg 4. In: *Initial Reports of the Deep Sea
    Drilling Program*, R.G. Bader, R.D. Gerard et al. (Eds.), Washington
    (U.S.Government Printing Office), Vol.4, pp. 577-643.

Bolli, H.M. & Bermudez, P.J. (1965). Zonation based on planktonic foraminifera of Middle
    Miocene to Pliocene warm-water sediments. *Boletin Informativo, Asociación
    Venezolana de Geologia, Mineria y Petroleo*. Vol. 8, pp. 119-149.

Bolli, H.M. & Saunders, J.B. (1985). Oligocene to Holocene low latitude planktic
    foraminifera. In: *Plankton Stratigraphy*, H.M. Bolli, J.B. Saunders, K. Perch-Nielsen
    (Eds.), pp. 155-262, Cambridge University Press, Cambridge, United Kingdom.

Boni, A.; Boni, P.; Peloso, G. F. & Gervasoni, S. (1985). Nuove osservazioni e considerazioni
    sui lembi pliocenici dal confine di stato a Capo Sant'Ampelio (Alpi Marittime). *Atti
    dell'Istituto Geologico dell'Università di Pavia*, Vol.30, pp. 246-309.

Boni, A.; Mosna, S. & Peloso, G. F. (1987). Considerazioni su una successione stratigrafica
    mio-pliocenica del Monregalese. *Atti Ticinensi di Scienze della Terra*, Vol.31, pp. 165-
    175.

Bonsignore, G.; Bortolami, G. C.; Elter, G.; Montrasio, A.; Petrucci, F.; Ragni, U.; Sacchi, R.;
    Sturani, C. & Zanella, E., (1969). *Note Illustrative della Carta Geologica d'Italia alla scala
    1:100.000, Fogli 56 e 57 Torino - Vercelli*, IIª ed., 96 pp., Servizio Geologico Italiano,
    Roma, Italy.

Bove Forgiot, L.; Trenkwalder, S. & Violanti, D. (2005). Le associazioni plioceniche a foraminiferi ed ostracodi di Verrua Savoia (Piemonte, Italia NW): analisi biostratigrafica e paleoambientale. *Rendiconti della Società Paleontologica Italiana*, Vol.2, pp. 9-27.

Breda, A.; Mellere, D.; Massari, F. & Asioli, A. (2009). Vertically stacked Gilbert-type deltas of Ventimiglia (NW Italy): The Pliocene record of an overfilled Messinian incised valley. *Sedimentary Geology*, Vol.219, pp. 58-76.

Brocchi, G. (1814). *Conchiologia fossile subapennina*. Vol. 1 & 2, 712 pp., Stamperia Reale, Milano, Italy.

Cande, S.C. & Kent, D.V. (1992). A new geomagnetic polarity time scale for the Late Cretaceous and Cenozoic. *Journal of Geophysical Research*, Vol.97 (B10), pp. 13917-13951.

Cande, S.C. & Kent, D.V. (1995). Revised calibration of geomagnetic polarity time scale for the Late Cretaceous and Cenozoic. *Journal of Geophysical Research*. Vol.100 (B4), pp. 6093-6095.

Capozzi, R. & Picotti, V. (2003). Pliocene sequence stratigraphy, climatic trends and sapropel formation in the northern Apennines (Italy). *Palaeogeography, Palaeoclimatology, Palaeoecology*, Vol.90, pp. 349-371.

Casnedi, R. (1971). Contributo alle conoscenze del Pliocene del Bacino Terziario Piemontese. *Atti dell'Istituto Geologico dell'Università di Pavia*, Vol.22, pp. 110-120.

Castradori, D. (1998). Calcareous nannofossils in the basal Zanclean of the Eastern Mediterranean Sea: remarks on palaeoceanography and sapropel formation. In *Proceedings of the Ocean Drilling Program*, Scientific Results, Vol.160, A.H.F. Robertson, K.C. Emeis, C. Richter and A. Camerlenghi (Eds.), pp.113 123, College Station, Texas, U.S.A.

Castradori, D.; Rio, D.; Hilgen, F.J. & Lourens, L.J. (1998). The Global Standard Stratotype section and Point (GSSP) of the Piacenzian Stage (Middle Pliocene*). Episodes*, Vol.21, No.2:, pp. 88-93.

Catalano, R. & Sprovieri, R. (1969). Stratigrafia e micropaleontologia dell'intervallo tripolaceo di Torrente Rossi (Enna). *Atti dell'Accademia Gioenia di Scienze Naturali*, Ser.7, Vol.1, pp. 513-527.

Catalano, R. & Sprovieri, R. (1972). Biostratigrafia di alcune serie Saheliane (Messiniano inferiore) in Sicilia. In: *Proceedings of the II Planktonic Conference Roma 1970*, M. Farinacci (Ed.), Vol.1, pp. 211-249.

Chaisson, W.P. & Pearson, P.N. (1997). Planktonic foraminifer biostratigraphy at Site 925: middle Miocene-Pleistocene. In: *Proceedings of the Ocean Drilling Program, Scientific Results*, N.J. Shackleton, W.B. Curry, C. Richter & T.J. Bralower (Eds.), Vol.154, pp. 3-32, College Station, Texas, U.S.A.

Channell, J.E.T.; Poli, M.S.; Rio, D.; Sprovieri R. & Villa, G. (1994). Magnetic stratigraphy and biostratigraphy of Pliocene "argille azzurre" (Northern Apennines, Italy). *Palaeogeography, Palaeoclimatology, Palaeoecology*, Vol.110, pp. 83-102.

Cita, M.B. (1973). Pliocene biostratigraphy and chronostratigraphy. In: *Initial Reports of the Deep Sea Drilling Program*, W.B.F. Ryan, K.J. Hsü et al. (Eds.), Vol.13, pp. 1343-1379, U.S.Government Printing Office, Washington, U.S.A.

Cita, M.B. (1975a). Studi sul Pliocene e gli strati di passaggio dal Miocene al Pliocene. VII. Planktonic foraminiferal biozonation of the Mediterranean Pliocene deep-sea record. A revision. *Rivista Italiana di Paleontologia e Stratigrafia*, Vol. 81, No. 4, pp. 527-544.

Cita, M.B. (1975b). The Miocene- Pliocene boundary: history and definition. In: *Late Neogene Epoch Boundaries*, T. Saito & L.D. Burckle, (Eds.), Special Publication Vol. 1, pp. 1-30, Micropaleontology Press, New York, U.S.A.

Cita, M.B. (2008). Summary of Italian marine stages of the Quaternary. *Episodes*, Vol.29: 107-114.

Cita, M.B. (2009). Mediterranean Neogene stratigraphy: development and evolution through the centuries. *Sedimentology*, Vol.56, pp. 43-62.

Cita, M.B. & Gartner, S. (1973). Studi sul Pliocene e gli strati di passaggio dal Miocene al Pliocene. IV. The stratotype Zanclean foraminiferal and nannofossil biostratigraphy. *Rivista Italiana di Paleontologia e Stratigrafia*, Vol.79, No.4, pp. 503-558.

Cita, M.B.; Rio, D.; Hilgen, F.J.; Castradori, D.; Lourens, L.J. & Vergerio, P. (1996). Proposal of the Global boundary Standard Stratotype section and Point (GSSP) of the Piacenzian (Middle Pliocene). *Neogene Newsletter*, Vol.3, pp. 20-46.

Cita, M. B.; Racchetti, S.; Brambilla, R.; Negri, M.; Colombaroli, D.; Morelli, L.; Ritter, M.; Rovira, E.; Sala, P.; Bertarini, L. & Sanvito, S. (1999). Changes in sedimentation rates in all the Mediterranean drillsites document basin evolution and support starved basin conditions after early Zanclean flood. *Memorie della Società Geologica Italiana*, Vol.54, pp.145-159.

Clari, P. & Polino, R. (2003). *Carta Geologica d'Italia alla scala 1:50.000. Foglio 157 Trino*. APAT & Dipartimento di Difesa del Suolo, Roma, Italy.

Coimbra, J.C.; Carreño, A.L. & de Santana dos Anjos-Zerfass, G. (2009). Biostratigraphy and paleoceanographical significance of the Neogene planktonic foraminifera from Pelotas Basin, southernmost Brazil. *Revue de Micropaléontologie*, Vol.52, pp. 1-14.

Colalongo, M.L.; Dondi, L.; D'Onofrio, S. & Iaccarino, S. (1982). Schema biostratigrafico a Foraminiferi per il Pliocene e il basso Pleistocene nell'Appennino settentrionale e nella Pianura Padana. In: *Guida alla geologia del margine appenninico padano*. G. Cremonini & F. Ricci Lucchi (Eds.), pp. 121-122.

Corselli C., Cremaschi M. & Violanti D. (1985) - Il Canyon messiniano di Malnate (Varese); pedogenesi tardo-miocenica ed ingressione marina pliocenica al margine meridionale delle Alpi. *Rivista Italiana di Paleontologia Stratigrafica*, Vol.91, No.2, pp. 259-286.

Dela Pierre, F.; Piana, F.; Fioraso, G.; Boano, P.; Bicchi, E.; Forno, M.G.; Violanti; D.; Balestro, G.; Clari, P.; d'Atri, A.; De Luca, D.; Morelli, M. & Ruffini, R. (2003). *Note illustrative della Carta Geologica d'Italia alla scala 1:50.000, Foglio 157 Trino*. 147 pp., APAT & Dipartimento di Difesa del Suolo, Roma, Italy.

Dervieux, E. (1892). Studio sui foraminiferi pliocenici di Villalvernia. *Atti della Regia Accademia delle Scienze di Torino*. Vol. 27, pp. 1-6.

Di Napoli Alliata, E. (1952). Nuove specie di foraminiferi nel Pliocene e nel Pleistocene della zona di Castell'Arquato (Piacenza). *Rivista Italiana di Paleontologia Stratigrafica*, Vol.58, No.3, pp. 95-110.

Di Stefano, E.; Sprovieri, R. & Scarantino, S. (1996). Chronology of biostratigraphic events at the base of the Pliocene. *Paleopelagos*, Vol.6, pp. 401-414.

Doderlein, P. (1870-1872). Note illustrative della Carta Geologica del Modenese e del Reggiano 1870. *Atti della Regia Accademia delle Scienze, Lettere e Arti di Modena*, Vol. 9, No. 1-2 (1870), pp. 1-114, No. 3 (1872), pp. 1-74.

Giammarino, S.; Sprovieri, R. & Di Stefano, I. (1984). La sezione pliocenica di Castel d'Appio (Ventimiglia). *Memorie della Società Toscana di Scienze Naturali*, Ser.A, Vol. 91, pp. 1-26.

Giammarino, S. & Tedeschi, D. (1980). Le microfaune a foraminiferi del Pliocene di Borzoli (Genova) e il loro significato paleoambientale. *Annali dell'Università degli Studi di Ferrara*, Ser.9, Vol.6, pp. 73-92.

Gibbard, P.L.; Head, M.J.; Walker, M.J.C & Subcommission on Quaternary Stratigraphy. (2010). Formal ratification of the Quaternary System/Period and the Pleistocene Series/Epoch with a base at 2.58 Ma. *Journal of Quaternary Science*, Vol. 25, No.2, pp. 96-102.

Gradstein, F.; Ogg, J. & Smith, A. (2004). *A Geologic Time Scale 2004*. 589 pp., Cambridge University Press, Cambridge. United Kingdom.

Hasegawa, S.; Sprovieri, R. & Poluzzi, A. (1990). Quantitative analysis of benthic foraminiferal assemblages from Plio-Pleistocene sequences in the Tyrrhenian Sea, ODP Leg 107. In: *Proceedings of the Ocean Drilling Program, Scientific Results*, K.A. Kastens, J. Mascle et al. (Eds.), Vol.107, pp. 461-478, College Station, Texas, U.S.A.

Haynes, J.R. (1981). *Foraminifera*. 433 pp., John Wiley and Sons, New York, U.S.A.

Hilgen, F.J (1991). Extension of the astronomically calibrated (polarity) time scale to the Miocene/Pliocene boundary. *Earth Planetary Science Letters*, Vol.107, pp. 349-368.

Hilgen, F.J & Langereis, C.G. (1988). The age of the Miocene-Pliocene boundary in the Capo Rossello area (Sicily). *Earth and Planetary Science Letters*, Vol. 91, pp. 214-222.

Hsü, K.J.; Cita, M.B. & Ryan, W.B.F. (1973). The origin of the Mediterranean evaporites. In: *Initial Reports of the Deep Sea Drilling Program*, W.B.F. Ryan, K.J. Hsü et al. (Eds.), Vol.13, pp. 1203-1231, U.S. Government Printing Office, Washington, U.S.A.

Kastens, K.A. Mascle J. et al. (Eds.) (1990). *Proceedings of the Ocean Drilling Program, Scientific Results*, Vol.107, 772 pp. College Station, Texas, U.S.A. doi:10.2973/odp.proc.sr.107.1990

Kennett J.P. & Srinivasan M.S. (1983). *Neogene Planktonic Foraminifera - A phylogenetic atlas*, 265 pp, Hutchinson Ross Publishing Company, Stroudsburg, Pennsylvania, U.S.A.

Kouwenhoven T.J.; Seidenkrantz, M.-S. & Van der Zwaan, G.J. (1999): Deep-water changes: The near-synchronous disappearance of a group of benthic foraminifera from the late Miocene Mediterranean. *Palaeogeography, Palaeoclimatology, Palaeoecology*, Vol.152, pp. 259-281.

Iaccarino, S. (1985). Mediterranean Miocene and Pliocene planktic foraminifera. In: *Plankton Stratigraphy*, H.M Bolli, J.B. Saunders & K. Perch-Nielsen (Eds.), pp. 283-314, Cambridge University Press, Cambridge, United Kingdom.

Iaccarino, S.; Cita, M.B.; Gaboardi, S. & Grappini, G.M. (1999a). High Resolution biostratigraphy at the Miocene/Pliocene boundary in Holes 974B and 975B, Western Mediterranean. In: *Proceedings ot the Ocean Drilling Program*, Scientific Results, R. Zahn, M.C. Comas & A. Klaus (Eds.), Vol.161, pp. 197-221, College Station, Texas, U.S.A.

Iaccarino, S.; Castradori, D.; Cita, M.B.; Di Stefano, E.; Gaboardi, S.; Mc Kenzie, J.A.; Spezzaferri, S. & Sprovieri, R. (1999b). The Miocene-Pliocene boundary and the significance of the earliest Pliocene flooding in the Mediterranean. *Memorie della Società Geologica Italiana*, Vol.54, pp. 109-131.

Langereis, C.G. & Hilgen, F.J. (1991). The Rossello composite: a Mediterranean and global reference section for the Early to early Late Pliocene. *Earth and Planetary Science Letters*, Vol. 104, pp. 211-225.

Loeblich, A.R. & Tappan, H. (1988). *Foraminiferal genera and their classification.* pp. 1-970, pls. 1-847, Van Nostrand Reinhold, New York, U.S.A.

Lourens, L.J; Antonarakou, A.; Hilgen, F.J.; Van Hoof, A.A.M.; Vergnaud-Grazzini, C. & Zachariasse, W. (1996). Evaluation of the Plio-Pleistocene astronomical timescale. *Paleoceanography*, Vol.11, pp. 391-413.

Lourens, L.J.; Hilgen, F.J.; Shackleton, N.J.; Laskar, J & Wilson, D. (2004). The Neogene Period. In: *A Geologic Time Scale 2004*, F. Gradstein, J. Ogg & A. Smith (Eds.), pp. 409-440., Cambridge University Press, Cambridge. United Kingdom.

Lozar, F.; Violanti, D.; Dela Pierre, F.; Bernardi, E.; Cavagna, S.; Clari, P.; Irace, A.; Martinetto, E. & Trenkwalder, S. (2010). Calcareous nannofossils and foraminifers herald the Messinian salinity crisis: the Pollenzo section (Alba, Cuneo; NW Italy). *Geobios*, Vol.43, pp. 21--32, ISSN: 0016-6995.

Lauldi, A. (1981). Il Pliocene di Folla d'Induno (Varese): indagine faunistica su campioni del sottosuolo. *Atti dell'Istituto Geologico dell'Università di Pavia*, Vol.29, pp. 115-119.

Lyell, C. (1833). *Principles of Geology.* Vol.3, 398 pp., John Murray, London, United Kingdom.

Martinis, B. (1950). La microfauna dell'affioramento pliocenico di Casanova Lanza (Como). *Rivista Italiana di Paleontologia e Stratigrafia*, Vol.56, No.2, pp. 55-64.

Martinis, B. (1954). Ricerche stratigrafiche e micropaleontologiche sul Pliocene piemontese. *Rivista Italiana di Paleontologia e Stratigrafia*, Vol.60, No.2-3, pp. 45-114/125-194.

Mayer-Eymar, K. (1858). Versuch einer neuen Klassifikation der Tertiär-Gebilde Europas. *Verhandlungen der Allgemeinen Schweizerischen Gesellschaft fur die gesammten Naturwissenschaften,* Vol.42 (1857), pp. 165-199.

Mayer-Eymar, K. (1868). *Tableau synchronistique des terrains tertiaires supérieurs.* 4th ed. Manz, Zurich, Switzerland (table only).

Montefameglio, L.; Pavia, G. & Rosa, D.A. (1979). Associazione a molluschi del Tabianiano del Basso Monferrato (Alba, Italia NW). *Bollettino della Società Paleontologica Italiana*, Vol.18, No.2, pp. 173-199.

Negri, A.; Morigi, C. & Giunta, S. (2003). Are productivity and stratification important to sapropel deposition? Microfossil evidence from late Pliocene insolation cycle 180 at Vrica, Calabria. *Palaeogeography, Palaeoclimatology, Palaeoecology*, Vol. 190, pp. 243-255.

Pavia, G.; Chiambretto, L. & Oreggia, G. (1989). Paleocomunità a molluschi nel Pliocene inferiore di Breolungi (Mondovì, Italia NW). In: *Atti del 3° Simposio di Ecologia e Paleoecologia delle Comunità Bentoniche* (1985), S.I. Di Geronimo (Ed.), pp. 521-569, Catania, Italy.

Piana, F.& Polino, R., (1995). Tertiary structural relationships between Alps and Apennines: The critical Torino Hill and Monferrato area, northwestern Italy: *Terra Nova*, Vol.7, pp. 138-143.

Pierre, C.; Caruso, A.; Blanc-Valleron, M.-M.; Rouchy, J.M., & Orzsag-Sperber, F. (2006). Reconstruction of the paleoenvironmental changes around the Miocene-Pliocene boundary along a West-East transect across the Mediteranean. *Sedimentary Geology*, Vol.188/189, pp. 319-340.

Premoli Silva I. 1964. Le microfaune del Pliocene di Balerna (Canton Ticino). *Eclogae Geologicae Helveticae*, Vol.57, No.2, pp. 731-742.

Raffi, S.; Rio, D.; Sprovieri, R.; Valleri, G.; Monegatti, P.; Raffi, I. & Barrier, P. (1989). New stratigraphic data on the Piacenzian stratotype. *Bollettino della Società Geologica Italiana*, Vol.108, pp. 183-196.

Raymo, M.E; Ruddiman, W.F.; Backman, J.; Clement, B.M. & Martinson, D.G. (1989). Late Pliocene variation in northern hemisphere ice sheets and North Atlantic Deep Water circulation. *Paleoceanography*, Vol.4,pp. 413-446.

Rio, D.; Sprovieri, R. & Raffi, I. (1984). Calcareous plankton biostratigraphy and biocronology of the Pliocene-Lower Pleistocene succession of the Capo Rossello area. *Marine Micropaleontology* Vol.9, pp. 135-180.

Rio, D.; Sprovieri, R.; Raffi, I. & Valleri, G. (1988). Biostratigrafia e paleoecologia della sezione stratotipica del Piacenziano. *Bollettino della Società Paleontologica Italiana*, Vol.27, No.2, pp. 213-238.

Rio, D.; Raffi, I. & Villa, G. (1990). Pliocene-Pleistocene calcareous nannofossil distribution patterns in the Western Mediterranean. In: *Proceedings of the Ocean Drilling Program, Scientific Results*, K.A. Kastens, J. Mascle et al., (Eds.), Vol.107, pp. 513-533, College Station, Texas, U.S.A.

Rio, D.; Sprovieri, R. & Thunell, R. (1991). Pliocene-lower Pleistocene chronostratigraphy: A re-evaluation of Mediterranean type sections. *Geological Society of America Bulletin*, Vol.103, pp. 1049-1058.

Rio, D.; Sprovieri, R. & Di Stefano, E. (1994). The Gelasian Stage: a proposal of a new chronostratigraphic unit of the Pliocene series. *Rivista Italiana di Paleontologia Stratigrafica*, Vol.100, No.1, pp. 103-124.

Rio, D.; Channell, J.E.T.; Bertoldi, R.; Poli, M.S.; Vergerio, P.P.; Raffi, I.; Sprovieri, R. & Thunell, R.C. (1997). Pliocene sapropels in the northern Adriatic area: chronology and paleoenvironmental significance. *Palaeogeography, Palaeoclimatology, Palaeoecology*, Vol.135, pp. 1-25.

Rio, D.; Sprovieri, R.; Castradori, D. & Di Stefano, E. (1998). The Gelasian Stage (Upper Pliocene): a new unit of the Global Standard Chronostratigraphic Scale. *Episodes*, Vol.21, pp. 82-87.

Robertson, A.H.F.; Emeis, K.-C.; Richter C. & Camerlenghi A. (Eds.) (1998). *Proceedings ot the Ocean Drilling Program, Scientific Results*, Vol.160, 817 pp., College Station, Texas, U.S.A. doi:10.2973/odp.proc.sr.160.1998

Roveri, M, & Taviani, M. (2003). Calcarenite and sapropel deposition in the Mediterranean Pliocene: shallow- and deep-water record of astronomically driven climatic events. *Terra Nova*, Vol.15, pp. 279-286.

Ryan W.B.F., Hsü K.J. et al. (1973). *Initial Reports of the Deep Sea Drilling Program*, Leg 13, 1447 pp., Washington (U. S. Government Printing Office), doi:10.2973/dsdp.proc.13.1973

Sacco, F. (1887). Le Fossanien, nouvel étage du Pliocène d'Italie. *Bulletin de la Société Géologique de France*, Sér.3, Vol.15: 27-36.

Salvador, A. (1994). International Stratigraphic Guide. A guide to stratigraphic classification, terminology, and procedure. *The International Union of Geological Sciences and the Geological Society of America* (Eds), 214 pp.

Seguenza, G. (1868). La Formation Zancléenne, ou reserches sur une nouvelle formation tertiarie. *Bulletin de la Societé Géologique de France*, Sér. 2, Vol. 25, pp. 465-485.

Sgarrella, F.; Sprovieri, R.; Di Stefano, E. & Caruso, A. (1997). Paleoceanographic conditions at the base of the Pliocene in the southern Mediterranean basin. *Rivista Italiana di Paleontologia Stratigrafica*, Vol.103, No.2, pp. 207-220.

Sgarrella, F.; Sprovieri, R.; Di Stefano, E.; Caruso, A.; Sprovieri, M. & Bonaduce, G. (1999). The Capo Rossello bore-hole (Agrigento, Sicily) cyclostratigraphic and paleoceanographic recontruction from quantitative analyses of the Zanclean foraminiferal assemblages. *Rivista Italiana di Paleontologia e Stratigrafia*, Vol.105, No.2, pp. 303-322.

Shackleton, N.J.; Crowhurst, S.J.; Hagelberg, T.K.; Pisias, N.G. & Schneider, D.A. (1995). A new late Neogene time scale: application to Leg 138 sites. In: *Proceedings ot the Ocean Drilling Program, Scientific Results*, N.G. Pisias, L.A. Mayer, T.R. Janecek, A. Palmer-Julson & T.H. van Andel (Eds.), Vol.138, pp. 73-101, College Station, Texas, U.S.A.

Sierro, F.J.; Hernandez-Almeida, I.; Alonso-Garcia, M. & Flores, J.A. (2009). Data report: Pliocene-Pleistocene planktonic foraminifer bioevents at IODP Site U1313. In: *Proceedings ot the Integrated Ocean Drilling Program*, Channell J.E.T., Kanamatsu T., Sato T., Stein R, Alvarez Zarikian C.A., Malone M.J. and the Expedition 303/306 Scientists (Eds.), Vol.303/306, pp. 1-11, College Station, Texas, U.S.A., doi:10.2204/iodp.proc.303306.205.2009

Spaak, P. (1983). Accuracy in correlation and ecological aspects of the planktonic foraminiferal zonation of the Mediterranean Pliocene. *Utrecht Micropaleontological Bulletin*, Vol.28, pp. 1-160.

Spezzaferri, S.; Cita, M.B. & McKenzie, J.A. (1998). The Miocene/Pliocene boundary in the Eastern Mediterranean: Results from Sites 967 and 969. In: A.H.F. Robertson, K.-C. Emeis, C. Richter & A. Camerlenghi (Eds.), *Proceedings ot the Ocean Drilling Program, Scientific Results*, Vol.160, pp. 9-28, College Station, Texas, U.S.A.

Sprovieri, R. (1978). I foraminiferi benthonici della sezione pliocenica di Capo Rossello (Agrigento). *Bollettino della Società Paleontologica Italiana*, Vol.17, No.1, pp. 68-97.

Sprovieri, R. (1986). Paleotemperature changes and speciation among benthic foraminifera in the Mediterranean Pliocene. *Bollettino della Società Paleontologica Italiana*, Vol.24, No.1, pp. 13-21.

Sprovieri, R. (1992). Mediterranean Pliocene biochronology: a high resolution record based on quantitative planktonic foraminifera distribution. *Rivista Italiana di Paleontologia e Stratigrafia*, Vol.98, No.1, pp. 61-100.

Sprovieri, R. (1993). Pliocene-Early Pleistocene astronomically forced planktonic foraminifera abundance fluctuations and chronology of Mediterranean calcareous plankton bio-events. *Rivista Italiana di Paleontologia e Stratigrafia*, Vol.99, No.3, pp. 371-414.

Sprovieri, R.; Di Stefano, E.; Howell, M.; Sakamoto, T.; Di Stefano, A, & Marino, M. (1998). Integrated calcareous biostratigraphy and cyclostratigraphy at Site 964. In: A.H.F. Robertson, Emeis, K.-C. Richter C. & A. Camerlenghi (Eds.), *Proceedings ot the Ocean Drilling Program*, Scientific Results, 160, pp. 155-165.

Sprovieri, R.; Sprovieri, M.; caruso, A.; Pelosi, N.; Bonomo, S. & Ferraro, L. (2006). Astronomic forcing on the planktonic foraminifera assemblage in the Piacenzian Punta Piccola section (southern Italy). *Paleoceanography*, Vol.21, pp. 1-21.

Stainforth, R.M; Lamb, J.L.; Luterbacher, H.; Beard, J.H. & Jeffords, R.M. (1975). Cenozoic planktonic foraminiferal zonation and characterists of index forms. *The University of Kansas Paleontological Contributions*, Vol.62, pp. 1- 162e, Appendix, pp. 1-425.

Sturani, C. (1978). Messinian facies in the Piedmont Basin. *Memorie della Società Geologica Italiana*, Vol.16 (1976), pp. 11-25.

Suc, J.-P.; Clauzon, G. & Gautier, F. (1997). The Miocene/Pliocene boundary: present and future. In: A. Montanari, G.S Odin & R. Coccioni (Eds.), *Miocene stratigraphy: an integrated approach*. Elsevier, Amsterdam, Netherland, pp. 149-154.

Trenkwalder, S.; Violanti, D.; d'Atri, A.; Lozar, F.; Dela Pierre, F. & Irace, A.. (2008). The Miocene/Pliocene boundary and the Early Pliocene micropalaeontological record: new data from the Tertiary Piedmont Basin (Moncucco quarry, Torino Hill, Northwestern Italy). *Bollettino della Società Paleontologica Italiana*, Vol.47, No.2, pp. 87-103.

Thunell, R.C.; Williams, D.F. & Belyea, P.R. (1984). Anoxic events in the Mediterranean Sea in relation to the evolution of Late Neogene climates. *Marine Geology*, Vol.59, pp. 105-134.

Vai, G.B. (1997). Twisting or stable Quaternary boundary? A perspective on the Glacial Late Pliocene concept. *Quaternary international*, Vol.40, pp. 11-22.

Van Couvering, J.A.; Castradori, D.; Cita, M.B. & Hilgen, F.J. (2000). The base of the Zanclean Stage and of the Pliocene Series. *Episodes*, Vol.23, No.3, pp. 179-187.

Van Morkhoven, F.P.C.M.; Berggren, W.A. & Edwards, A.S. (1986). Cenozoic Cosmopolitan Deep-Water Benthic Foraminifera. *Centres Recherches Exploration-Production Elf-Aquitaine Bulletin*, Vol.11, 421 pp.

Violanti, D. (1987). Analisi paleoambientali e tassonomiche di associazioni a Foraminiferi del Pliocene ligure (Rio Torsero). *Bollettino del Museo Regionale di Scienze Naturali di Torino*, Vol.5, No.1, pp. 239-293.

Violanti, D. (1989). Foraminiferi plio-pleistocenici del versante settentrionale dei monti Peloritani: analisi biostratigrafica e paleoambientale. *Rivista Italiana di Paleontologia e Stratigrafia*, Vol. 95, No.2, pp. 173-216.

Violanti, D. (1994). Pliocene mud-dwelling benthic foraminiferal assemblages dominated by *Globobulimina affinis* (Castel di Sotto, Canton Ticino, Switzerland). *Bollettino della Società Paleontologica Italiana*, Vol. Spec.2, pp. 365-379.

Violanti, D. (2005). Pliocene Foraminifera of Piedmont (north-western Italy): a synthesis of recent data. *Annali dell'Università degli Studi di Ferrara – Museologia Scientifica e Naturalistica*, Vol. Spec. 2005, pp. 75-88.

Violanti, D. & Giraud, V. (1992). Contributi allo studio del Neogene delle Langhe sud-occidentali (Mondovì). *Rivista Italiana di Paleontologia e Stratigrafia*, Vol.97, No.3-4 (1991), pp. 639-660.

Violanti, D. & Sassone, P. (2008). Il Pliocene del sottosuolo di Casale Monferrato (Piemonte, Italia Nord-occidentale): dati preliminari. *Atti Museo Civico di Storia Naturale di Trieste*, Vol.53 (2006), Suppl, pp. 233-264.

Violanti, D.; Trenkwalder, S.; Lozar, F.& Gallo, L.M. (2009). Micropalaeontological analyses of the Narzole core: biostratigraphy and palaeoenvironment of the late Messinian and early Zanclean of Piedmont (Northwestern Italy). *Bollettino della Società Paleontologica Italiana*, Vol.48, No.3, pp. 167-181.

Violanti, D.; Dela Pierre, F.; Trenkwalder, S.; Lozar, F.; Clari, P.; Irace, A. & d'Atri, A. (2011). Biostratigraphic and palaeoenvironmental analyses of the Messinian/Zanclean boundary and Zanclean succession in the Moncucco quarry (Piedmont, Northwestern Italy). *Bulletin de la Société Géologique de France*, Vol.182, No.2, pp. 149-162.

Vismara Schilling, A. & Stradner, H. (1977). I "Trubi" di Buonfornello (Sicilia). Biostratigrafia e tentativo di valutazione paleoclimatica. *Rivista Italiana di Paleontologia e Stratigrafia*, Vol.83, No.4, pp. 869-896.

Walker, J.D. & Geissman, J.W. (2009). *2009 Geologic Time Scale*. Geological Society of America, doi: 10.1130/2009.CTS004R2C

Wright, R. (1978). Neogene paleobathymetry or the Mediterranean based on benthic foraminifers from DSDP Leg 42A. In: *Initial Reports of the Deep Sea Drilling Program*, R. B. Kidd & P. J. Worstell (Eds.), Vol. 42, pp. 837-846, U.S. Government Printing Office, Washington, U.S.A.

Zachariasse, W.J. (1975). Planktonic foraminiferal biostratigraphy of the Late Neogene of Crete (Greece). *Utrecht Micropaleontological Bulletin*, Vol.11, 171 pp.

Zachariasse, W.J.; Riedel, W.R.; Sanfilippo, A.; Schmidt, R.R.; Brolsma, M.J.; Schrader, H.J.; Gersonde, R.; Drooger, M.M. & Broekman, J.A. (1978). Micropaleontological counting methods and techniques - an exercise on an eight metres section of the lower Pliocene of Capo Rossello, Sicily. *Utrecht Micropaleontological Bulletin*, Vol. 17, 265 pp.

Zachariasse, W.J. & Spaak, P. (1983). Middle Miocene to Pliocene paleoenvironmental reconstructions of the Mediterranean and adjacent Atlantic Ocean: planktonic foraminiferal records of southern Italy. *Utrecht Micropaleontological Bulletin*, Vol. 30: 91-110.

Zahn, R.; Comas, M.C. & Klaus A. (Eds.) (1999). *Proceedings ot the Ocean Drilling Program, Scientific Results*, Vol.161, College Station, Texas, U.S.A. doi:10.2973/odp.proc.sr.161.1999

Zappi, L. (1961). Il Pliocene di Castel Verrua. *Atti della Società Italiana di Scienze Naturali*, Vol.100, No.1-2, pp. 73-204.

Zijderveld, J.D.A.; Hilgen F.J.; Langereis, C.G.; Verhallen, P.J.J.M. & Zachariasse, W.J. (1991). Integrated magnetostratigraphy and biostratigraphy of the upper Pliocene-lower Pleistocene from the Monte Singa and Crotone areas in Calabria (Italy). *Earth and Planetary Science Letters*, Vol.107, pp. 6997-714.

8

Late Silurian-Middle Devonian Miospores

Adnan M. Hassan Kermandji
Department of Biology and Ecology, Faculty of Nature and Life,
University of Mentoury-Constantine,
Algeria

1. Introduction

Recent palynological researches on late Silurian to Middle Devonian terrestrial palynomorph have culminate on cryptospores and miospores zonation system (e.g., Richardson and McGregor 1986; Streel *et al.*1987; Massa and Moreau-Benoît 1976; Melo and Loboziak 2003; Hassan Kermandji *et al.*2008). Although stratigraphic, ecologic and geographic importance of these fossils grasp great effort, but much more work is required on their biostratigraphy and taxonomy before their complete potential can be understand. Nevertheless, palynological research is creating a major contribution to enable an overview of most important characteristics of cryptospores and miospores (Devonian miospore from Algeria, Tunisia and Libya, e.g., Magloire 1967; Moreau-Benoît *et al.*1993; Loboziak *et al.* 1992; Spina and Vecoli 2009) also Silurian and Devonian miospore from Bolivia, Spain United Kingdom and Saudi Arabia (e.g., McGregor, 1984; Richardson *et al.*2001; Wellman and Richardson 1993, 1996; Wellman *et al.*1998; Steemans *et al.*2007). Sporomorphs are more abundant and wide spread than cryptospores where decreasing in number and variety from older to the younger sediments (e.g., Rubinstein and Steemans 2002). Miospores by their structural complexity and augmentation they may provide more reliable evidence for formal zonal schemes to be proposed. However, uncertainty also exist the precise definition of the binomial species regarding their lateral distribution on local and regional scale, some may have restricted distribution pattern (e.g., Hassan Kermandji 2007), this is also may be due to the timing of their appearance and development and place of their parent plant evolutionary events (e.g., Edwards and Richardson 1996) and ecological niches.

Local or regional variations in plant distribution during late Silurian-Lower Devonian are due to: close contact of the Rheic Ocean during the early Lochkovian, or moderate distance separating Gondwana and Laurussia, or close proximity of the Avalonia Maguma and Aquitaine areas to the northern border of Gondwana during Přídolí times and or a possible land plant migration route between Laurussia and Gondwana (Rubinstein and Steemans 2002; Steemans and Lakova 2004; Hassan Kermandji 2007; Richardson *et al.* 2001; Spina and Vecoli 2009). Since the land invaded by plants, palaeoenvironment constrictions are possible. It is, however, significant in demonstrating the influence of Gondwanan parent flora on the miospore populations of the southern regions of Laurussia. This is may serve for providing a link between Laurussian and Gondwanan palynofloras.

Late Silurian-Lower Devonian stratigraphy is rather complicated where different miospore biozones are recognized. Correlations of North Africa Gondwanan miospore biozones remain hard to be compared precisely with Euramerica well established spore zonation but may be resolved by studies based on phytoplankton, chitinozoa, conodonts and graptolites. Primary biostratigraphic subdivision within the late Silurian-Lower Devonian has been done by recognition of progressive change in terrestrial miospores. Subdivision is based on new incoming of well established miospores. Euramerican stages containing six Spore Zones and two Opple Zones where they provide bases for local and regional correlations. Although, some miospore assemblage similarities exist between Gondwana and Euramerica, there are significant differences as well. For instance species characteristics of Lochkovian (e.g., *Streelispora newportensis* s.s., *Emphanisporites zavallatus*, *E. micrornatus* s.s) are absent or rare in most parts of Gondwana (Melo and Loboziak 2003; Steemans *et al.* 2008), in Euramerica appear to be geographically well established which may suggests that they are apparently regionally confined. On the other hand, other biozonal species characteristics of late Silurian-Lower Devonian are found in Europe, North America and Gondwana (e.g., *Chelinospora cassicula*, *C. hemiesferica*, *Leonispora argovejae*, *Perotrilites microbaculatus*, *Dictyotriletes emsiensis*, *Verrucosisporites polygonalis*, *D. subgranifer*, *E. annulatus*, *D. echinaceus*, *Calyptosporites velatus*, *Rhabdosporites langii*). These palynological similarities might point out possible common constituents of their parent plants producing these miospores.

Late Silurian and Lower Devonian times are thus a key period for the developing of miospore zonation with confident suggestion in biostratigraphic, phytogeographic and palaeoenvironmental studies.

This chapter therefore will concentrate on late Silurian and early Devonian miospores biozones and systematic issues also will deal with early Middle Devonian miospores regarding their biozonation consequence and evolutionary significance.

2. Sporomorphs evolution

Richardson (1996) reviewed in brief development record of terrestrial sporomorphs. Stated that the problematic organic-walled microfossils occur in Vendian rocks and named by Volkova (1976) and 'are associated with leiosphaerid acritarchs' probably derived from brown alga. Also he believes that the first occurrence of cryptospores and miospores is approximate. The earliest terrestrial sporomorphs resembling miospores record is outside the scope of this chapter. However, it is extremely interesting to record the development of sporomorphs concerning the study.

The sporomorph events described below are based on data from type sequences. These will provide minimum age range for the events stated by, (Hassan Kermandji 2007; Hassan Kermandji *et al.* 2008) which agrees with (Richardson and McGregor 1986; Streel *et al.* 1987; Richardson *et al.*2001) detection. Detailed review of the many palynological studies of Silurian and Devonian deposits of Euramerica and western Godwana, revels obvious coincidence of many of significant palynological events.

In both Euramerica and western Gondwana, possibly Llanvirn-Caradoc cryptospore population are characterised by distinctive elements (permanent tetrads, pseudodyads and monads), they posses nearly identical structure and sculptural ornaments. Many studies (Gray 1992; Richardson 1988; Wellman *et al.* 1998; Steemans 1999a, b) reported possible cryptospores from Middle and late Ordovician. The recorded possible cryptospore assemblages from deposits older than Caradoc are either naked or enclosed within a thin, smooth or variously ornamented envelops. Consequently, Richardson (1988) considers that pseudotetrads, pseudodyads and monads envelop possibly developed prior to Caradoc. Cryptospore bearing plants were becoming more widespread and numerous during late Ordovician and early Silurian (Gray 1985, 1988; Burgess 1991; Burgess and Richardson 1991). They were reported from terrestrial and near shore environments. Many studies (e.g.; Vaverdova 1988) believe that high abundance of cryptospores may be facies controlled. Others (e.g.; Richardson 1992) relate high abundance to the upper Ashgill glacial period which played a major role in the extinction of faunas and floras and not affected the cryptospores of the high latitude. This is based on Burgess's study (1991) on Llandovery deposits, south Wales that there is only minor differences between Ashgill and Llandovery cryptospore assemblages.

Although, there is no specific record for the earliest cryptospore creator, it is very hard to analyse that the cryptospores reflects true diversity. Because, different cryptospore morphotypes have similar envelope (taking in consideration that many envelops may be lost or changed during preparation). Furthermore, many of these forms, morphologically and structurally, are simple. Such simple forms could have been produced by a number of plant types or by the same parent plant (Richardson 1988, 1992; Gray 1991; Strother 1991). Similarly, Fanning et al. (1988) refer to morphologically identical plant may produce different spore forms; they reflect it to the rapid evolution of the spore morphology. Wellman et al. (1998) came to the comparable conclusion that superficially morphologically similar miospores were produced by different plant types. This is all because the unavailability of studies denotes the in situ relationship between the parent plants and dispersal cryptospores and miospores.

Studies of (Gray et al.1982; Gray 1985; Vaverdova 1984; Burgess and Edwards 1991; Gensel et al.1991; Wellman 1995; Edwards and Wellman 1996; Hassan 1982; Hassan Kermandji and Khelifi Touhami 2008) show that the enigmatic dispersal phytodebris and land plants of the late Lower and Middle Palaeozoic sediments probably, are the core stone of the plants produced dispersal cryptospores and miospores. They show that the dispersal fragments consist of ornamented and laevigate types of cuticles, tubular structures, tissues and filamentous, occurs isolated or commonly in complex form mainly in continental deposits. The oldest known record of permanent tetrad and dyads is from the early Llanvirn of Bohemia by Vaverdova (1984) and Vaverdova's (1990) record for Prague basin. Although, cryptospore spreading declined upward, it exists in its diversity until to the early Devonian. Early cryptospores, permanent tetrads, dyads and monads, though it is not well documented, believed that they are of terrestrial origin (non marine sediments) and near shore environment (Strother and Traverse 1979; Wellman and Richardson 1993; Richardson 1996) and that 'cryptospores are more abundant than miospores in inshore sediments' Richardson and Rasul (1990). Many of these cryptospores (dyads and tetrads) were derived from plants with bifurcating axes and sporangia Wellman et al. (1998). It appears that monads, dyads and tetrads frequently have similar envelops, this tend to suggest that they closely related, probably derived from the same species (Richardson 1988, 1992; Johnson 1984). Whereas trilete miospores dissociated from immature tetrads which are match forms among existing embryophytes.

The cryptospore assemblages decline in its taxonomic diversity in Homerian and younger sediments perhaps is environmental effect (probably facies controlled) Richardson (1988) and or an evolutionary indication (Hassan Kermandji et al. in preparation).

The earliest recorded embryophyte miospores was by Vaverdova (1984) from early Llanvirn of Bohemia; also identical miospore assemblages were reported from mid Ordovician to early Silurian by: Richardson (1988) Gray (1985, 1991), Burgess (1991) Strother (1991),

Strother *et al.*(1996), Wellman (1996), Steemans (1999a). However, (Wellman and Richardson 1993; Strother and Traverse 1979; Richardson 1992; Gray *et al.* 1992; Wellman and Gray 2000; Steemans 1999b) believe that the structure and morphology of these early miospores are atypical comparing with existing miospores of younger ages, therefore some times, embryophyte miospores been used as cryptospores. This is due to that the widely spread vegetation was of restricted diversity and diminutive evolutionary change.

Most studies reveal that major changes occur to the miospore nature during late Lower Silurian. The mature miospores of existing embryophytes become dominant and more abundant than cryptospores by Ludlow Wellman and Gray (2000, fig. 1). Their diversity and abundance increased throughout the late Silurian. Steemans *et al.* (1996), Wellman, Higgs and Steemans (2000) studies show that trilete miospores and hilate monads of northeast Gondwana appear during late Ordovician and early Silurian. Sculptured forms of both trilete miospore and hilate monads flourished throughout late Silurian and early Devonian Burgess and Richardson (1995); Steemans (1999b). The hilate monads number decline in latest Silurian and earliest Devonian, and cryptospores exist as a minor constituents in miospore assemblages through Lower Devonian and extinct at the end of Emsian Hassan Kermandji *et al.* (In the press).

Considerable variations in dispersed miospore record exist through late Silurian and early Devonian, (mainly Přídolí and Lochkovian). This creates problems when trying to recognize phytogeographic variations (Rubinstein and Steemans 2002; Richardson *et al.* 2001; Hassan Kermandji 2007). Sequences of spore assemblages from Western Europe are similar in composition except some minor differences, this is indicate that it contain a flora representing a single palaeophytogeographic province. Although miospore assemblages sequences of northern Gondwana sharing some elements but display striking differences, signifying that they belong to a different palaeophytogeographic province. Comparison between Lochkovian miospore assemblages of the two provinces is difficult; probably they belong to different parent plants. The differences become less obvious during late Emsian and Eifelian.

Although, the *in situ* miospore records for early land plant megafossils are rare, it is more common in higher sediments. This is due to nature of land plant preservation, where they preserved as coalified compressions. Though, suitable preservations (uncompressed coalifications with splendid cellular detail) have been recorded in Lochkovian localities of south Wales (Edwards and Richardson 1996; Edwards 1996). The record of early land plants show that many miospores (*Ambitisporites, Synorisporites,* and *Streelispora*) have been recovered from the sporangia of rhyniophyte *Cooksonia* (Funning *et al.* 1988; Richardson 1996) which is a factual tracheophyte Edwards *et al.* (1992) and most trilete miospores appear to have consisted of plants with bifurcating axes/sporangia and more frequently stomata (Edwards 1995; Wellman *et al.* 1998). On the other hand, some dispersed miospore assemblages such as patinate species and *Emphanisporites* are abundant, but there *in situ* forms are extremely rare. This is probably facies and/or preservational effect.

The main evolutionary performance divided into events based on dispersed palynomorphs (text fig. 1), nearly all events described below are based on information from type sequences.

2.1 Crassitate, distally laevigate miospore event

The Aeronian and early Telychian miospores are distally laevigate and equatorially crassitate and belong to the genus *Ambitisporites avitus, Amitisporits dilutes.* Also appear some tetrads and

System	Stage/Series	Palynological events	Main Characters
D E V O N I A N	Eifelian	Zonate-pseudosaccate and coarse bifurcate spinose miospore	Incoming of considrable significant finely sculptured pseudosaccate miospores (*C.velatus and R. langii*). Waning of *Acinosporites* and patinate taxa.
	Eifelian	Monopseudosacciti and Bifurcate-tipped appendages miospore	Incoming of bifurcate appendages (*Hystricosporites* species).Persistent of *Grandispora, Acinosporites, Geminospora* and *Grandispora* species. Disappearance of *Scylaspora* and cryptospores.
	Emsian	Bizonate, proximally cristate and distally annulate miospore	Proliferation of zonate and pseudosaccate (*C. Sextantii*), *Geminospora* forms Appearance of proximally radially ribbed with distinct distal annulae (*E. annulatus*) and distally convoluted murornate species (*Acinosporites*).
	Emsian	Diversification of proximal prominent stout radial muri miospore	Incoming of retusoid miospores with verrucate or murornate sculpture (*D. subgranifer, V. polygonalis*) . Waning of cryptospores.
	Pragian	Coarse biform apiculae with cingulicavati and foveolae miospore	Incoming of *Dibolisporites,* distally with spines, and biform elements and retusoid *Apiculiretusispora*. Diversification of cingulate species of *Camptozonotrileres* and *Clivosispora,*. Waning of *Scylaspora* species
	Pragian	Distally murornate, proximally radially ribbed miospore	Incoming of *E. spinaeformis* and *D. emsiensis*; persistent of hilate cryptospore monads and dyads. Diversification of apiculate monad cryptospores and laevigate, apiculate tetrads.
	Lochkovian	Diversification of distally cingulate,patinate miospore	Incoming of annulate, cingulate rugulate and patinate spores. Persistant of perinate and crassitate species. Appearance of *A. miserabilis* and *C.proteus*
	Lochkovian	Proximally slightly ribbed, distally apiculate crassitate, perinate miospore	Persistent of apiculate, microrugulate, crassitae miospores, incoming of perinate and patinate species. Proliferation of laevigate dyads, monads and tetrads. Appearance of *S .tidikeltense* and *P. microbaculatus.*
S I L U R I A N	Přídolí	Diversification of crassitae, disto-equatorially multi sculptured miospore	Diversification and proliferation of murornate equatorially crassitate miospores, incoming of apiculate curvaturate forms and hilate dydes. Appearance of *S. radiata* and *A. synoria.*
	Přídolí	Murornate, apiculate patinate miospore	Diversifications of distally reticulate patinate and foveolate sculptured miospores. Incoming of tripapillate and equatorially crassitate murornate forms and hilate dyads. Appearance of *C. hemiesferica*
	Lud-fordian	Patinate, proximally hilate with distal faint muri miospore	Diversification of varially sculptured patinate, crassitate and cingulate and radially ribbed sporomorphs. Appearance of *C. sanpetrensis* and *C. triangulatus.*
	Grostian	Apiculate, patinate with faint radial muri miospore	Incoming of distally murornate miospores with radially ribed patinate miospores. Consistent of hilate cryptospores and tetrads.
	Homerian	Granulate,apiculate, crassitate miospore	Incoming of sporomorphs with proximal radial muri and equatorial radial thickening and distal granules and verrucae. Appearance of *Cheilotetras caledonica*
	Homerian	Murinate,verrucate crassitate,patinate miospore	Proliferation of verrucate, muromate emphanoid forms and patinates miospores with distal radial muri. Also, include verrucate cryprospores. Appearance of *S. vetusta* and *S. kozlica.*
	Homerian	Murornate,verrucate miospore	Domination of distally apiculate and verrucate crassitate and cingulate miospores. Persistent of hilate cryptospores. Appearance of *Hispanaediscus verrucatus.*
	Shienw-oodiaan	Hilate monad miospore	Incoming of patinate miospores, also including dyads and tetrads, persistent of cryptospores.
	Shienw-oodiaan	Patinate, proximaly hilate laevigate miospore	Incoming of hilate miospores, *A. chulus/nanus* complex, equatorially crassitate miospores and tetrads and dyads
	Telychia n	Crassitate,distally laevigate miospore	Incoming of laevigate crassitate miospores and some tetrads and dyads and persistent of cryptospores.

Fig. 1. Summery of principle Silurian-Devonian sporomorphs evolution.

some dyads and cryptospores are present (Hassan Kermandji *et al.* in preparation).The age is based on acritarchs and chitinozoans in the type Llndovery area Burgess and Richardson (1991).

2.2 Patinate, proximaly hilate laevigate miospore event

The exact stratigraphic range where equatorially crassitate, patinate miospores, tetrads and dyads occur is uncertain, possibly they appear in rocks older than Homerian probably Upper Aeronian (Richardson 1988, 1996), they include *Archaeozonotriletes chulus/nanus* complex where associated with graptolite zone (*turriculatus* or *cripus*) (Richardson 1996). The *A. chulus/nanus* compound occurs in Libyan sediments of Middle and Upper Telychian (Al-Ameri 1980 in Richardson 1996). The assemblages are probably of Middle and Upper Telychian, they may extend to late Homrian.

2.3 Hilate monad miospore event

These palynomorphs include *Laevolancis divellomedia*, recorded in marine sediments as *Tasmanites avelinoi* by Al-Ameri (1980), ranging from Landovery to Lower Devonian. This is also including pseudodyads, dyads, tetrads and halite cryptospore monad. They occur possibly in Telychian to become predominant in much younger sediments.

2.4 Murornate, verrucate miospore event

Appearance of sporomorphs with varied distal sculpture including verrucate, murornate *Synorisporites* and emphanoid forms. Both cryptospores and miospores were appeared with distal verrucae and proximal radial muri. Cryptospores were first appeared with verrucate-murinate distal sculpture and proximal radial muri in south Wales by Burgess and Richardson (1991). This event is also, include patinate miospore with distal radial muri. Telychian to Sheinwoodian.

2.5 Murinate, verrucate crassitate miospore event

Appearance of murinate, verrucate, crassitate, confined proximal radial muri radiating from curvaturae perfectae trilete Upper Homerian miospores including *Emphanisporites protophanus*. In addition, it is including, distally verrucate, murinate hilate alete monads and some cryptospores with radial proximal muri (Richardson and McGregor 1986; Richardson *et al.* 2001; Hassan Kermandji 2007).

2.6 Granulate, apiculate, crassitate miospore event

It is characterized by miospores with equatorial radial thickening "curvatural crassitude" becoming indistinct towards the proximal pole, proximally with radial anastomosing muri and/or regulae, distally apiculate granulat and minutely verrucate of *Scylaspora* and *Synorisporites* species. In addition it is including cryptospores Burgess and Richardson (1995). Early and late Homerian.

2.7 Apiculate, patinate with faint radial muri miospore event

Appearance of granulate, micro rugulate curvatural proximal sculupture, distally with low narrow murornate, verrucate patinae and narrow hilum *Chelinospora*, *Scylaspora* species.

This is also including cryptospores (Burgess and Richardson 1995; Richardson *et al.* 2001; Hassan Kermandji *et al.* in preparation). Grostian to Lower Ludfordian.

2.8 Patinate, proximally hilate with distal faint muri miospore event

Characterised by the diversification of verrrucate, apiculate patinate, domination of crassitate and cingulated and proximally radially ribbed forms of *Chelinospora, Cymbosporites, Concentricosisporites and Emphanisporites* miospores (Richardson and Ioannides 1973; Richardson et al. 2001; Hassan Kermandji 2007) and cryptospores. Early and late Ludfordian

2.9 Murornate, apiculate patinate miospore event

In coming of distally reticulate, patinate miospores with thin contact areas (*Chelinospora*) Richardson *et al.* (2001). Appearance of species with foveolate sculpture (*Brochotriletes*) Richardson and McGregor (1986). Persistent of murornate, apiculate species of the previous zones Hassan Kermandji (2007).

2.10 Diversification of crassitae, disto-equatorially multi sculptured miospore event

Characterized by diversification of equatorially crassitate and proximally tripapillate, distally murornate, apiculate patinate miospores of *Scylaspora, Synorisporites, Cymbosporites* and *Chelinospora* (Richardson and Lister 1969; Burgess and Richardson 1995). Proliferation of verrucate, retusoid, apiculate and radially ribbed, foveolate and murornate miospores of *Apiculiretusispora* and *Dictyotriletes*. Permanent tetrads and cryptospores also occur, Hassan Kermandji (2007).

2.11 Proximally slightly ribbed, distally apiculate, crassitate, perinate miospore event

Proliferation of apiculate and proximally radially ribbed miospores (*Apiculiretusispora* and *Emphanisporites*). Persistence of apiculate, proximo-equatorial microgranulate crassitate, patinate and perinate miospores of (*Scylaspora, Perotilites* and *Cymbosporites*) Hassanan Kermandji *et al.* (2008). Early (but not earliest) to early late Lochkovian.

2.12 Diversification of distally cingulate, patinate miospore event

Appearance of cingulated, annulate and patinate miospores of *Amocosporites* and *Cybosporites*. Persistent of perinate and crassitate species of the older strata. Late early and early late Lochkovian.

2.13 Distally murornate, proximally radially ribbed miospore event

Proliferation and diversification of proximally retusoid, distally apiculate, reticulate *Dictyotriletes*. Increasing of the variety of distally granulate-apiculate-spinose, proximally radially ribbed *Emphanisporites* and incoming of bizonate, proximally with highly irregular'scalloped' folds and distally with distinct annular thickening *Breconsporites* Richardson *et al.* (1982). Latest Lochkovian and earliest Pragian.

2.14 Coarse biform apiculae with cingulicavati and foveolae miospore event

In coming of *Dibolisporites* species with prominent spines, tubercles and biform elements. Diversification of *Camptozonotriletes* and *Clivosispora* and *Apiculiretusispora* species. Late early to early late Pragian.

2.15 Diversification of proximal prominent stout radial muri miospore event

Appearance of retusoid reticulate, non reticulate and foveolate species of *Dictyotriletes* and *Brochotriletes*. Proliferation of regularly verrucate taxa of *Verrucosisporites*. Persistent of tuberculornati and cingulicavati species from the previous zone. Latest Pragian and earliest Emsian.

2.16 Bizonate, proximally cristate and distally annulate miospore event

First appearance and diversification of proximally radially ribbed and distinct delimited distally annulate species of *Emphanisporites* and distally convoluted murornate species of *Acinosporites*. Abundance of biform sculptured forms of *Dibolisporites*. First appearance of stout, prominent-spined species of *Acanthotriletes*. Proliferation of distally patinate species of *Tholisporites* and *Chelinospora*. Diversification of zonate and pseudosaccates forms of *Camarozonotriletes* and *Geminospora* species. Disappearance of most reticulate and tripapillate miospores (Richardson and McGregor 1986; Streel *et al.* 1987; Hassan Kermandji *et al.* 2008). Early and early late Emsian.

2.17 Bifurcate-tipped appendages and monopseudosacciti miospore event

First appearance of prominent bifurcate appendages forms of *Hystricosporites* species. Proliferation and diversification of *Grandispora* forms. Propagation and persistence of *Dibolisporites, Acinosporites* and *Geminospora* species. Persistence of many species from the previous zone Hassan Kermandji *et al.* (2008). Latest Emsian and earliest Eifelian.

2.18 Zonate-pseudosaccate and coarse bifurcate spinose miospore event

Characterized by incoming of finely-sculptured zonate-pseudoccate miospores of *Calyptosporites* and *Rhabdosporites* species. Proliferation of *Samarisporites, Corystisporites* and *Ancyrospora* species. Gradually waning of *Acinosporites* and patinate taxa (Richardson and McGregor 1986; Streel *et al.* 1987; Hassan Kermandji *et al.* 2008). Early Eifelian.

The correlation between the Algerian miospore assemblage biozones and the standard Euramerican North Gondwanan miospore biozones is shown in Figure 2. All these data originated from the Tidikelt Plateau, Triassic Province, Oued Saoura and Illizi regions, west and east Sahara Algerian Desert synclines, North Africa, Old Red Sandstone continent and adjacent regions, but other regions may contribute for improvement of the results. For instance, I have used in the legend of the figure additional correlation data from south Wales.

3. Stratigraphic miospore distribution

Significant progress has been made in areas regarding morphologic events of the sporomorphs stratigraphic distribution Richardson (1996) and Zonal concepts were used by many workers;

some described it as local events (Loboziak *et al.* 1992; Rubinstein and Steemans 2002; Richardson and Edwards 1989; Wellman and Richardson 1996; Jardiné and Yapaudjian 1968; Melo and Loboziak 2003; Moreau-Benoît *et al.* 1993; Hassan Kermandji *et al.* 2008; Richardson and Ioannides 1973, Massa and Moreau-Benoît, 1976), others are used the concept in more wider sense "on global scale" (e.g. Richardson and McGregor 1986; Streel *et al.* 1987). Richardson and McGregor (1986) concept for Devonian miospore zonation, based on spore assemblage biozones to describe miospore distributions of the Old Red Sandstone Continent and adjacent regions, whereas Streel *et al.* (1987) employed the scheme based on spore Oppel zones and Interval zones recognized in the progressive changes from shallow to more deeper marine sediments of the Ardenne-Rhenish region. Both schemes have been utilized beyond their regions where they established (e.g. McGregor and Playford 1992, Loboziak and Streel 1995). Fundamentally both schemes based on the appearance of well recognized, laterally extensively distributed spore taxa. Although the Richardson and McGregor's magnificent zonation concept is loose (Streel and Loboziak, 1996), it is practical, based on assemblages of characteristic taxa allow correlations on great scale. The branded Interval zone concept of Streel *et al.* (1987) is build up within a single great phytogeographic province and on distinctive miospore taxa, to some extent, many of their typical taxa hard to occur in another region (Wellman 2006; Breuer *et al.* 2007; Steemans *et al.* 2008; Hassan Kermandji *et al.* 2008). Nevertheless, Streel and Loboziak (1996) are employed Western Gondwana and southern Euramerica as a single major phytogeographic province.

There are significant difficulties concerned correlations between Upper Silurian-early Lower Devonian miospore assemblage sequences of north Africa (Sahara Algeria, Libya and Tunisia) and those of Old Red Sandstone and adjacent regions by Richardson and McGregor (1986) and the Ardenne-Rhenish regions by Streel *et al.* (1987). The major differences between miospores sequences of Moreau-Benoît *et al.* (1993), Melo and Loboziak (2003), Steelmans *et al.* (2008), Rubinstein and Steemans (2002), Steemans *et al.* (2007), Hassan Kermandji (2007), Hassan Kermandji *et al.* (2008), Spina and Vecoli (2009) and many others are that: they contain groups of palynomorphs of different stratigraphic significance. They are different particularly, in terms of the characteristic taxa and the absence of common index species from the zones. The degree of similarities in terms of the general characteristics and composition between Euramerican and Gondwana miospores were overestimated by many workers: Loboziak and Streel 1989; Loboziak *et al.* 1992, and many others. However, some similarities exist between Euramerica and Western Gondwana but many regions of northern Gondwana contain limited numbers of Euramerican characteristics zonal miopores within assemblages of different composition Wellman (2006). Therefore, Upper Silurian-early Lower Devonian Euramerican miospore biozones are hard to be in use in the Western Gondwana province.

Comparable zonation schemes for the Upper Silurian and Lower Devonian of North Africa are unexpectedly few despite a long history of investigations that have concentrated on phytoplankton and land plant miospore assemblages of deep drillings. No Proposal exists for a formal zonation. Nevertheless, the Silurian and Devonian biostratigraphic studies neither for the Western Libyan deep drilling deposits by: Massa and Moreau Benoit 1976; Richardson *et al.* 1981; Rubinstein and Steemans 2002; Le Hérissé 2002; Spina and Vecoli 2009; nor those of Mid Palaeozoic Algerian Sahara petroleum sediments by Magloire 1967; Jardiné and Yapaudjian 1968; Moreau-Benoît *et al.* 1993; Abdesselam-Rouighi, 1986, 1996,

2003. They are preliminary and do not permit exact correlation with standard zonations of Euramerican palynozones by Richardson and McGregor (1986), Streel *et al.* (1987) and Richardson *et al.* (2001).However, detail proposal do exist for the Tidikelt and Oued Myia deposits in the central and south western Algerian Sahara synclines (Hassan Kermandji 2007, Hassan Kermandji *et al.* 2008) which are equivalent to the Přídolí, Lochkovian, Pragian, Emsian and Eifelian deposits of Euramerica. They establish well-defined ten miospore assemblage biozones and one interval zone extending from the Homerian to early Eifelian. Zonal index taxa were recognized and the general pattern of changes in miospore distribution through Homerian and early (but not earliest) Emsian to Lower Eifelian, more or less, comparable to those changes reported in Euramerica region. While the pattern of changes in miospore distribution through Ludfordian, Přídolían, Lochkovian, Pragian and Lower Emsian does not reflects changes comparable to those stated in Euramerican province. They are differing in the selection of zonal index taxa, zonal miospore composition and vertical stratigraphic occurrence.

The most diverse and distinctive, moderately well preserved Silurian and Devonian sporomorphs assemblages occur in shallow marine and terrestrial strata of the Triassic Province and Tidikelt plateau were studied biostratigraphically. This is well coincide with Richardson (1984) finding that 'sporomorph assemblages occur in rocks deposited either on land or in marine environments'.

4. Sporomorphs distribution development

Detailed studies of many local and regional palynological zonations of the middle to late Silurian and early to early Middle Devonian deposits in the Euramerica and Western Gondwana reveal many difficulties, probably due to the time of closure (Lochkovian or Pragian-Emsian) of the Rheic Ocean which may cause climate and palaeophytogeographic isolation. On the other hand, this may lead to widespread dispersal ability and climatic tolerance of late Silutian and early Devonian land plants Raymond *et al.* (2006). This palaeophytogeographic differentiation of macroflora does not imitate spore distributional pattern. On the basis of dispersal miospores distribution patterns Wellman and Gray (2000) and Edwards and Wellman (2001) distinguished a Euramerican miospore biogeographic unit from a Gondwanan unit for late Silurian times. Though palaeophytogeographic differentiation does not well reflects miospores distribution patterns for this period Richardson *et al.* (2001).

On the basis of sporomorphs classification which is based on morphology, appears that earliest cryptospores with apparent thick curvatural crassitude may have close relation with initial *Ambitisporites*. This is may indicate that plants producing cryptospores may be predecessors of those producing *Ambitisporites* miospores Richardson (1996). Morphological features show that there are close relation between some cryptosopre tetrads with retusoid miospores bearing distinct contact areas, curvaturae perfectae not confined to the equator and distally patinate with thin proximal wall (*Ambitisporites* and *Archaeozonotriletes* which occur a little latter). The progressive advantages in complexity of morphological structure occur in sequence of parallel changes in both closely related cryptospores and miospores. The appearance of some types of ornaments was not always synchronous in both groups, for instance proximal papillae first occur on miospores in early Přídolí, whereas it appears on cryptospores in early Lochkovian. This is probably related to the environmental responses with two closely related groups of plants responding to the same motivation.

With reference to the main spromorphs evolution events, their distribution patterns are discussed. However, the appearance or the disappearance of structural group or a single genus and/or whole complex of forms provides helpful indicators in establishing local or regional miospore zonations.

4.1 Cryptospores

Cryptospores are usually more abundant and diversified than miospores in early sediments, mainly occur in inshore environments, and decline in abundance offshore more than miospores. They decrease in abundance number in younger sediments and taxonomically become more varied. This is may indicate that their parent plants possibly were living on sediments of ephemeral environment. The high abundance and diversity of cryptospores suggesting that their parent plant flourishing and occurred together with miospores producing plants, probably of vascular type Wellman and Richardson (1996). The earliest geological record of cryptospores was in Caradoc, late Ordovician, but probably they appear much earlier.

The cryptospore diversifications are varied from place to place. The variations are due to the nature of depositional environment. Laevigate taxa are more diverse than sculptured forms. Some granulate; apiculate cryptospore types are found *in situ* (Funning *et al.* 1991).

Cryptospores 'generally' are represented by long ranging taxa such as, *Tetrahedraletes medienensis* (Pl.1, Fig.2, Pl.2, Fig.6), *Laevolancis divellomedia* (Pl.1, Figs. 8, 13), *Cheilotetras caledonica* (Pl.1, Fig.1), *Cymbohilates horidus*, (Pl.1 Figs. 3, 5, 6, 11) and *Cymbohilates allenii* var. *allenii* (Pl.1, Fig.14), *Cymbohilates allenii* var. *magnus* (Pl.1, Fig.4), *Hispanaediscus verrucatus* (Pl.1, Figs.7, 12), *Artemopyra? scalariformis* (Pl. 1, Fig. 9) and *Acontotetras inconspicuis* (Pl.1, Fig.10). All these taxa have been recorded in many late early Silurian to Lochkovian sedimentary sequence, from many Euramerican and Gondwanan areas.

On the basis of proliferation and diversification of cryptospores from the Type Wenlock and Towy Anticline, Burgess and Richardson (1991, 1995) confirmed the first occurrence of seven zonally significant species in Sheinwoodian to Ludfordian and graptolite bearing strata of Euramerican province. Most of these sub-zones are regionally confined and have not been observed in North Africa. Out of two spore assemblage biozones, two spore biozones, two sub-biozones and one spore interval biozone of Ludfordian-Lochkovian from the Cantabrian Mountains, NW Spain by Richardson *et al.* (2001), only *Scylaspora vetusta-Scylaspora kozlica* (Dufka) Spore Assemblage Biozone and the spore interval biozone correspond to just two Přidolí spore interval biozone (*hemiesferica*) and the *Scylaspora radiate-Apiculiretusispora synoria* Miospore Assemblage Biozone of western and central Algeria by Hassan Kermandji (2007, Fig.3). The endemism of some miospore species caused many terrestrial plants to appear earlier in Gondwana, also infrequency of plants producing these palynomorphs, palynologically unfavorable types of sediments, incomplete study of available rocks and inappropriate environments may originate these variations.

4.2 Miospores

The earliest recorded laevigate retusoid, equatorially thickened miospores is specimens resemble *Ambitisporites* appear little earlier than true *Ambitisporites* Hoffmeister (1959) Pl. 2,

Figs. 2, 4, of latest Aeronian. Whereas laevigate, distally patinate, proximally generally hilate, *Archaeozonotriletes* (Naumova) Allen (1965), appear in Telychian. Both taxa persist into Lower Devonian. This is indicating that land flora producing these taxa are uniformly distributed over wide regions. Some retusoid forms are found *in situ*, for instance *Retusotriletes coronadus* found in the lower Downton Group, also some sporangia have yielded verrucate crassitate *Synorisporites verrucatus* and some others has yielded, crassitate papillate *Synorisporites tripapillatus*. Whereas, no laevigate patinate *Archaeozonotriletes* has been found *in situ*, this is may indicate that their parent plants lived in ephemeral environment Richardson (1996). Nevertheless, equally crassitate and patinate taxa are remarkably abundant in offshore environment. Richardson and McGregor (1986) in their zonal concept used *A. avitus-A. dilutus* for Aeronian and *A. chulus-A. nanus* for Sheinwoodian as cosmopolitan assemblage zones. The two assemblage zones are considered by McGregor and Playford (1992) and Streel and Loboziak (1996) as loose biozones.

There are some familiarities in miospore assemblage composition throughout Homerian-Přídolían sediment reported from Euramerican and Gondwanan phytogeographic provencies. The common zonal species are represented by: *S. kozlica* Pl.2, Fig.3; *S. vetusta* Pl.2, Fig. 11; *C. hemiesferica* Pl.2, Fig. 5; *C. sanpetrensis* Pl.4, Fig.12, Pl.5, Fig. 13; *C. (Lophozonotriletes?) poecilomorpha* Pl.2, Figs. 7, 8). The first two species form nominal taxa of the Middle Homerian sediments, whereas, the rest are nominal taxa for Ludfordian and Přídolían strata (Burgess and Richardson 1995, Richardson *et al.* 2001, Hassan Kermandji 2007).

Lower Devonian sediments of the Gondwanan province includes many miospore species known elsewhere, though striking assemblages of the same stratigraphic range are difficult to be found in other studied regions. For instance *S. tidikeltense* Pl.2, Fig. 10 of early Lochkovian and *C. triangulates* Pl.2, Fig. 9 of Ludfordian are recorded only in Gondwanan province. Whereas, *S. newportensis* ss, *E. micrornatus* ss, *E. zavallatus* ss are recorded in Lockovian sediments of Euramerican phytogeographic province. Some other characteristic species such as *C. cassicula, L. argovejae* have different stratigraphic occurrence in the two provinces. This is probably are due to insufficient favorable sediment, nature or rarity of plants producing these miospores, unsuitable environments, the absence of an effective physiographic barrier (Steemans 1999b; Steemans *et al.*2007, 2008; Edwards and Richardson 2004; Richardson 2007).

To illustrate these differences between the two phytogeographic provinces, the miospore zonal scheme of text figure (2) demonstrate these differences. A comparison with the Old Red Sandstone Continent zonation shows that only two Western Sahara Algeria syncline (*annulatus-sextantii* and *velatus-langii*) is correspond to just two Emsian to early Middle Eifelian assemblage zones of Richardson and McGregor (1986). Nearly 6 Interval zones (Po-AB) of the Ardenne-Rhine regions zonation of Streel *et al.* (1987) correspond to just two Middle and Upper Pragian miospore assemblage biozones (*arenorugosa-caperatus* and *polygonalis-subgranifer*) of Western Sahara Algeria syncline of Hassan Kermandji *et al.* (2008, Fig.4). Despite this limited matching, they contain many more identical characteristic species of Lochkovian to early Eifelian but of different occurrence and stratigraphic range. The Innovation contrast with miospore assemblages is by their cosmopolitanism culminating in the Emsian *Emnanisporites annulatus*. This species began in early but not earliest Emsian and well expanded during Emsian and early Eifelian and collapsed at latest Devonian, allowing accurate correlations throughout the world.

System	Series	Stage	Richardson & McGregor 1986	Streel et al. 1987		Richardson et al. 2001	Rubinstein & Steemans 2002	Hassan Kermandji 2007	Hassan Kermandji et al. 2008, 2009
D E V O N I A N	MIDDLE	EIFELIAN	velatus-langii	AD	Mac				velatus-langii
					Vel				
	LOWER	EMSIAN	douglastownense-eurypterota	AP	net.				microancyreus-protea
					Pro.				
					ked.				
			annulatus-sextantii	FD	Car				annulatus-sextantii
					Min.				
					Pra.				
					Fov.				
		PRAGIAN	polygonalis-emsiensis	PoW	AB				polygonalis-subgranifer
					Su				arenorugosa-caperatus
					Pa				
					W				emsiensis-spinaeforms
					Po				
		LOCHKOVIAN	breconensis-zavallatus	BZ	E				tidikeltense-microbaculatus
				MN	Z	micrornatus-newportensis	micrornatus-newportensis		
					G				
			micrornatus-newportensis		Si				
					M				
					R				
					N				
			tripapillatus-spicula			elegans-cantabrica	inframurinata-inframurinata	radiata-synoria	
S I L U R I A N	PRIDOLI	DOWNTONIAN				hemiesferica	Biozone A	hemiesferica	
							tripapillatus-spicula		
	LUDLOW	LUDFORDIAN	libycus-poecilomorphus			reticulata-sanpetrensis	Apiculiretusispora	sanpetrensis-triangulatus	
							libycus-poecilimorphus		
		GROSTIAN	cf. protophanus-verrucatus			brevicosta-verrucata	cf. protophanus-verrucatus		
		HOMERIAN				vetusta-kozlika		vetusta-kozlica	
	WENLOCK	SHEINWOODIAN	chulus-nanus						

Fig. 2. Correlation between miospore zonation in the Lower and Middle Silurian and Devonian of Euramerica and Western Gondwana.

PLATE 1

All figures x 1000 where stated otherwise.

Figure 1. *Cheilotetras caledonica* Wellman and Richardson, 1993. (NL: samp. 1/21, sl. 3/1) Sheinwoodian, ECF-1 borehole, Illizi Basin, Sahara Algeria.

Figure 2. *Tetrahedraletes medinensis* (Strother and Traverse, 1979) Wellman and Richardson, 1996. (NL: samp. 1/14, sl. 2/1), Upper Homerian, NGS-1 borehole, Triassic Province, Sahara Algeria.

Figures 3, 5, 6&11. *Cymbohilates horridus* Richardson, 1996. Figs. 3 & 6 (NL: samp. 21082.5m, sl. 224), figs. 5 &11 (NL: dep. 2082.5m, sl. 223), Lower Lochkovian, GMD-3 & ISS-1 boreholes respectively, Tedikelt Plateau, Sahara Algeria. Figs. 5, 6 x500.

Figure 4. *Cymbohilates allenii* var. *magnus* Richardson, 1996. (NL: dep. 2082.5m, sl. 223), Lower Lochkovian, ISS-1 boreholes, Tedikelt Plateau, Sahara Algeria.

Figure 7 & 12. *Hispanaediscus verrucatus* (Cramer) Burgess and Richardson, 1991. (NL: samp. 7/20, sl. 15/2), Homerian, ECF -1 borehole, Illizi Basin, Sahara Algeria.

Figures 8&13. *Laevolancis divellomedia* (Chibrikova) Burgress and Richardson, 1991. (NL: samp. 2/20, sl. 2/2), Sheinwoodian, ECF -1 borehole, Illizi Basin, Sahara Algeria.

Figure 9. *Artemopyra? scalariformis* Richardson, 1996. (NL: dep. 2082.5m, sl. 223) Lower Lochkovian, ISS-1 borehole, Tedikelt Plateau, Sahara Algeria.

Figure 10. *Acontotetras inconspicuis* Richardson, 1996. (NL: dep. 2082.5m, sl. 223)Lower Lochkovian, ISS1-borehole, Tedikelt Plateau, Sahara Algeria.

Figure 14. *Cymbohilates allenii* var. *allenii* Richardson, 1996. (NL: samp. 1/14, sl. 2/1), Homerian, NGS-1 borehole, Triassic Province, Sahara algeria.

PLATE 1

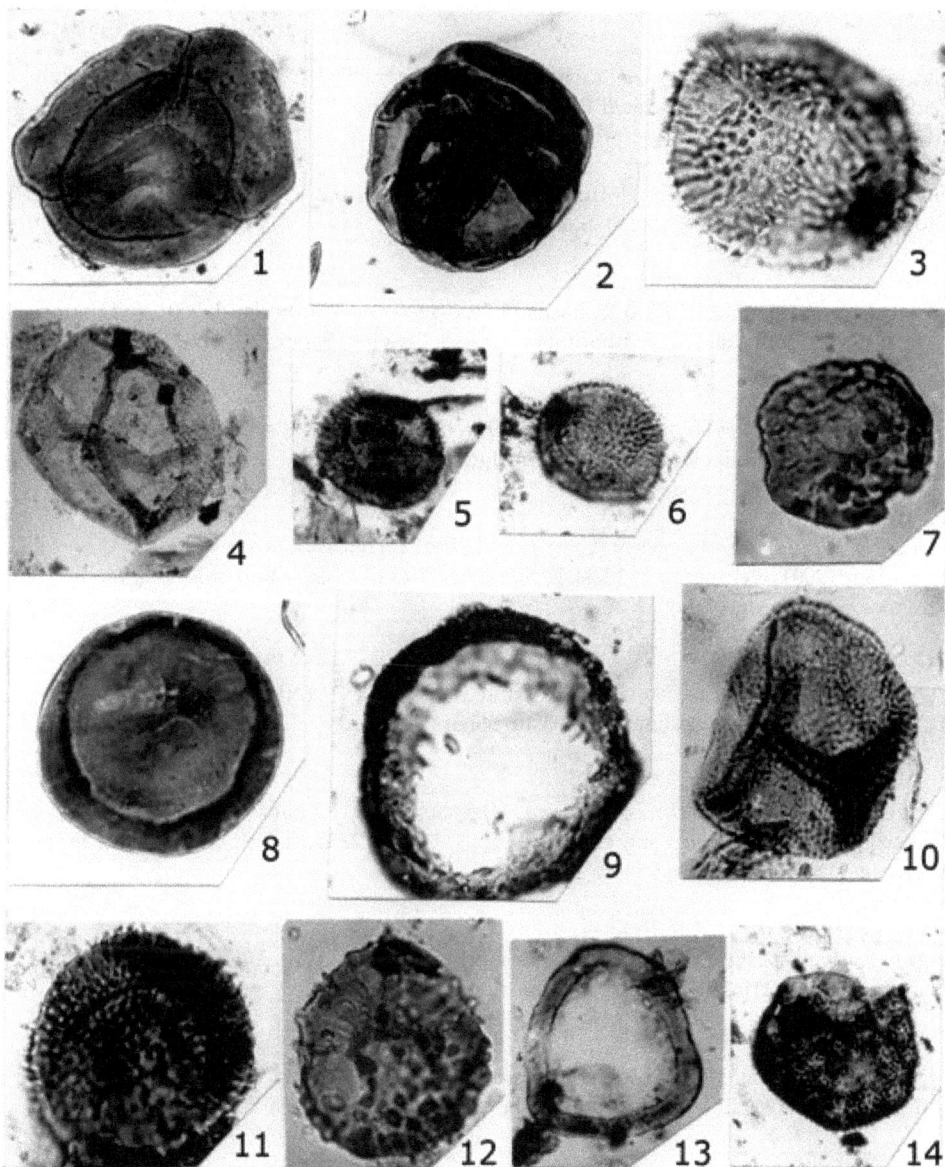

PLATE 2

All figures x 1000 where stated otherwise.

Figure 1. *Apiculiretusispora spicula* Richardson and Lister, 1969. (NL: dep. 2082.5m, sl.222), Přídolí , GMD-3 borehole, Tidikelt Plateau, Sahara Algeria.

Figure 2. *Ambitisporites dilutus* Hoffmeister, 1959. (NL: samp.2/11, sl. 9/2), Homerian, GMD-2 borehole, Tidikelt Plateau, Sahara Algeria.

Figs.3. *Scylaspora kozlica* (Dufka) Richardson, Rodriguez and Sutherland, 2001. (NL: samp. 1/14, sl. 2/1), Homerian, NGS-1 borehole, Triassic Provine, Sahara Algeria.

Figure 4. *Ambitisporites avitus* Hoffmeister, 1959. (NL: samp.4/11, sl. 16/4), Sheinwoodian, GMD-2 borehole, Tidikelt Plateau, Sahara Algeria.

Figure 5. *Chelinospora hemiesferica* (Cramer and Diez) Richardson, Rodriguez and Sutherland, 2001. (NL: samp.5/13, sl. 7/5), Lower Přídolí, NGS-1 borehole, Triassic Province, Sahara Algeria.

Figure 6. *Tetrahedraletes mediensis* Strother and Traveres, 1979. (NL: samp.2/21, sl. 5/2) ECF-1 borehole, Sheinwoodian, Illizi Basin, Sahara Algeria..

Figures 7&8. *Chelinospora (Lophozonotriletes?) poecilomorpha* (Richardson and Ioannides) Richardson, Rodriguez and Sutherland, 2001. (NL: samp.2/11, sl.8/2), Ludfordian, NGS-1 borehole, Triassic Province, Sahara Algeria.

Figure 9. *Cymbosporites triangulatus* Hassan Kermandji, 2007. (NL: 3/13, 5/3), Ludfordian, NGS-1 borehole, Triassic Province, Sahara Algeria.

Figure 10. *Scylaspora tidikeltense* Hassan Kermandji 2008. (NL: dep. 2082.5m, sl. 223), Lower (but not lowermost) Lochkovian, GMD-3 borehole, Tidikelt Plateau, Sahara Algeria.

Figure 11. *Scylaspora vetusta* (Rodriguez) Richardson, Rodriguez and Sutherland, 2001. (NL: samp.2/11, sl.5/2), Homerian, NGS-1 borehole, Triassic Province, Sahara Algeria.

PLATE 2

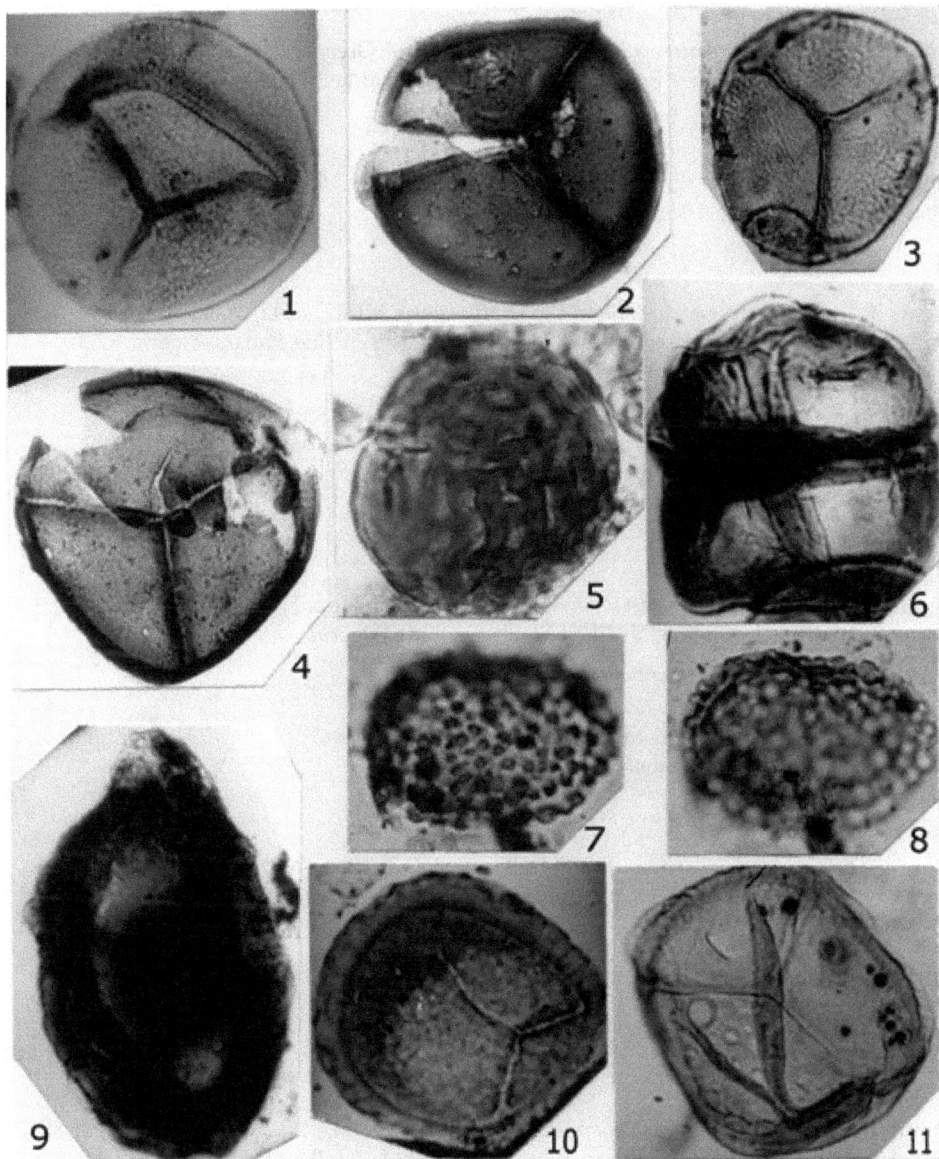

PLATE 3

All figures x 1000 where stated otherwise.

Figures 1&8. *Verrucosisporites polygonalis* (Lanninger) McGregor, 1973. (NL: dep. 1012.7m., sl. 419), late Pragian-earliest Emsian, MSR-1 borehole, Tidikelt Plateau, Sahara Algeria.

Figure 2. *Emphanisporites neglectus* Vigran, 1964. (NL: samp.2/11, sl.8/2), Ludfordian, NGS-1 borehole, Triassic Province, Sahara Algeria.

Figure 3. *Cymbosporites cyathus* Allen, 1965. (NL: samp. MD42, sl. 42/3), late Pragian-earliest Emsian, Djabel el Kahla, Oued Saoura, Western syncline, Sahara Algeria. x700.

Figures 4, 7&10. *Perotrilites microbaculatus* Richardson and Lister, 1969. (NL: dep.2082.5m, sl. 224), Lower (but not lowermost) Lochkovian, GMD-3 borehole, Tidikelt Plateau, Sahara Algeria.

Figures 5&11. *Amocosporites miserabilis* Cramer, 1966. (NL: dep. 1183.0m, sl. 102), Lower Lochkovian, MSR-1 borehole, Tedikelt Plateau, Sahara Algeria.

Figure 6. *Brochotiletes libyensis* Moreau-Benoît, 1979. (NL: dep.1012.7m, sl. 419), Pragian, MSR-1 borehole, Tidikelt Plateau, Sahara Algeria.

Figure 9. *Emphanisporites spinaeformis* Schultz, 1968. (NL: samp. MD21, sl. 21/5), late Lochkovian, Moungar Debad Km 30, Oued Souara, Western syncline, Sahara Algeria.

Figure 12. *Dibolisporites echinaceus* (Eisenack) Richardson, 1965. (NL: dep. 1911.0m, sl. 511), late Pragian-earliest Emsian, ISS-1 borehole, Tidikelt Plateau, Sahara Algeria.

Figure 13. ?*Cymbosporites* cf. *proteus* McGregor and Camfield, 1976. (NL: dep. 1077.1m, sl. 310), latest Lochkovian-earliest Pragian, MSR-1 borehole, Tidikelt Plateau, Sahara Algeria.

PLATE 3

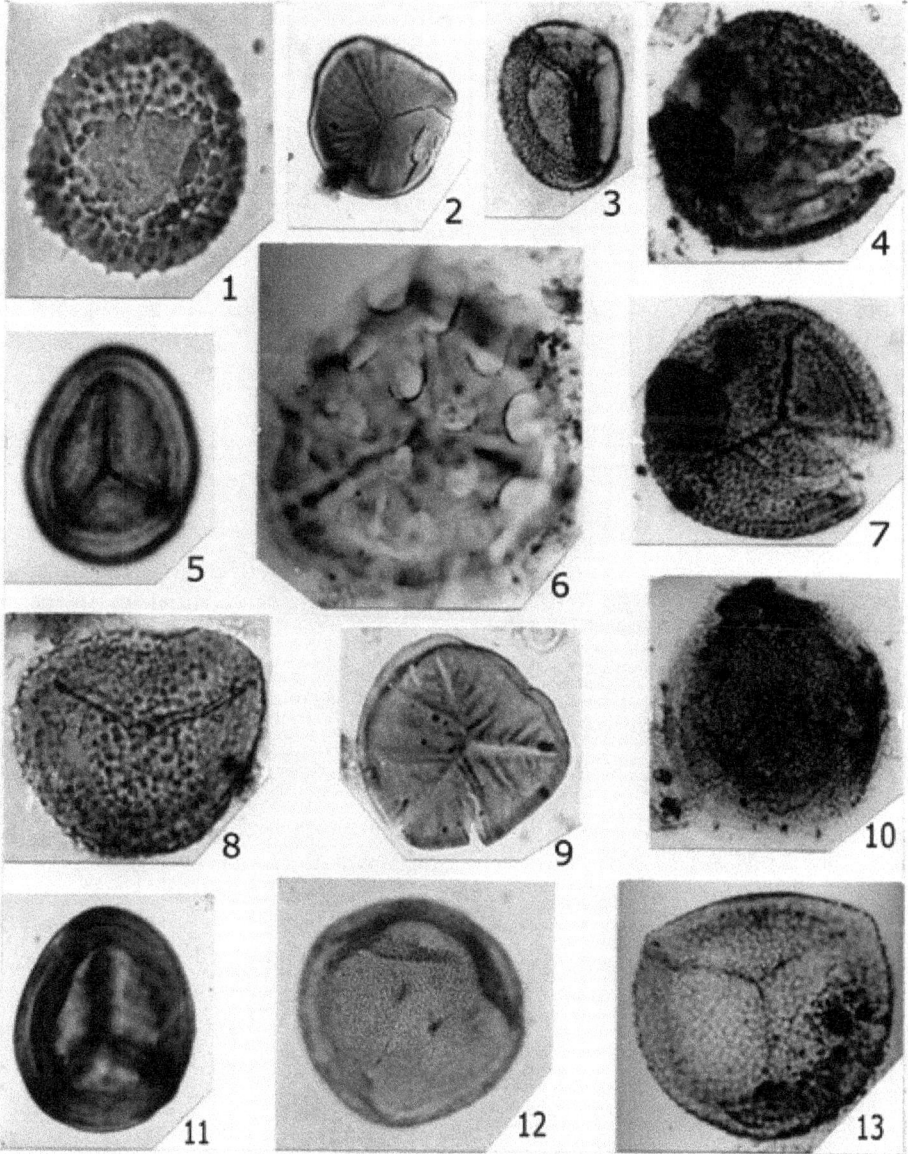

PLATE 4

All figures x 1000 where stated otherwise.

Figures 1 & 4. *Clivosispora verrucata* var. *convoluta* McGregor and Camfield, 1976. (NL: dep. 1012. 7m, sl. 419), Middle Pragian, MSR-1 borehole, Tidikelt Plateau, Sahara Algeria.

Figure 2. *Corystisporites* cf. *multispinosus* Richardson, 1965. (NL : samp. MD56, sl. 56/5) latest Emsian, Moungar Debad Km 30, Oued Saoura, Western syncline, Sahara Algeria.

Figures 3 & 6. *Samarisporites orcadiensis* Richarson, 1965. (NL: dep. 955.0 m, sl. 405), Eifelian, MSR-1 borehole, Tidikelt Plateau, Sahara Algeria.

Figures 5 & 9. *Camptozonotriletes caperatus* McGregor, 1973. (NL: dep. 1012. 7m, sl. 419), Middle Pragian, MSR-1 borehole, Tidikelt Plateau, Sahara Algeria.

Figure 7. *Samarisporites mediconus* (Richardson) Richardson, 1965. (NL: samp. MD62, sl. 62/3), Eifelian, Moungar Debad Km 30, Oued Saoura, Western syncline, Sahara Algeria.

Figure 8. *Geminospora* cf. *treverica* Riegel, 1973. (NL: dep. 955.0 m, sl. 405), Eifelian, MSR-1 borehole, Tidikelt Plateau, Sahara Algeria.

Figures 10 & 11. *Dictyotriletes emsiensis* (Allen) McGregor 1973. (NL: dep. 1012. 7m, sl. 419), Middle Pragian, MSR-1 borehole, Tidikelt Plateau, Sahara Algeria.

Figures 12.*Chelinospora sanpetrensis* (Rodriguez) Richardson, Rodriguez, Sutherland, 2001. (NL: samp. OA, sl. 7/2), Ludfordian, Gurziem, Oued Saoura, Western syncline, Sahara Algeria.

Figure 13. *Dictyotriletes subgranifer* McGregor 1973. (NL: dep. 1027.5m, sl. 405), latest Pragian, MSR-1 borehole, Tidikelt Plateau, Sahara Algeria.

PLATE 4

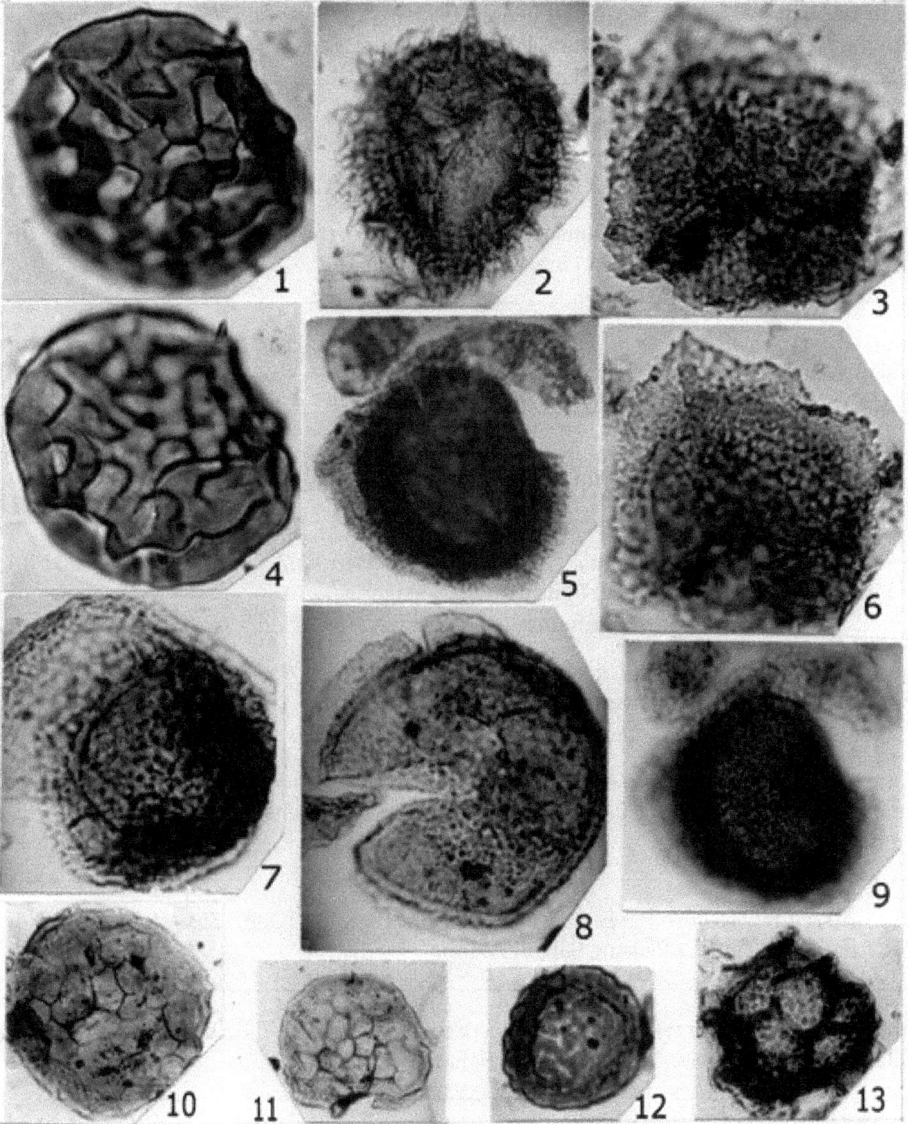

PLATE 5

All figures x 1000 where stated otherwise.

Figure 1. *Grandispora libyensis* Moreau Benoît, 1980. (NL: DM 62, sl.62/5), latest Eifelian, Moungar Debad Km 30, Oued Saoura, Western syncline, Sahara Algeria..

Figure 2. *Grandispora inculta* Allen, 1965. (NL: dep. 955.0m, sl. 405), Eifelian, MSR-1 borehole, Tidikelt Plateau, Sahara Algeria.

Figure 3. *Rhabdosporites langii* (Eisenack) Richardson 1960. (NL: dep. 955.0m, sl. 312), Eifelian, MSR-1 borehole, Tidikelt Plateau, Sahara Algeria.

Figure 4. *.Apiculiretusispora arenorugosa* McGregor, 1973. (NL: dep. 1055.7m, sl. 400), Middle Pragian, MSR-1 borehole, Tidikelt Plateau, Sahara Algeria.

Figures 5. *Grandispora protea* (Naumova) Allen, 1965. (NL: samp. MD56, sl. 56/9), latest Emsian, Moungar Debad Km 30, Oued Saoura, Western syncline, Sahara Algeria.

Figures 6 & 7. *Calyptosporites velatus* (Eisenack) Richardson 1962. (NL: dep. 955.0m, sl. 310), Eifelian MSR-1 borehole, Tidikelt Plateau, Sahara Algeria.

Figure 8. *Emphanisporites annulatus* McGregor, 1961. (NL: dep. 955.0m, sl. 425), Emsian, MSR-1 borehole, Tidikelt Plateau, Sahara Algeria.

Figure 9 & 12. *Camarozonotriletes filatoffi* Breur, Al-Ghazi, Al-Ruwaili, Higgs, Steemans and Wellman, 2007. (NL: dep. 955.0m, sl. 427), Emsian, MSR-1 borehole, Tidikelt Plateau. Algeria.

Figures 10 & 14. *Rhabdosporites parvulus* Richardson, 1965. (NL: dep. 955.0, sl. 312), Eifelian, MSR-1 borehole, Tidikelt Plateau, Sahara Algeria..

Figure 11. *Ancyrospora ancyrea* var. *brevispinosa* Richardson 1962. (NL: dep. 955.0m, sl. 311), Eifelian, MSR-1 borehole, Tidikelt Plateau. Sahara Algeria.

Figures 13. *Chelinospora sanpetrensis* (Rodriguez) Richardson, Rodriguez, Sutherland, 2001. (NL: samp. OA, sl. 7/2), Ludfordian, NGS-1borehole, Triassic Province, Sahara Algeria.

PLATE 5

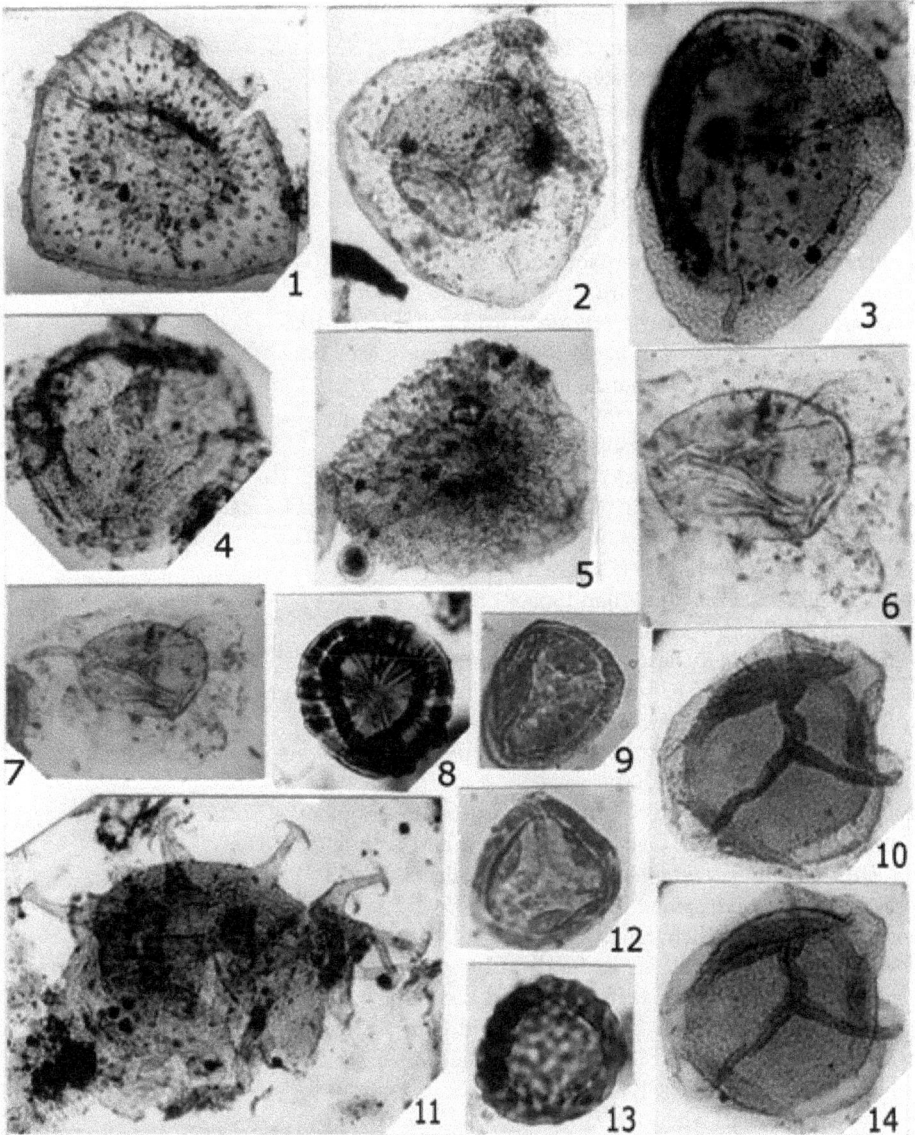

5. Appendix 1 List of species

Acanthotriletes raptus Allen 1965
Acinosporites verrucatus Streel 1967
Acinosporites conatus Hassann Kermandji 2008
Acontotetras inconspicuis Richardson 1996
Ambitisporites avitus Hoffmeister 1959
Ambitisporites dilutus (Hoffmeister) Richardson & Lister 1969
Ambitisporites tripapillatus Moreau-Benoît, 1976
Amocosporites miserabilis Cramer and Diez 1975
Ancyrospora ancyrea var. *brevispinosa* Richardson 1962
Apiculiretusispora arenorugosa McGregor 1973
Apiculiretusispora plicata (Allen) Streel 1967
Apiculiretusispora spicula Richardson and Lister, 1969
Apiculiretusispora synoria Richardson & Lister 1969
Archaeozonotriletes chulus var. *chulus* Richardson and Lister 1969
Archaeozonotriletes chulus var. *nanus* Richardson and Lister 1969
Artemopyra ?scalariformis Richardson 1996
Brochotriletes foveolatus ? Naumova 1953
Brochotriletes libyensis Moreau-Benoît 1979
Calyptosporites velatus (Eisenack) Richardson 1965
Camarozonotriletes filatoffi Breur, Al-Ghazi, Al-Ruwaili, Higges, Steemans and Wellman 2007
Camarozonotriletes sextantii McGregor and Camfield 1976
Camptozonotriletes caperatus McGregor 1973
Camptozonotriletes aliquantus Allen 1965
Cheilotetras caledonica Wellman and Richardson 1993
Chelinospora cassicula Richardson & Lister 1969
Chelinospora hemiesferica (Cramer & Diez) Richardson, Rodriguez and Sutherland 2001
Chelinospora perforata Allen 1965
Chelinospora (*Lophozonotriletes* ?) *poecilomorpha* (Richardson & Ioannides) Richardson, Rodriguez and Sutherland 2001
Chelinospora sanpetrensis (Rodriguez) Richardson, Rodriguez and Sutherland 2001
Clivosispora verrucata var. *Convoluta* McGregor and Camfield 1976
Concentricosisporites sagittarius (Rodriguez) Rodriguez 1983
Corystisporites cf. *multispinosus* Richardson 1965
Cymbohilates horridus Richardson 1996
Cymbohilates allenii var. *allenii*, Richardson 1996
Cymbosporites cf. *dittonensis* Richardson and Lister 1969
Cymbosporites proteus McGregor and Camfield 1976
Cymbosporites cf. *proteus* McGregor and Camfield 1976
Cymbosporites catillus Allen 1965
Cymbosporites cyathus Allen 1965
Cymbosporites triangulatus Hassan Kermandji 2007
Dibolisporites saharansis Hassan Kermandji 2008
Dibolisporites echinaceus (Eisenack) Richardson 1965
Dibolisporites cf. *gibberosus* (Naumova) var. *major* (Kedo) Richardson 1965

Dictyotriletes emsiensis (Allen) McGregor 1973
Dictyotriletes subgranifer McGregor 1973
Emphanisporites annulatus McGregor 1973
Emphanisporites decoratus Allen 1965
Emphanisporites epicautus Richardson and Lister 1969
Emphanisporites cf. *micrornatus* Richardson and Lister 1969
Emphanisporites neglectus Vigran 1964
Emphanisporites protophanus Richardson and Ioannides 1973
Emphanisporites spinaeformis Schultz 1968
Emphanisporites splendens (Richardson and Ioannides) Richardson and Ioannides 1979
Geminospora cf. *spinosa* Allen 1965
Geminospora svalbardiae (Vigran) Allen 1965
Geminospora cf. *treverica* Reigel 1973
Grandispora diamphida Allen 1965
Grandispora inculta Allen 1965
Grandispora libyensis Moreau-Benoît, 1980
Grandispora protea (Naumova) Allen 1965
Hispanaediscus lamontii Wellman 1993
?Hystricosporites cf. *corystus* Richardson 1962
Hystricosporites microancyreus Reigel 1973
Laevolancis divellomedia (Chibrikova) Burgess and Richardson 1991
Leonispora argovejea Cramer and Diez 1975
Lophozonotriletes curvatus Naumova 1953
Perotrilites microbaculatus Richardson and Lister 1969
Retusotriletes abundo Rodriguez 1978
Retusotriletes actinomorphus Chibrikova 1962
Retusotriletes cf. *frivolus* Chibrikova 1959
Retusotriletes triangulatus (Streel) Streel 1967
Rhabdosporites langii (Eisenack) Richardson 1960
Rhabdosporites parvulus, Richardson 1965
Samarisporites mediconus (Richardson) Richardson 1965
Samarisporites orcadiensis Richardson 1965
Scylaspora cymba Hassan Kermandji 2007
Scylaspora distincta Hassan Kermandji 2007
Scylaspora kozlica (Dufka) Richardson, Rodriguez and Sutherland 2001
Scylaspora radiata Hassan Kermandji 2007
Scylaspora undulata Hassan Kermandji 2007
Scylaspora vetusta (Rodriguez) Richardson, Rodriguez and Sutherland 2001
Scylaspora tidikeltense Hassan Kermandji 2008
Synorisporites tipapillatus Richardson and Lister 1969
Synorisporites verrucatus Richardson and Lister 1969
Stenozonotriletes furtivus Allen 1965
Tetrahedralets medinensis (Strother and Traverse) Wellman and Richardson 1993
Tholisporites ancylus Allen 1965
Verrucosisporites polygonalis (Lanninger) McGregor 1973

6. References

Abdesselam-Rouighi, F. 1986. Premiers résultats biostratigraphiques (miospores, acritarches et chitinozoaires) concernant le Dévonien moyen et supérieur de la mole d'Ahara (bassin d'Illizi, Algérie). *Revue Micropaléontologie* 29, 87-92.

Abdesselam-Rouighi, F. 1996. Biostratigraphie des spores du Dévonien de la synéclise Illizi-Ghadamès, Algérie. *Serves Géologique Algérie Bulletin* 7, 171-209.

Abdesselam-Rouighi, F. 2003. Biostratigraphie des spores, acritarches et chitinozoaires du Dévonien Moyen et Superieur du Bassin d' Illizi (Algérie). *Serves Géologique Algérie Bulletin* 14, 97-117.

Al-Ameri, T.K., 1980. Palynology, biostratigraphy and palaeoecology of subsurface Mid-Palaeozoic strata from the Ghadames Basin, Libya. *Unpublished Ph.D. thesis*, University of London, London.

Breuer, P., Al-Ghazi, A., Al-Ruwaili, M., Higgs, K.T., Steemans, P. and Wellman, C.H. 2007. Early to Middle Devonian miospores from northern Saudi Arabia. *Revue de Micropaleontologie* 50, 27-57.

Burgess, N.D., 1991. Silurian Cryptospores and Miospores from the Type Llandovery Area. south-west Wales. *Palaeontology* 34, 565-599.

Burgess, N.D. and Edwards, E. 1991. Classification of uppermost Ordovician to Lower Devonian tubular and filamentous macerals from the Anglo-Welsh Basin. *Botanical Journal of Linen Society* 106, 41-66.

Burgess, N.D. and Richardson, J.B. 1991. Silurian cryptospores and miospores from the wenlock area, Shropsire, England. *Palaeontology* 34,601-628.

Burgess, N.D. and Richardson, J.B., 1995. Late Wenlock to Early Přídolí cryptospores and miospores from south and south-west Wales-Great Britain. *Palaeontographica. B 236*, 1-44.

Edwards, D. 1996. New insights into early land ecosystems: a glimpse of a Liliputian world. *Review of Palaeobotany and palynology* 90, 159-174.

Edwards, D. Davies, K. L. and Axe, L. 1992. A vascular conducting strand in the early land plants. *Cooksonia. Nature* 357, 683-685.

Edwards, D. and Richardson, J.B. 1996. Review of *in situ* spores in early land plants; in: Jansonius, J. and McGregor, D. C. (ed), *Palynology: principles and applications*; American Association of Stratigraphic Palynologists Foundation, v.2, p. 391-407.

Edwards, D. and Richardson, J.B. 2004. Silurian to Lower Devonian plant assemblages from the Anglo-Walesh Basin: a palaeobotanical and palynological synthesis. *Geological Journal* 39,375-402.

Edwards, D. and Wellman, C. H. 1996. Older plant macerals (excluding spores); in: Jansonius, J. and McGregor, D. C. (ed), *Palynology: principles and applications*; American Association of Stratigraphic Palynologists Foundation, v.1, p. 383-387.

Edwards, D. and Wellman, C. H. 2001. Embryophytes on land: the observation to lochkovian (Lower Devonian) Record; in: Gensel, P.G. and Edwards, D. (ed.), Plants invade the Land. Columbo University Press, New York, p.3-28.

Fanning, U., Richardson, J. B. and Edwards, E. 1988. Cryptic evolution in an early land plant. *Evolutionary trends in plants* 2, 13-24.

Fanning, U., Richardson, J. B. and Edwards, E. 1991. A review of *in situ* spores in Silurian land plants; in: Mlackmore,S. and Barnes, S. H. (ed.),Pollen and spores, pattern of diversification; *Systematic Association Special Volume* 44:25-47.

Gensel, P. G., Johnson, N. G. and Srother, P. K. 1991. Early land plant debries (Hooker's waifs and strays'?). *Palaios* 5, 520-547.

Gray, J. 1985. The microfossil record of early land plants: advances in understanding of early terrestrialization, 1970-1984. *Philosophical Transaction of the Royal Society of London, Series B* 309, 167-195.

Gray, J. 1988. Land plant spores and the Ordovician and Silurian boundary. *Bulletin of the British Museum, Natural History (Geology)* 43: 351-358.

Gray, J. 1991. Tetrahedraletes, Nodospora and the 'cross' tetrad: an accretion of myth.49-87; in Blackmore, S and Barnes, S. H. (ed.) *Pollen et spores.* Systematic Association Special Volume, 44.Clarendon Press, Oxford, 391pp.

Gray, J., Boucot, A. J., Grahn, Y. and Himes, G. 1992. A new record of early Silurian land plant spores from the Parana Basin,Paraguay (Malvinokaffric Realm). *Geological Magazine* 129, 741-752.

Gray, J., Massa, D. and Boucot, A. J. 1982. Caradocian land plant microfossils from Libya. *Geology* 10, 197-201.

Hoffmeister, W. 1959. Lower Silurian plant spores from Libya. *Micropaleontology* 5, 331 - 334.

Hassan, A.M., 1982. *Palynology, stratigraphy and provenance of the Lower Old Red Sandstone of the Brecon Beacon, (Powys) and Black Mountains (Gwent and Powys), South Wales;* Ph.D. dissertation, University of London, King's College, v.1, 440pp; v.2, text figs. and plates. (Unpublished)

Hassan Kermandji, A.M. 2007. Silurian and Devonian miospores from the West and Central Algerian Synclines. *Revue de Micropaleontologie* 50, 109-128.

Hassan Kermandji, A.M., Kowalski, W.M. and Khelifi Touhami, F. 2008. Miospore stratigraphy of Lower and middle Devonian deposits from Tidikelt, Central Sahara, Algeria. *Geobios* 41, 227-251.

Hassan Kermandji, A.M. and Khelifi Touhami, F. 2008. Plant microfossils from Lochkovian and Pragian sequence of Brecon Beacons and Black Mountains, south Wales, United Kingdom; in: El-Sayed A. A. Y. (ed.), *Proceedings, The Third International Conference on the Geology of Tethys;* The Tethys Geological Society Cairo, Egypt, p.149-170.

Hassan Kermandji A.M., Khelifi Touhami, F., Kowalski, W. M., Ben Abbés, S., Boularak, M., Chabour, N., Laifa E.L and Bel Hannachi, H. 2009. Stratigraphie du Dévonien Inférieur du Plateau du Tidikelt d'In Salah (Sahara Central Algérie). *Comunicações Geológicas*, 96, 67-82.

Jardiné, S., and Yapaudjian, L., 1968. Lithostratigraphie et palynologie du Dévonian-Gothlandien grésseux du Bassin de Polignec (Sahara). *Revue de l'Institut de la France du Pétrole*, 23, 439-469.

Johnson, N. G. 1984. Early Silurian palynomorphs from the Tuscarora Formation in central Pennsylvania and their paleobotanical and geological significance. *Review of Palaeobotany and Palynology*, 45, 307-360.

Le Hérissé, A., 2002. Palaeoecology, biostratigraphy and biogeography of late Silurian to early Devonian acritarchs and prasinophycean phycomata in WellA1-61Western Libya, North Africa. *Review of Palaeobotany and Palynology* 118, 359-395.

Loboziak, S. and Streel, M. 1989. Middle-Upper Devonian miospores from the Ghadamis Basin (Tunisia-Libya): systematic and stratigraphy. *Review of Palaeobotany and Palynology* 58,173-196.

Loboziak, S. and Streel, M. 1995: West Gondwanan aspect of the Middle and Upper Devonian miospore zonation in the North African and Brazil. *Review of Palaeobotany and Palynology* 86, 147-155.

Loboziak, S. Streel, M., Caputo, M. V. and Melo, J. H. G., 1992: Middle Devonian to Lower Carboniferous miospore stratigraphy in the central Parnaiba Basin (Brazil). *Annales de la Société Géologique de Belgique*, 115, p. 215-226.

Magloire, I. 1967. Etude stratigraphique, par la palynologie, des dépôts argilo - gréseux du Silurien et du Dévonien inférieur dans la région du Grand Erg Occidental (Sahara Algérien); in: Oswald, D. H. (ed). *International Symposium of Devonian System*, Alberta, Calgary, p. 473-495.

Massa, D., Moreau-Benoît, A. 1976. Essai de synthèse stratigraphique et palynologique du système Dévonien en Libye Occidentale. *Revue de l'Institut de la France du Pétrole* 31, 287-333.

McGregor, D.C. 1984. Late Silurian and Devonian spores from Bolivia. *Academia National Cien. Cordoba, Misce* 69,57p.

McGregor, DC and Playford, G., 1992. Morphology and distribution of the miospore *Teichertospora torquata* comb.nov. in the Upper Devonian of Euramerica and Australia. *Palynology* 14, 7-18.

Melo, J. H. G. and Loboziak, S. 2003. Devonian and Early Carboniferous miospore biostratigraphy of the Amazon Basin, northern Brazil. *Review of Palaeobotany and Palynology* 124, 131-202.

Moreau-Benoît, A. Coquel, R. and Latreche, S. 1993. Étude palynologique du Dévonien du Bassin d'Illizi (Sahara Oriental Algérien). Approche Biotratigraphique. *Géobios*, 26, 3-31.

Raymond, A., Gensel, P. and Stein, W. E. 2006. Palynostratigraphy of Late Silurian macrofossils. *Review of Palaeobotany and Palynology* 142,165-192.

Richardson, J. B. 1984. Mid Palaeozoic palynology facies and correlation; proceeding of the 27th International Geological Congress, Musco, 1,Stratigraphy;VNU Science Press, Utrecht, p.341-365

Richardson, J. B. 1988. Late Ordovician and Early Silurian cryptospores and miospores from northeast Libya. In: El-Arenauti, A. Owens, B. and Thusu, B. (ed.), *Subsurface palynostratigraphy of Northeast of Libya*. Benghazi, Garyounis University, p.89-109.

Richardson, J.B. 1992. Origin and Evolution of the earliest land plants. In; Schopf, J. W. (ed.), *Major events in the history of life;* Jones and Bartlett, Bosten, p. 95-118.

Richardson, J. B. 1996. Lower and Middle Palaeozoic records of terrestrial palynomorphs; in: Jansonius, J. and McGregor, D. C. (ed.), *Palynology: principles and applications;* American Association of Stratigraphic Palynologists Foundation, vol.2, p. 555-574

Richardson, J. B. 2007. Cryptospores and miospores, their distribution patterns in the Lower Old Red Sandstone of the Anglo-Welsh Basin, and the habitat of their parent plants. *Bulletin of Geosciences* 82,355-364.

Richardson, J.B. and Edwards, D. 1989: Sporomorphs and plant megafossils. In: Holland, C. H., Bassaett, M. G.(ed.), *A Global standard for the Silurian System. National Museum of Wales Geological Series, Cardiff,* p. 216-226.

Richardson, J. B. and Ioannides, N., 1973. Silurian palynomorphs from the Tanezzuft and Acucus Formations. Tripolitania, North Africa, *Micropaleontology* 19, 257–307.

Richardson, J. B. and Lister, T. R., 1969. Upper Silurian and Lower Devonian spore assemblage from the Welsh Borderland and South Wales. *Palaeontology* 12, 201–252.

Richardson, J. B. and McGregor, D.C., 1986. Silurian and Devonian spore zones of the Old Red Sandstone Continent and adjacent regions. *Geological Survey of Canad. Bulletin* 364. 79pp.

Richardson, J. B. and Rasul, S. M. 1990. Palynofacies in a late Silurian regressive sequence in the Welsh Borderland and Wales. *Journal of Geological Society of London* 147, 675-685.

Richardson, J. B., Rasul, S. M. and Al-Ameri, T. 1981. Acritarchs, miospores and correlation of the Ludlovian-Downtonian and Silurian-Devonian boundaries. *Review of Palaeobotany and Palynology* 34, 209-224.

Richardson, J. B., Rodriguez, R. M., Sutherland, S. J. S. 2001. Palynological zonation of Mid-Palaeozoic sequences from the Cantabrian Mountains, NW Spain: implications for inter-regional and interfacies correlation of the Ludford/Přídolí and Silurian/Divonian boundaries, and plant dispersal patterns. *Bulletin of the British Museum, Natural History (Geology)* 57, 115-162.

Richardson, J. B. Streel, M., Hassan, A. M., and Steemans, P., 1982: A new spore assemblage to correlate between the Breconian (British Isles) and the Gedinnian (Belgium). *Annales de la Société Géologique de Belgique* 105, 135-143.

Rubinstein, C. and Steemans P., 2002. Miospore assemblages from the Silurian Devonian boundary, in borehole A1-61, Ghadames Basin, Libya. *Review of Palaeobotany and Palynology* 118, 397-421.

Spina, A. and Vecoli, M. 2009. Palynostratigraphy and vegetational changes in the Siluro-Devonian of the Ghadamis Basin, North Africa. *Palaeogeography, Palaeoclimatology, Palaeoecology* 282, 1-18.

Steemans, P. 1999a. Paléodiversification des spores et des cryptospores de l'Ordovicien au Dévonien inferieur. *Geobios* 32, 341-352.

Steemans, P. 1999b. Cryptospores and spores from the Ordovician 566 to the Llandovery. A review. *Acta Universitatis Carolinae Geologica* 43, 271-273.

Steemans P., Le Hérissé, A. and Bozdogan, N. 1996. Ordovician and Silurian cryptospores and miospores from southern Turkey. *Review of palaeobotany and Palynology* 93, 35-76.

Steemans, P. and Lakova, I. 2004. The Moesian terrane during the lochkokvian a new palaeogeographic and phytogeographic hypothesis based on miospore assemblage. *Palaeogeography, Palaeoclimatology, Palaeoecology* 208, 225-233.

Steemans, P., Rubinstein, C. and Melo, J. H. G. 2008. Siluro-Devonian biostratigraphy of the Urubu River area, western Amazon Basin, northern Brazil. *Geobios* 41, 263-282.

Steemans, P., Wellman, C.H. and Filatoff, J. 2007. Palaeophytogeographical and palaeoecological implications of a miospore assemblage of earlist Devonian (Lochkovian) age from Saudi Arabia. *Palaeogeography, Palaeoclimatology, Palaeocology* 250, 237-254.

Streel, M., Higgs, K., Loboziak, S, Reigel, W. and Steemans, P. 1987. Spore stratigraphy and correlation with faunas and floras in the type marine Devonian of the Ardenne-Rhenish regions. *Review of palaeobotany and Palynology* 50, 211–229.

Streel, M. and Loboziak, S. 1996. Middle and Upper Devonian miospores; in Jansonius, J and McGregor, D.C. (ed.) *Palynology: principles and applications*; American Association of Stratigraohic Palynologists Foundation, vol. 2, p.575-587.

Strother, P. K. 1991. A classification schema for the cryptospores. Palynology 15, 219-236.

Strother, P.K. and Traverse, A. 1979. Plant microfossils from Llandoverian and Wenlokian rocks of Pennsylvania. *Palynology* 3, 1-21.

Strother, P.K., Al-Hajari, S. and Traverse, A. 1996. New evidence for land plants from the Lower Middle Ordovician of Saudi Arabia. *Geology* 24, 55-58.

Vaverdová, M.1984. Some plant microfossils of possible terrestrial origin from the Ordovician of Central Bohima. *Věstink Ústředniho Ústavu Geologikého* 3,165-170.

Vaverdová, M. 1988. Further acritarchs and terrestrial plant remains from the Late Ordovician at Hiásná Třebaň (Czechoslovakia). *Časopis pro Miniralogii a Geologii* 33, 1-10.

Vaverdová, M. 1990a. Early Ordovician acritarchs from the locality Myto near Rokycany (late Arenig, Czechoslovakia) *Časopis pro Miniralogii a Geologii* 35, 239- 250.

Vaverdová, M. 1990b. Coenobial acritarchs and other palynomorphs from the Arenig/Llanvirn boundary, Prague basin. *Věstink Ústředniho Ústavu Geologikého* 65, 237-242.

Volkova, N. A. 1976. On finds of Precambrian spores with a tetrad scar. In: Sokolov,B.S., Gekker, R.F. *et al.* (ed), *paleontology, marine geology;* International Geological Congress, XXV session, Reports of Soviet Geologists; Akademiia Nauk SSSR, Nauka, Moscow, p.14-18 (In Russian.)

Wellman, C. H. 1995. 'Phytodebris' from Scottish Silurian and Lower Devonian continental deposits. *Review of Palaeobotany and Palynology* 84, 255-279.

Wellman, C. H. 1996. Cryptospores from the type area of the Caradoc Series in southern Britain. *Special paper in palaeontology* 55,103-136.

Wellman, C. H. 2006. Spore assemblages from the Lower Devonian 'Lower Old Red Sandstone' deposits of the Rhynie outlier, Scotland. *Transaction of the Royal Society of Edinburgh, Earth Sciences* 97, 167-211.

Wellman, C. H. Edwards, D. and Axe, L. 1998. Permanent dyads in sporangia and spore masses from the Lower Devonian of the Welsh Borderland. *Botanical Journal of Linen Society* 127, 117-147.

Wellman, C. H. and Gray, J. 2000. The microfossil record of early land plants. *Philosophical Transaction of the Royal Society of London, Series B* 355, 717-732.

Wellman, C. H. Higgs, K. T. and Steemans, P. 2000. Spore Assemblages in the Silurian sequence in borehole HWYH-151 from Saudi Arabia. *Special GeoArabia Publication* no. 1, p. 116-133. Bahrain: Gulf PetroLink.

Wellman, C. H. and Richardson, J. B. 1993. Terresterial plant microfossils from Silurian inliers of the Midland Valley of Scotland. *Palaeontology* 36, 155-193.

Wellman, C. H. and Richardson, J. B. 1996. Sporomorph assemblages from the 'Lower Old Red Sandstone' of Lorne, Scotland. *Special Papers in Palaeontology* 55, 41-101.

Wellman, C. H., Thomas, R. G., Edwards, D. and Kenrick, P. 1998. The Coshestone Group (Lower Old Red Sandstone) in southwest Wales: age, correlation and palaeobotanical significance. *Geological Magazine* 135, 397-412.

Wellman, C. H. and Steemans, P. 2000. Spore assemblages from a Silurian sequences in borehole Hawiyah 151 from Saudi Arabia In: Al-Hajri, S. and Owens, B. (ed.), *Stratigraphic Palynology of the Palaeozoic of Saudi Arabia*; GeoArabia Special Publication. 1, p116-133.

Section 3

Sequence Stratigraphy

9

Paleocene Stratigraphy in Aqra and Bekhme Areas, Northern Iraq

Nabil Y. Al-Banna, Majid M. Al-Mutwali and Zaid A. Malak

Mosul University,
Iraq

1. Introduction

The Paleocene Kolosh Formation of Iraq is comprised of flysch deposit of sandstones, marls, shales, intraformational conglomerates and thin beds of arenaceous limestone, deposited in subduction trench, parallel to the suture zone formed by closing of the southern Neotethyan ocean and finally collision between the Arabian, Anatolian and Iranian plates, extended North West-South East, from Mushorah (NW) to Kashti (SW) (van Bellen *et al.*, 1950 and Jassim and Buday, 2006) (Figure 1). Kolosh Formation is coeval with several other formations in other parts of Iraq. These formations are diachronous and define according to lithology. In north Iraq, it is pass to or inter-tongue with algal reef limestone (Sinjar Formation) and reef - back reef deposit (Khurmala Formation) (van Bellen *et al.*, 1959), moreover, a different set of paleocene formations are used in central, western and southern Iraq: Aaliji, Akashat and Umm Er Radhumma formations. The complex lithostratigraphy relationships between these units are not suitable for correlation and best resolved by biostratigraphy and sequence stratigraphy.

The Kolosh Formation was firstly described by Dunnington (1952; in van Bellen *et al.*, 1959) in Kolosh area northern Koi Sanjaq City, northern Iraq. It consists of shales and fine sandstones, composed of fragments of various grain size of green rock, chert and radiolarite, their age extended Paleocene - lower Eocene, while Al-Omari *et al.* (1993) and Al-Mutwali (2001) reported early Paleocene – early Eocene age to Kolosh Formation in Shaqlawa City north Iraq, Al-Mutwali and Al-Wazan (2010) recorded early – late Paleocene age to Kolosh succession in Duhok area (Figure 1).

The present study based on a (172) m-thick Kolosh and Sinjar formations surface section in the south limb of the Barat Anticline, Bekhme area, the mid point of the section at 36° 39ó 48ó N, 44° 14ó 42ó E and a (42) m-thick surface section in the north limb of Aqra anticline, the mid point of the section at 36° 46ó 39ó N, 43° 58ó 10ó E (Figure 1) (Figure 2 and 3). Sixty five samples are collected from both surface sections. The study involved a detailed field lithological description and identification of foraminifera assemblages from Bekhme surface section are interpreted as early Paleocene.

Based on paleontological and sedimentological attributes facies are recognized and used to interpret the sequence stratigraphy of Kolosh and Sinjar formations.

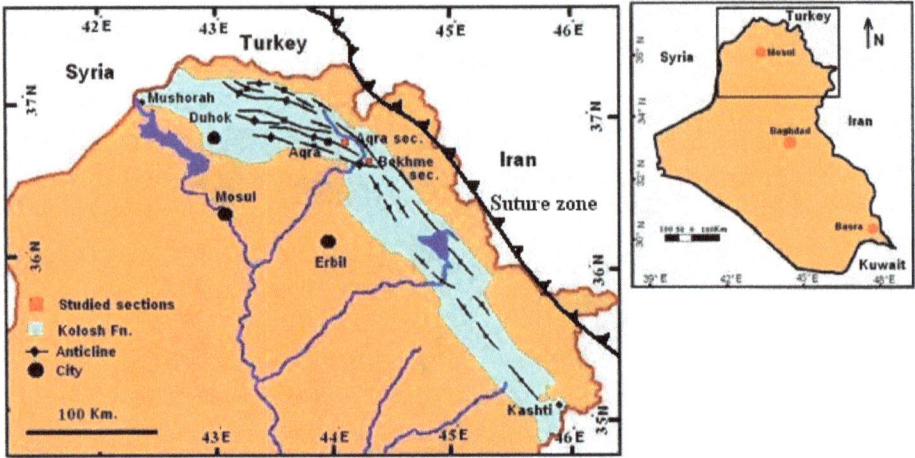

Fig. 1. Paleogeographic distribution of the Kolosh Formation and locations of the studied sections in the northern Iraq.

2. Petrography

Modal analysis was carried out for the sandstones of Kolosh Formation by point counting of an average of 300 point per thin section using the Gazzi-Dickinson method (Ingersoll *et al.*, 1984). The composition of sandstone (Table 1) indicated that they are lithic arenite – lithic graywacke (Figures 4). The mineralogical constituent are:

2.1 Quartz

Monocrystalline (Qm) and Polycrystalline (Qp) quartz occur through out the sandstones of the Kolosh Formation, in which monocrystaline has higher percentage than polycrystalline. Qm is commonly angular to subangular and parallel extinction with inclusion of zircon (Plate 1-1) and rutile suggesting a plutonic origin (Folk, 1974). Most Qp grain consist of >3 crystals (Plate 1-2), the contact between the subgrains are straight to suture and their grain size reach to 0.5mm, with subangular shape and wavy extinction suggesting metamorphic origin (Folk, 1974; Yan *et al.*, 2006).

2.2 Feldspar

K-feldspar (K) and plagioclase (P) representing the second rank of the total content of the studied sandstone. Orthoclase grains are subrounded with wavy extinction and perthitic texture (Plate 1-3), while microcline grains represented by cross twining texture (Plate 1-4). The plagioclase grains are less common, their grain size reach to 0.8 mm, with subrounded and contain poly twining. The provenance rock of feldspar are metamorphic and plutonic (Pettijohn *et al.*, 1973).

2.3 Rock fragments

The most abundant constituents are the carbonate rock fragments (Plate 1-5), generally, they are subangular with grain size reach to 0.5mm, represented micritic grains and bioclasts of

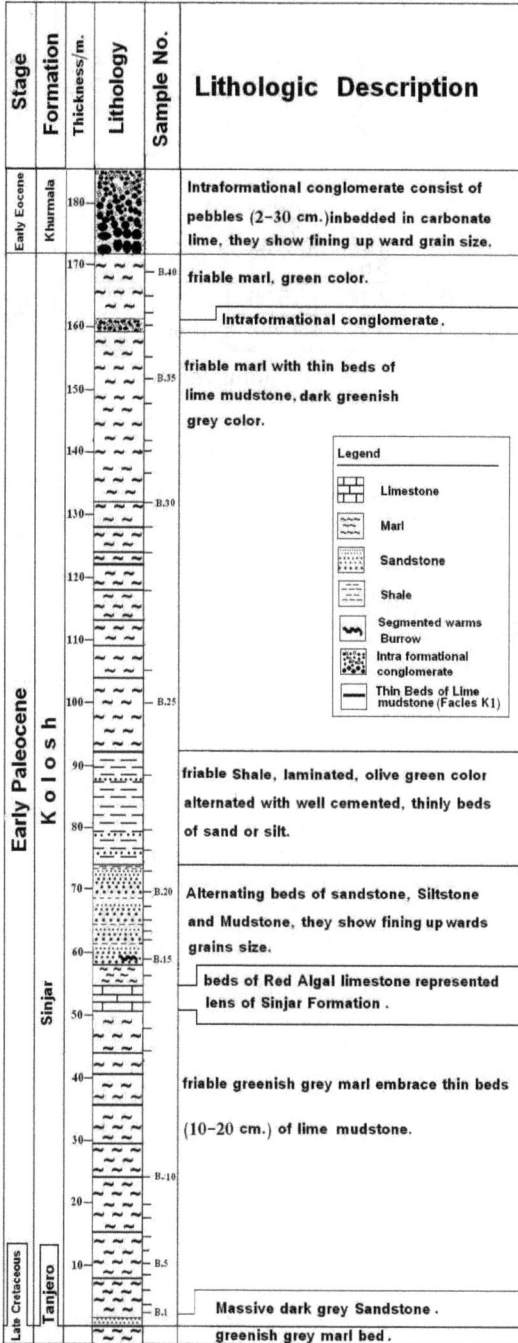

Fig. 2. Lithological description of Kolosh Formation in Bekhme section.

Sample No.	Monocrys. Quartz	Polycrys. Quartz	Plagioclase	K.Feldspar	Chert R. F.	Carbonate R.F.	Ign., Meta. & Other R.F.	Cement	Heavy Minerals	Matrix
B.15	7.56	1.20	0.48	1.20	4.20	24.00	3.36	15.00	3.00	40.00
B.16	11.30	2.06	0.35	3.22	8.34	17.10	16.94	24.75	5.64	10.30
B.18	7.68	0.80	1.04	1.04	2.08	44.24	5.84	12.80	4.48	20.00
B.19	10.26	3.42	1.36	5.64	15.93	25.65	4.27	15.04	3.93	14.50
B.20	9.54	5.04	0.45	3.24	15.30	27.54	11.07	16.47	1.35	10.00
B.21	3.82	1.16	0.34	0.92	1.74	31.90	0.920	16.04	1.16	42.00

Table 1. Modal analysis of the various petrographic constituents of Bekhme section.

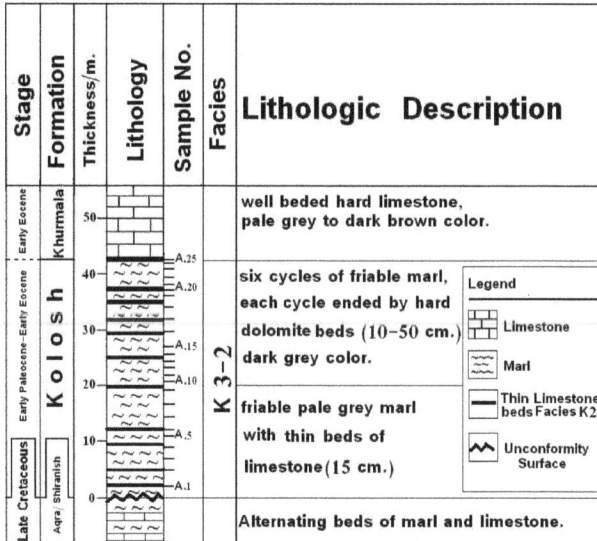

Fig. 3. Lithological description of Kolosh Formation in Aqra section.

pelecypod and gastropod. The source of the carbonate rock fragments is believe to be from the underlaying Mesozoic carbonate rock of the Arabian shelf, which are less resistant, accordingly they can not sustain long distances of transportation. The large content of carbonate rock fragments requires high relief and rapid erosion (Pettijohn, 1975).

Chert rock fragments are microcrystalline and cryptocrystalline (Plate 1-6), the grains are mostly angular with sharp to subsharp edges indicated to nearby source, the source rocks should be the radiolarian cherts of the Cretaceous Qulqula series and the carbonate formations include chert nodules.

The metamorphic (Plate 1-7) and igneous rock fragments show low content, indicating to the paucity of these rocks in the source areas, accordingly heavy minerals found with less amount.

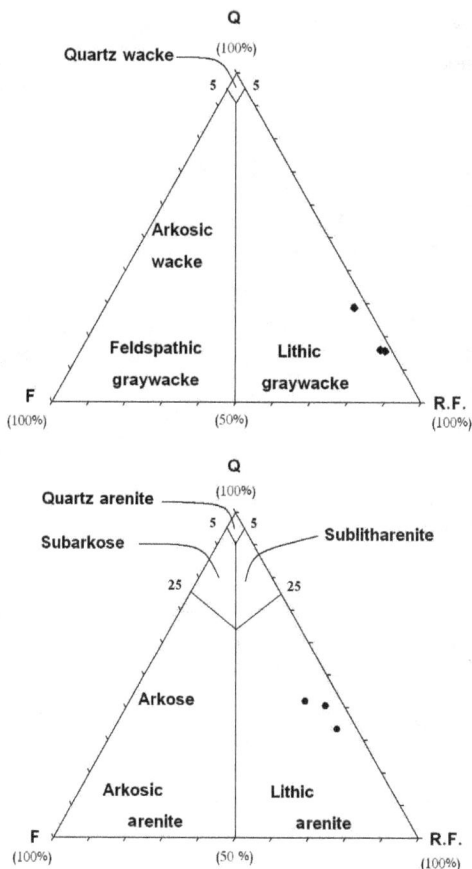

Fig. 4. Position of Kolosh Formation sandstones on the classification of Dott (1964).

2.4 Matrix and cement

The carbonate cement form the abundant type of cement in all sandstones of Kolosh Formation, it occurs as microspar and sparry mosaic cement (Plate 1-8). The matrix consist of clayey – silty grains that filled the interpartical vugs (Plate 1-9), it embrace clay minerals in addition to quartz, feldspar and mica minerals.

3. Sedimentary facies

The studied Paleocene succession in Bekhme section consist of Kolosh and lens of Sinjar formations, while in Aqra section only Kolosh Formation was recognized.

3.1 Facies and depositional setting of Kolosh Formation

The Kolosh Formation contains six facies, donated K1 to K6, for which sedimentological and biological evidences are used to determine their depositional environment and bathymetry.

1- *Inclusion of Zircon in Monocrystalline Quartz.*
2- *Polycrystalline Quartz.*
3- *Perthitic Texture of Orthoclase.*
4- *Cross Twinning of Microcline.*
5- *Carbonate Rock Fragment in Sandstone.*
6- *Angular to Subangular Chert in Sandstone.*
7- *metamorphic rock fragments in Sandstone.*
8- *Calcite Cement in Sandstone.*
9- *Clay Matrix in Sandstone.*

Plate 1.

3.1.1 Lime mudstone facies K1

The facies represent thin bedded limestone with thickness ranging between 5 – 10 cm, dark grey color. The facies has a mud supported texture and allochems make up to 10% of the total contain (Plate 2-1). The predominant allochems are planktonic foraminifera represented by *Parasubbotina pseudobulloides*, *Praemurica pseudoinconstans* and *Subbotina triloculinoides*. The few miliolid, pelecypod and gastropod are recorded. Extraclasts of quartz and chert are record. The matrix consist of micrite and microspar. Carbonate cement fill the chambers of fossils. The sedimentological and paleontological evidences indicated to deep shelf margin environment of the facies (Wilson, 1975; Flugel, 2004).

3.1.2 Pelloidal lime wackestone facies K2

The facies is characteristically as grey dolomitic limestone with abundance of pelloids amount to 40 % of the total contain. The pelloids are rounded to subrounded shape consist of micrite, it have good preservation (Plate 2-2). The few generally affected by selective dolomitization that forming pelloidal fabric, according to Randazzo and Zachos, (1984), the origin of this type fabric is pelloidal lime wackestone. All the attributes indications subtidal – intertidal of restricted marine shelf (Wilson, 1975; Flugel, 2004).

3.1.3 Marl facies K3

This facies consists of green to dark grey marl, soft beds with thickness ranging between 3 – 15 m, embracing thin beds of hard limestone (Plate 2-3). The insoluble residue analysis indicates to carbonate contain of the facies that ranging between (35 – 60%). The allochems percentage ranging between (1 – 45%). These wide range of allochems content lead to subdivision the facies in to two subfacies:

3.1.3.1 Highly fossiliferous marl subfacies K3-1

The hard part of the subfacies included to the presence of allochems ranging between 25 – 45% of total subfacies content, represent by Paleocene planktonic foraminifera, the washing of 10 gm of soft marl samples indicated to the present of 400 – 600 specimen of planktonic foraminifera, the paleontological evidence point to outer shelf – upper bathyal environments of the subfacies, with depth ranging between 150 – 225 m (Gibson, 1989). The inter-tongue of thin beds of hard planktonic foraminiferal limestone with soft marls indicated to the base absent Bouma sequence which is dominated in deep marine basin (van Vliet, 1978).

3.1.3.2 Poorly fossiliferous marl subfacies K3-2

The allochems of this subfacies represent less than 10% of the subfacies content. Also the washing samples show less than 100 specimens of Paleocene planktonic foraminifera. The subfacies of soft marl embracing thin beds of limestone and dolomitic limestone. The paleontological and sedimentological evidences indicate shallow marine environment extended from inner to middle shelf with water depth ranging between 40 – 100 m. (Gibsom, 1989).

3.1.4 Sandstone facies K4

The facies is typically soft greenish grey sandstones, it consist of fine – medium sandstone with thickness ranging between 2 - 5m. The grain size decreased upward, it started with structureless massive sandstone contain low percentage gravel grains (2 – 3mm in diameter) ranging between 1.00 - 2.26% of the total facies contain, these layer followed by laminated silty sandstone and the succession ended by thin bed of mudstone. Vertical burrows were recognized in some bedding planes and segmental warms burrows were recorded at level 60 m (Plate 2-4). Skeletal allochems are dominated by planktonic foraminifera and few benthonic foraminifera of shallow marine environment were recorded in the facies.

The sedimentological and paleontological evidences indicated to the lower part of the submarine fans environment of the facies (Emery and Myers, 2006) or lower part of Bouma sequence (Bouma, 2000; Stelting et al., 2000 and Nichols, 2004).

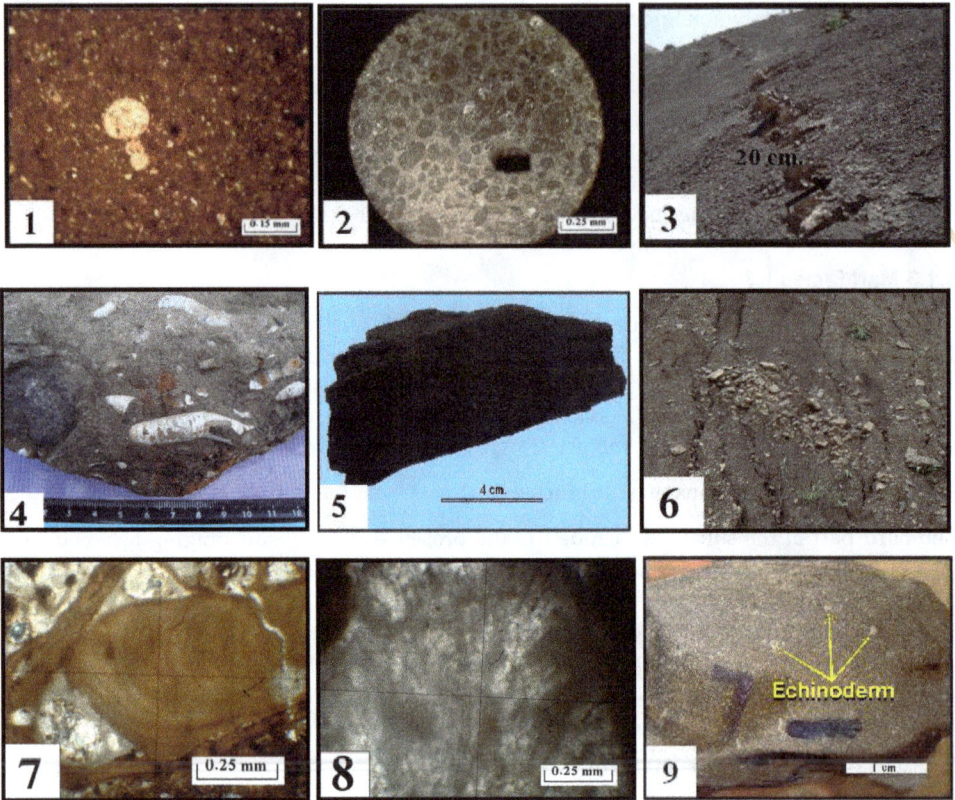

1- Lime Mudstone Facies (K1) in Kolosh Formation.
2- Pelloidal Lime Wackestone Facies (K2) in Kolosh Formation.
3- Hard Thin Limestone Bed Within Marl Facies (K3).
4- Segmented Warms Burrow in Sandstone Facies (K4).
5- Lamination Structure in Shale Facies (K5).
6- Intraformational Conglomerate Facies (K6) in Kolosh Formation.
7- Lithophyllum operculatum Red Algae in Boundstone Facies (S1) of Sinjar Formation.
8- Parachaetetes sp. Red Algae in Boundstone Facies (S1) of Sinjar Formation.
9- Bioclastic and Lime intraclasts in Rudstone Facies (S2) of Sinjar Formation.

Plate 2.

3.1.5 Shale facies K5

The facies is characterized by soft greenish grey shale located at the middle part of the succession of Bekhme section with 17m thick (Plate 2-5), in which allochems dominated by planktonic foraminifera. The facies included low percentage of silt grains, thin beds (>15cm) of hard shale contain high percentage of carbonate cement and lamination structure of alternated silt and clay laminas, which is produced by the oscillated velocity of transportation medium was recognized in the facies.

All the attributes point to levee deposit of submarine fan channels which dominated in the middle part of the submarine fan (Bouma, 2000; Stelting *et al.*, 2000 and Nichols, 2004).

3.1.6 Intraformational conglomerate facies K6

The facies has a total thickness of about 2 m at the top of Kolosh Formation in Bekhme section (Plate 2-6), it consist of random distribution intraclasts in calcareous clay matrix. The intraclasts diameter up to 25cm. the sedimentological evidence point to submarine channels depositional environment, in which they transported the sand and pebble grains to the deep marine (Al-Qayim and Salman, 1986; Boggs, 2006), generally these channels found at the upper part of the submarine fans(Nichols, 2004).

3.2 Facies and depositional setting of Sinjar Formation

The Sinjar Formation crop out at Bekhme section as a lens of carbonate facies with thickness ranging between (0.2m – 11m.) (Figure 2). Their facies donated S1 and S2, the sedimentological and biological evidences are used to determine their depositional environment and bathymetry.

3.2.1 Boundstone facies (S1)

The facies consist of two layer of white hard limestone of 0.6m. thick, it is grain – supported texture and allochems reach to 90% of the total content. The predominant allochems are red algae represented by *Lithothamnium ramosissimum, Lithophyllum sp., Lithothamnium androsovi, Lithothamnium operculatum* (Plate 2-7), *Parachaetetes sp.* (Plate 2-8), other allochems are miliolid, rotalid, echinoderm, pelecypod, and gastropod. Micrite fill the intraparticls vugs of the red algae . All the attributes indicated to organic buildup environments (Wilson, 1975 and Flugel, 2004).

3.2.2 Rudstone facies (S2)

The facies is typically hard grey to light brown limestone. Their thickness ranging between 0.3m. to 2.5m . The facies consist of bioclasts represented by lime intraclasts, echinoderm, red algae and coral up to 0.5cm in diameter (Plate 2-9), in addition to planktonic foraminifera and ostracoda. The allochems content ranging between 10 – 25% of the total facies content. The matrix embraces microspar produced by neomorphism. The facies affected by chemical and physical compaction, which indicated by fracturing of fossil and stylolite. Carbonate cement was recorded as granular, drusy and syntaxial cement. The sedimentological and biological evidences indicate to fore reef deposit within marine slope (Wilson, 1975 and Flugel).

4. Biostratigraphy

The studied rocks of Kolosh Formation exposed in Bekhme area are characterized by abundant well preserved planktonic foraminiferal assemblages, washed residues of Bekhme section were investigated from common diverse planktonic foraminifera to be used as biostratigraphic tools for investigating the marine setting of the formation. Twenty species belonging to six genera were identified in Kolosh Formation, the stratigraphic distribution of these species indicated two biozones and four subzone of the early Paleocene age (Danian) (Figure 5), these zones and subzones are correlated with similar ones established by other authors as shown in Table 2 and Table 3. The zonal scheme (P zones) followed in this study is that of Blow (1979). The zones and subzones are described in ascending orders as follows:

EPOCH	CRETACEOUS	PALEOCENE / EARLY
AGE	**MAASTR.**	**DANIAN**
Bolli 1966	Abathomphalus mayaroensis	Glt. eugubina | Glt. pseudobulloides | Glt. trinidadensis | Glt. uncinata
Toumarkine & Luterbacher 1985	Globigerina eugubina	Morozovella pseudobulloides | Morozovella trinidadensis | Morozovella uncinata
Berggren et al., 1995	Not Studied	Parveugubina–Pr.uncinata P1 / P2 — Pα & P0: Parv. eugubina & G. cretacea (P1a) Parv. eugubina–S.trioculinoides (P1b) S. trioculinoides–Gl. compressa (P1c) Gl. compressa–Pr.inconstans (P2) Pr. uncinata–M. angulata
Olsson et al., 1999	Not Studied	Prav. eugubina – Pr. uncinata P1 / P2 — Pα: c.cretacea (P1a) Parv. eugubina–S.trioculinoides (P1b) Parv. eugubina/S. trioculinoides (P1c) S.triloculinoides–Pr.inconstans / Gl.compressa/Pr.uncinata (P2) Pr. uncinata–M. angulata
Keller et al., 1995; Li & Keller 1998; Keller et al., 2007	Plummerita hantkeninoides	P0 / P1a(1): Parv eugubina | P1a(2): P1b | P1b | P1c(1): P.varianta | P1c(2): Pr.inconstans | P1d: Pr.trinidadensis
Arenillas et al., 2004; 2006 & Molina et al., 2005	Abathomphalus mayaroensis	Parv. eugubina / G.cretacea | Eoglobigerina trivialis | Subbotina triloculinoides | Gl. compressa / P. pseudobulloides
Berggren & Pearson 2005		Pα & P0: Parv. eugubina & G. cretacea | P1a: P. pseudobulloides | P1b: S.triloculinoides | P1c: Gl. compressa/Pr. inconstans | P2: Pr. uncinata
Present Study	Not studied	Parasubbotina pseudobulloides (P1) / Praemurica uncinata (P2) — Subbotina trivialis (P1a) | Subbotina triloculinoides (P1b) | Gl.compressa/Pr.inconstans (P1c) | Pr.trinidadensis (P1d) | Praemurica uncinata (P2)

Table 2. Correlation of planktonic foraminiferal biozones with the outside Iraq biozones.

	PALEOCENE		EPOCH
CRETACEOUS	EARLY		
MAAST	DANIAN		AGE

				Study
Hiatus	Planorotalia compressa	Grt. triloculinoides	Grt. inconstans	Grt.p. uncinata / Grt.p. uncinata / Grt.p.pseudo.sp. / Radiata	**Munim 1976**
Hiatus	Grt.pseudobulloides	Grt. trinidadensis		Grt. uncinata	**Abdel-Kareem 1986**
Hiatus	Grt. pseudobulloides	Grt. compressa	Grt. trinidadensis / Grt. inconstans	Grt. uncinata	**Al–Omari et al. 1993**
G. cretacea P0	P. eugubina– P & G. cretacea & P0	Pav. eugubina– S.triloculinoides P1a	S. triloculinoides – Pr. inconstans (P1b)	—	**Sharbazheri et al. 2009**
Hiatus	P.pseudobulloides (P1)	Subbotina triloculinoides (P1b)	G. compressa (P1c)	Praemurica uncinata (P2)	**Al–Mutwali & Al-Wazan 2010**
Not studied	Parasubbotina pseudobulloides (P1) / Subbotina trivialis (P1a)	Subbotina triloculinoides (P1b)	G. compressa/P. inconstans / P. trinidadensis (P1d)	Praemurica uncinata (P2)	**Present Study**

Table 3. Correlation of planktonic foraminiferal biozones with the inside Iraq biozones.

4.1 *Parasubbotina pseudobulloides* interval zone (P1)

Definition: Interval Zone of the nominate taxon (Plate 3, figs. 5-6).

Boundaries: the zone bounded at the base by the disappearance of the Late Cretaceous planktonic foraminifera and the disconformity with the underlying Tanjero Formation. The top of the zone is placed at the first appearance of the *Praemurica uncinata* (Bolli).

Age: early Paleocene (Danian)

Thickness: The zone is 100 m. thick.

Characteristics: The zone is characterized by the presence of cancel planktonic foraminiferal species in the lower part. The easy distinguished and dominated of (*Parasubbotina pseudobulloides*) make it favor in the biostratigraphic studies (Bolli, 1966, Toumarkine & Luterbacher, 1985, Molina *et al.*, 2005, Arenillas *et al.* 2004; 2006). Accordingly the biozone was subdivided into four subzones, these subzones are from base to top:

4.1.1 *Subbotina trivialis* Partial range subzone (P1a)

Definition: Partial range subzone of the nominate taxon (Plate 3, figs.14-15).

Boundaries: The subzone bounded at the base by the first appearance of the *Parasubbotina pseudobulloides* (Plummer). The top of the subzone is placed at the first appearance of the *Subbotina triloculinoides* (Plummer).

Age: early Paleocene (early Danian).

Thickness: The subzone is 5 m. thick.

Characteristics: In addition to the dominated of the nominate taxon, the following species have been recognized *Globanomalina archeocompressa, Eoglobigerina eobulloides, Subbotina trivialis, Parasubbotina pseudobulloides, Praemurica pseudoinconstans* (Plate 3, figs. 16-17; 1-2; 14-15; 5-6; 26), *Eoglobigerina edita*.

This subzone is equivalent to the *Parvularugoglobigerina eugubina – Subbotina triloculinoides* subzone (P1a) of Berggren *et al.*, (1995) and Olsson *et al.* (1999). It also equivalent to early Paleocene (Early Danian) *Parasubbotina pseudobulloides* subzone (P1a) of Berggren and Pearson (2005) and *Eoglobigerina trivialis* subzone of Arenillas *et al.* (2004); Molina *et al.* (2005) and Arenillas *et al.* (2006) (Table 2). In Iraq the subzone equivalent to the *Grt. Pseudobulloides* subzone of Al-Omari *et al.* (1993) (Table 3) , Therefore, we considered the *Subbotina trivialis* subzone to represent the early Paleocene (early Danian).

4.1.2 *Subbotina triloculinoides* interval subzone (P1b)

Definition: interval subzone of the nominate taxon (Plate 3,figs. 12-13).

Boundaries: The subzone bounded by the first appearance of the nominated taxon and ended by the first appearance of the *Praemurica inconstans* (Subbotina) and *Globanomalina compressa* (Plummer).

Age: early Paleocene (early to middle Danian).

Thickness : The subzone is 12m. thick.

Characteristics: The subzone is characterized by the nominate taxon in addition to the first appearance of *Woodringina hornerstowensis* (Olsson) and *Woodringina claytonensis* (Loeblich and Tappan) in the lower part of the subzone. The following species have been recognized *Globanomalina archeocompressa* (Plate 3, figs. 16-17), *Eoglobigerina eobulloides* (Plate 3, figs.1-2), *Eoglobigerina trivialis*, *Parasubbotina pseudobulloides*, *Eoglobigerina edita*, *Praemurica pseudoinconstans*.

This subzone is identical to the *Subbotina triloculinoids* subzone (P1b) of Berggren and Pearson (2005); Molina *et al.* (2005); Arenillas *et al.* (2004); Arenillas *et al.* (2006); Sharbazheri *et al.* (2009) and Al-Mutwali and Al-Wazan (2010). All the previous studies point to the early – middle Danian age of the subzone.

4.1.3 *Globanomalina compressa / Praemurica inconstans* interval subzone (P1c)

Definition: interval subzone of the two nominated taxa (Plate 3, figs. 18-19; fig. 23).

Boundaries: The subzone bounded by the first occurrence of the nominated taxa and the initial appearance of the *Praemurica trinidadensis* (Bolli).

Age: early Paleocene (middle to late Danian).

Thickness: The subzone is 31m. thick.

Characteristics: The subzone characterized by well – preserved planktonic foraminifera, it is distinctive by presence of *Subbotina cancellata* (Blow)(Plate 3, figs. 8-9) and *Parasubbotina varianta* (subbotina) (Plate 3, fig. 7), In addition to the following species: *Eoglobigerina eobulloides*, *Eoglobigerina trivialis*, *Parasubbotina pseudobulloides*, *Eoglobigerina edita* , *Praemurica pseudoinconstans*, *Subbotina triloculinoides*.

The subzone is equivalent to *Globanomalina compressa / Praemurica inconstans* subzone (P1c) of Berggren *et al.* , (1995) and Berggren and Pearson (2005), also it is identical to *P. inconstans* subzone of Keller *et al.* (2007), *Grt. inconstans* subzone of Munim (1976), Al-Omari *et al.* (1993) and Al-Mutwali and Al-Wazan (2010), which are assigned to early Paleocene (middle – late Danian) age of the subzone.

4.1.4 *Praemurica trinidadensis* interval subzone (P1d)

Definition: interval subzone of the nominate taxon (Plate 3, figs. 27-28).

Boundaries: The lower boundary is defined by the first occurrence of the nominate taxon. The upper of this subzone is placed at the first appearance of *Praemurica uncinata* (Bolli).

Age: early Paleocene (late Danian).

Thickness: The subzone is 52m. thick.

Characteristics: the predominant planktonic foraminifera in this subzone are *Praemurica trinidadensis* (Bolli) and *Praemurica praecursoria* (Morozova)(Plate 3, figs. 24-25), this subzone characterized by the occurrence of *Globanomalina imitata* (Subbotina) (Plate 3, fig. 22), *Eoglobigerina spiralis* (Bolli) (Plate 3, figs. 3-4), *Globanomalina ehrenbergi* (Bolli) (Plate 3, figs. 20-21), *Subbotina triangularis* (White)(Plate, 3, figs. 10-11) in addition to *Eoglobigerina trivialis*, *Parasubbotina pseudobulloides*, *Eoglobigerina edita*, *Praemurica pseudoinconstans*, *Subbotina triloculinoides*, *Praemurica inconstants*, *Globanomalina compressa*, *Parasubbotina varianta*, *Subbotina cancellata* .

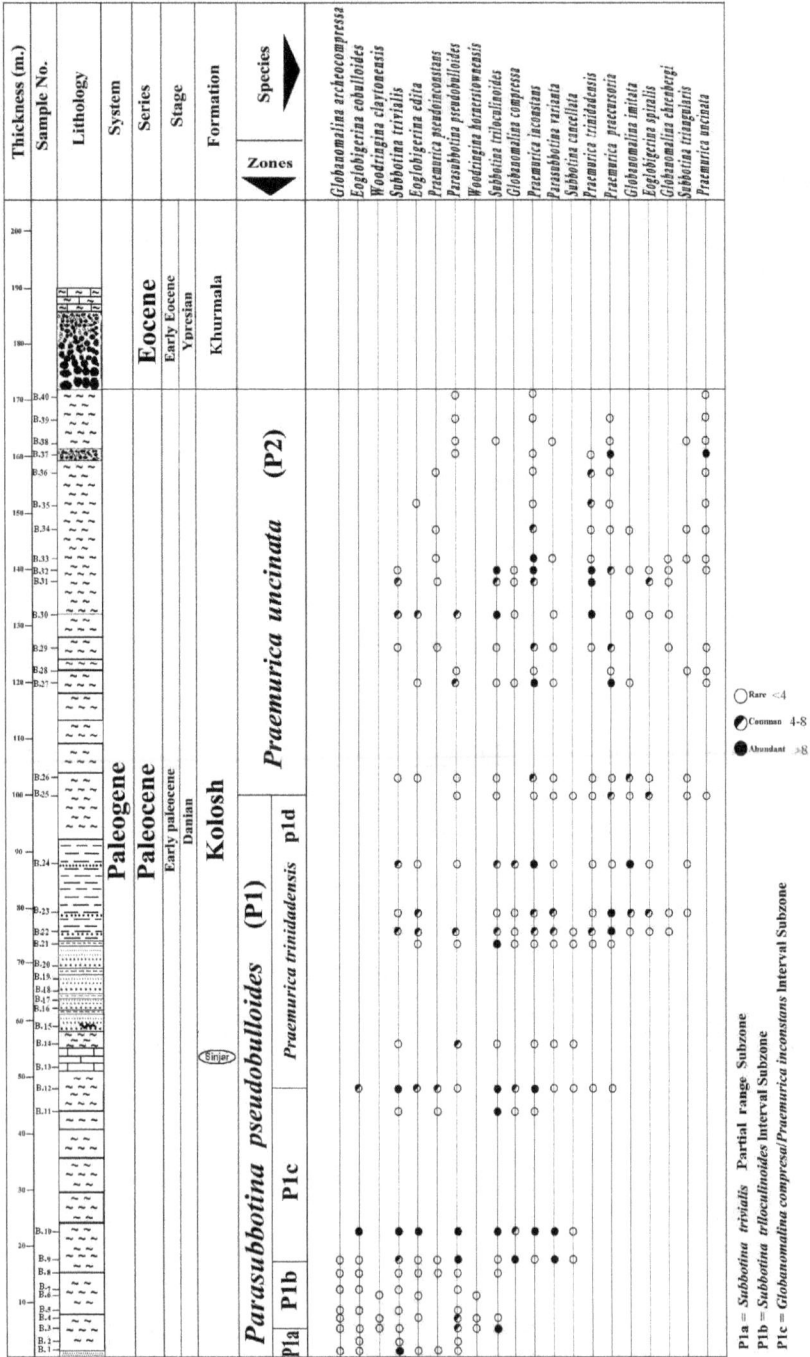

Fig. 5. Distribution of planktonic foraminifera In Kolosh Formation at Bekhme section.

The subzone is equivalent to *Praemurica trinidadensis* subzone (P1d) of Li and Keller (1998), Keller *et al.* (1995) and Keller *et al.* (2009), all of late Danian.

4.2 *Praemurica uncinata* interval zone (Part) (P2)

Definition: interval Zone (part) of the nominated taxon (Plate 3, figs. 29-30).

Boundaries: The base of the zone is defined by the first occurrence of the nominate taxon. The top of the zone is identified by the disappearance of the early Paleocene planktonic foraminifera at the disconformity with the overlying shallow marine carbonate rock of Khurmala Formation.

Age: late early Paleocene (late Danian).

Thickness : The zone is 72m. thick.

Characteristics: In addition to the nominate taxon this zone is characterized by the predominate of the following species: *Subbotina triloculinoides*, *Eoglobigerina trivialis*, *Globanomalina imitate* and *Praemurica trinidadensis*, and continuous occurrence of the following species: *Parasubbotina pseudobulloides*, *Eoglobigerina edita*, *Praemurica pseudoinconstans*, *Praemurica inconstant*, *Globanomalina compressa*, *Parasubbotina varianta*, *Subbotina cancellata*, *Eoglobigerina spiralis*, *Globanomalina ehrenbergi*, *Subbotina Triangularis* and *Praemurica praecursoria*.

This zone is equivalent to the *Praemurica uncinata* Zone (P2) of Kassab *et al.* (1986), Berggren *et al.* (1995), Olsson *et al.* (1999), Berggren and Pearson (2005), Chacon and Martin - Chivelet (2005), Bilotte *et al.* (2007) and Al-Mutwali and Al-Wazan (2010). The zone is also identical with *Grt. uncinata* Zone of Al-Omari (1995) and *Morozovella uncinata* of Abawi and Abdo (2001). All of late early Paleocene.

5. Sequence stratigraphy

The sequence stratigraphic analysis of Paleocene deposits (Kolosh and Sinjar formations) in Bekhme section interpreted four depositional sequence, designated Kolosh sequence 1 to 4 (Figure 6), The sequence, which vary in thickness from (13 – 66 m.) are described in terms of sequence boundaries (SB), lowstand system tract (TS), lowstand fan (LSF), transgressive system tract (TST), maximum flooding surfaces (MFS) and highstand system tracts (HST). As discussed in the final sections of this paper, correlations of depositional sequences and isochronous surface must be based on correlative biozones and follow the most recent conventions of the International commission on stratigraphy (ICS) as documented in Gradstein *et al.* (2004).

5.1 Kolosh sequence 1

The sequence 58 m. thick, begins with the deposition of (2 m.) of facies (K4), their content point to lowstand fans by turbidity currents. It is overlying the deep marine deposit of the Tanjero Formation (late Cretaceos age).

This is followed by marls (subfacies K3-2 and Lower part of K3-1) accompanied by an increased abundance and greater diversity of planktonic foraminifera that represent the TST. The MFS is positioned in the middle of subfacies K3-1, where the maximum number of species is recorded.

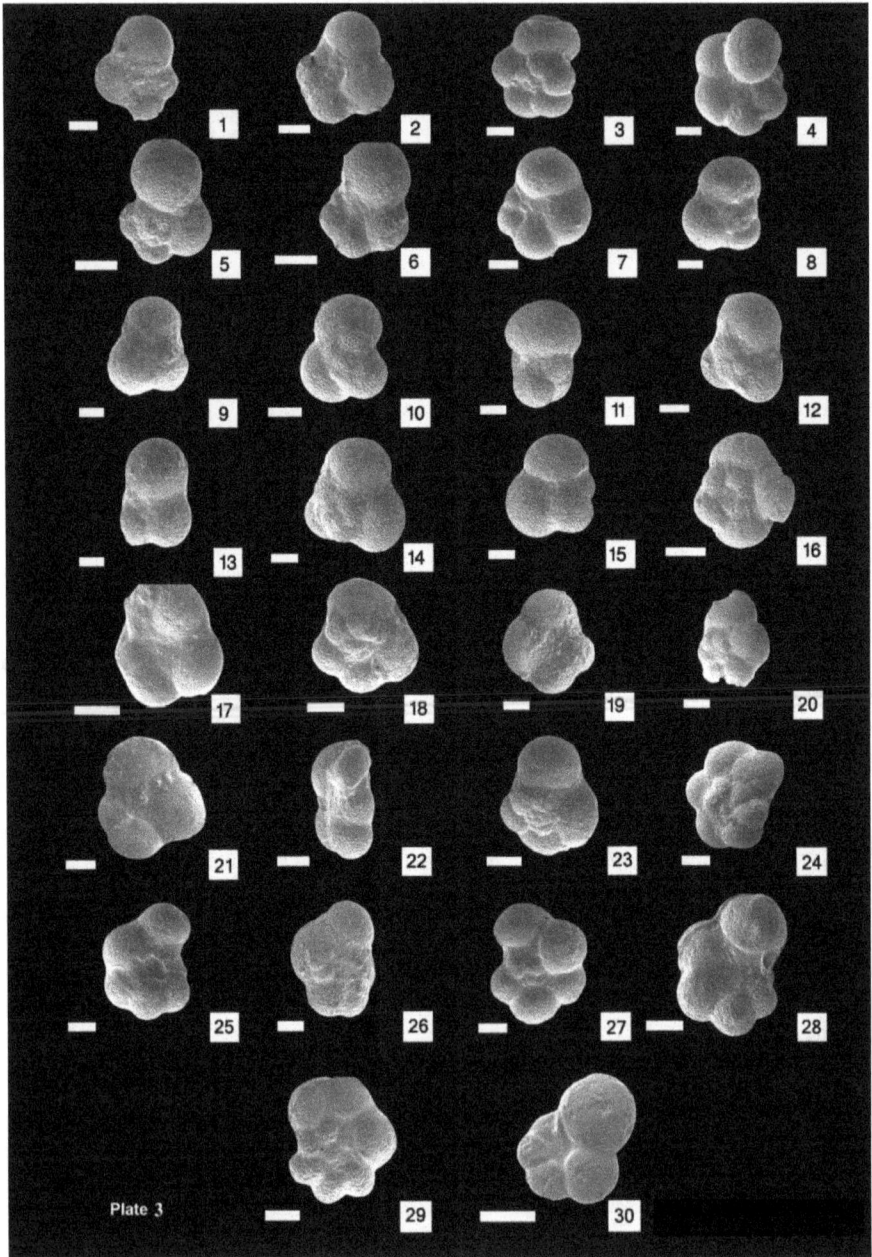

Plate 3. Scale bars 100 Micron

FIGURE 1 *Eoglobigerina eobulloides*, (Morozova), spiral side, Sample no. 1.
FIGURE 2 *Eoglobigerina eobulloides*, (Morozova), umbilical side, Sample no. 1.
FIGURE 3 *Eoglobigerina spiralis* (Bolli), spiral side, Sample no. 23.
FIGURE 4 *Eoglobigerina spiralis* (Bolli), umbilical side, Sample no. 23.
FIGURE 5 *Parasubbotina pseudobulloides* (Plummer), spiral side, Sample no. 1.
FIGURE 6 *Parasubbotina pseudobulloides* (Plummer), umbilical side, Sample no. 1.
FIGURE 7 *Parasubbotina varianta* (Subbotina), umbilical side, Sample no. 9.
FIGURE 8 *Subbotina cancellata* (Blow), spiral side, Sample no. 12.
FIGURE 9 *Subbotina cancellata* (Blow), umbilical side, Sample no. 12.
FIGURE 10 *Subbotina triangularis* (White), spiral side, Sample no. 26.
FIGURE 11 *Subbotina triangularis* (White), side view, Sample no. 26.
FIGURE 12 *Subbotina triloculinoides* (Plummer), spiral side, Sample no. 30.
FIGURE 13 *Subbotina triloculinoides* (Plummer), umbilical side, Sample no. 30.
FIGURE 14 *Subbotina trivialis* (Subbotina), spiral side, Sample no. 10.
FIGURE 15 *Subbotina trivialis* (Subbotina), umbilical side, Sample no. 10.
FIGURE 16 *Globanomalina archeocompressa* (Blow), spiral side, Sample no. 1.
FIGURE 17 *Globanomalina archeocompressa* (Blow), umbilical side, Sample no. 1.
FIGURE 18 *Globanomalina compressa* (Plummer), spiral side, Sample no. 9.
FIGURE 19 *Globanomalina compressa* (Plummer), umbilical side Sample no. 9.
FIGURE 20 *Globanomalina ehrenbergi* (Bolli), spiral side, , Sample no. 31.
FIGURE 21 *Globanomalina ehrenbergi* (Bolli), umbilical side, section, Sample no. 31.
FIGURE 22 *Globanomalina imitata* ((Subbotina),side view, Sample no. 24.
FIGURE 23 *Praemurica inconstans* (Subbotina), spiral side, Sample no.10.
FIGURE 24 *Praemurica praecursoria* (Morozova), spiral side, S.no.27.
FIGURE 25 *Praemurica praecursoria* (Morozova) umbilical side, S.no.27.
FIGURE 26 *Praemurica pseudoinconstans* (Blow), spiral side, Sample no. 9.
FIGURE 27 *Praemurica trinidadensis* (Bolli), spiral side, Sample no. 31.
FIGURE 28 *Praemurica trinidadensis* (Bolli), umbilical side, Sample no. 31.
FIGURE 29 *Praemurica uncinata* (Bolli), spiral side, Sample no. 33.
FIGURE 30 *Praemurica uncinata* (Bolli), umbilical side Sample no. 33.
All figures for Kolosh Fn. Bekhme section.

Plate 3. Figure listing

The HST is characterized by the upper part of subfacies K3-1 followed by subfacies K3-2 and capping limestone beds (Facies S1 and S2), both the lower and the upper contacts are interpreted as SB Type-1. The sequence spars about 2.8 my. The age of the MFS as calculated from biozone is (63.6 Ma) at interval (24 m.), according to Gradstein *et al.* (2004) (Figure 6).

5.2 Kolosh sequence 2

Sequence 2, 66m. thick, commence with the deposition of facies (K4) that deposited by gravity flow of the shallow marine sediment to accumulate in the deep marine as LSF. It is followed by facies (K5), their lower part showing increasing water depth upward indicated by the increasing of planktonic foraminifera percentage, it is represented the TST of the sequence. The middle of facies (K5) (at 80 m.) is picked as the MFS, where the planktonic foraminifera percentage reach maximum (35%) in the facies.

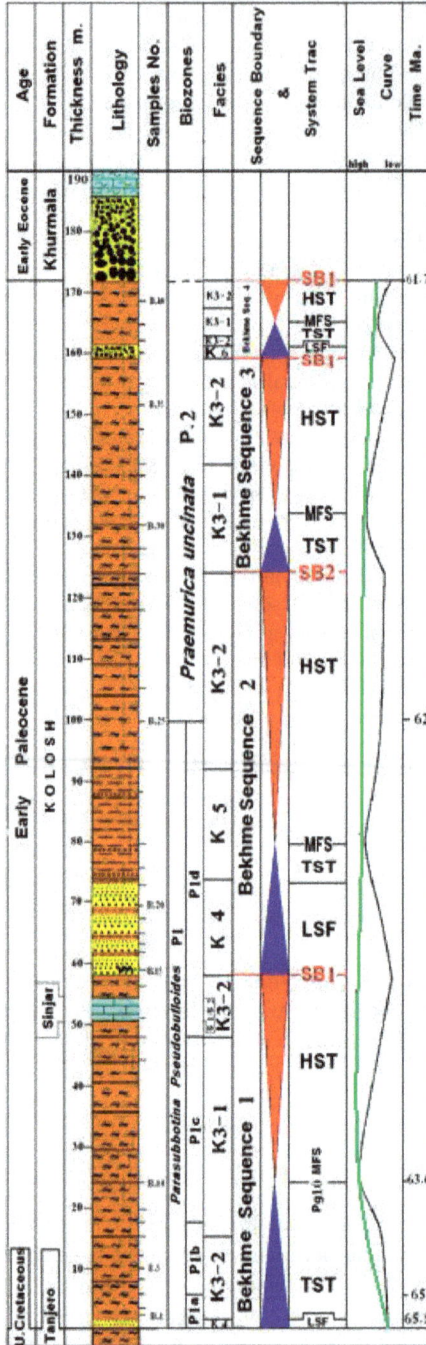

Fig. 6. Lithostratigraphy, biostratigraphy and sequence stratigraphy of the Bekhme section

The upper part of (K5) followed by subfacies (K3-2), both display a shallow environment upward, suggesting the HST of the sequence. The sequence spans less than (1 my), the upper SB is Type 2 (Figure 6).

5.3 Kolosh sequence 3

This sequence begins with the deposition of the facies (K 3-1), It is lower part show increasing of diversity and percentage of the planktonic foraminifera assigned to the TST.

The MFS is represented at the middle of subfacies (K3-1) where the planktonic foraminifera percentage decreasing and followed by subfacies (K3-2) as the HST, the upper SB is Type-1.

5.4 Kolosh sequence 4

Facies (K6) represents the LSF of this (13m.) thick sequence, (Figure 6). It is followed by subfacies (K3-2) and part of subfacies (K3-1) showing increasing water depth and the TST with a retrogradational stacking pattern. The MFS occurs at interval (163m.) within subfacies (K3-1), the upper part of the (K3-1) and the overlying subfacies (K3-2) reflect decreasing water depth and the HST with a progradational stacking pattern. The upper boundary is SB Type-1 below the shallow marine deposit of Khurmala Formation.

The four sequence of Paleocene form a third – order sequence with a total thickness of about, duration of approximately (3.8 my.) and MFS at interval (24m.), and (63.6 Ma).

6. Correlation to Aruma sequence

In central Arabia, Aruma Formation subdivided into three informal members Khanasir limestone, the Hajajah limestone and the Lina shale members (Phillip et al., 2002), They interpretated the khanasir member as Aruma sequence 1, hajajah member as Aruma sequence 2 and 3 and Lina member as Paleocene age Aruma sequence 4. the maximum flooding surface of Aruma sequence 4 may correlate with Kolosh third orders MFS.

7. Correlation to Arabian plate maximum flooding surface

Sharland et al. (2001) consider MFS Pg10 as late paleocene and positioned it in the base of Kolosh Formation; But in the global planktonic foraminiferal Biozone (P4), which is younger than the studied section in the time scale of Gradstein et al. (2004) MFS Pg10 has an age of 58 Ma, this age appear too younger to correlate with the third order MFS of Kolosh in Danian.

8. Correlation to global maximum flooding surface

The 3rd order Kolosh sequence can be correlated with European Danian transgressive surface of Gradstein et al. (2004), which lies within the planktonic foraminifera biozone (P1b), also it can tentatively correlated with Danian 3rd order sequence (TP1) of Blake Plateau, of the eastern united state (Schlager, 1992).

9. Depositional environment and sedimentary model

The sedimentary evidences of Kolosh Formation clastic deposit indicated to submarine fan environment that spilled over in the narrow Ne-Tethys onto the passive continental margin of the Arabian plate from the active margins of the Iranian and Anatolian plates (Numan, 1997).

The Kolosh Formation is widely distributed in the subsurface and at outcrop in parts of northern Iraq, and is coeval with several other formations in other part. These formations are diachronous and defined according to lithology; for examples: marine marl and marly limestone (Aaliji Formation), reef and back reef (Sinjar and Khumala formation) and clastics (Kolosh Formation) (Dunnington, 1958; van Bellen et al., 1959 and Jassim and Goff, 2006). These formations can pass below, above and inter-tongue with one another.

SLOPE MINI-BASIN & MUD RICH FINE-GRAINED SUBMARINE FANS

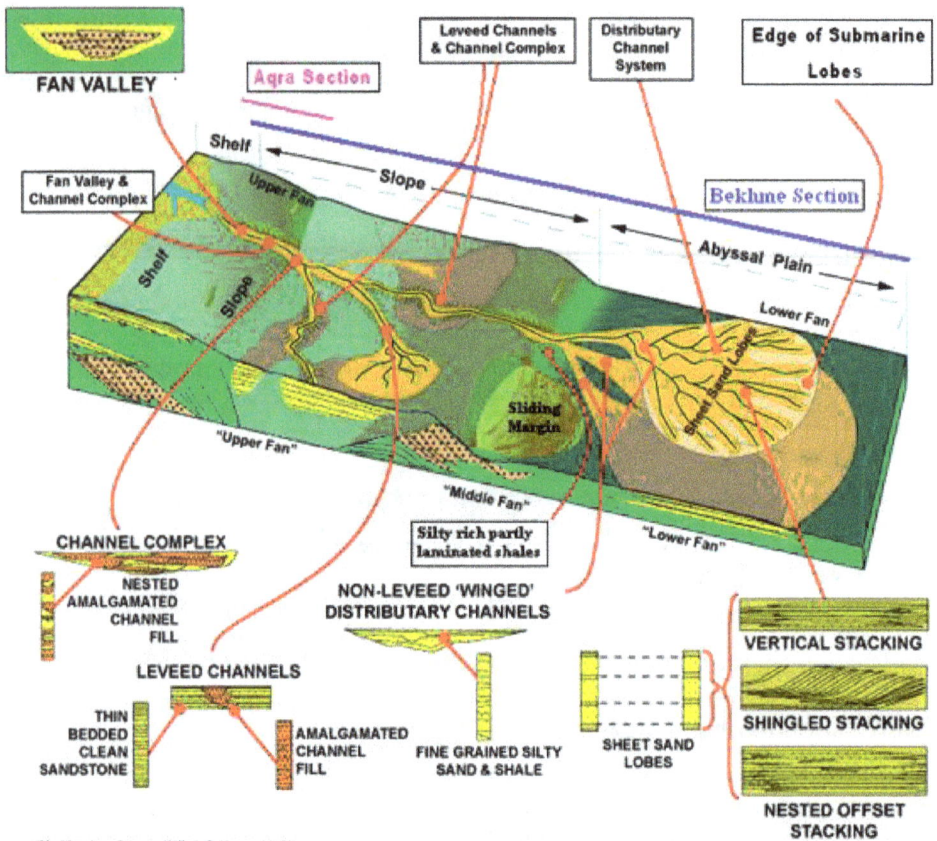

Fig. 7. Mud-rich, submarine fan model (modified from Bouma, 1997 and DeVay et al., 2000)

The lower boundary of Kolosh Formation in Bekhme section was mentioned as a disconformity K/T boundary with the underlying Tanjero Formation. It is followed by 2m thick of sandstones facies (K4) contains gravel grains of shallow marine deposit bypasses the middle and outer shelf and ended up in the outer fan imprinting the usual turbidite sole marks. (Al-Qayim and Salman, 1985 and Bouma, 2000).

Grey to olive green thick marl beds facies (K3) bearing fine silt and planktonic foraminifera deposited in continental slope at the edge of submarine fan lobes, where the turbidite current speed diminished and the foraminifera rainfall dominated, these unit continue until another influx of turbidite sandstone (Facies K4) is introduce or carbonate thin bed (Facies K1) where deposited as a result of short- lived phase of pelagic sedimentation dominated, when the sea level relatively stagnate and the turbidity current laterally switching to built a new lobs leaving the previous lobes area with carbonate production conditions. The cycles of marl and carbonate beds represent the lowest part of the outer fan (Figure 7).

Interfingering lens of Sinjar Formation was identify in the top of marl – limestone cycles, ranging in thickness between (0.2m – 11m), it consist of boundstone facies (S1) and rudstone facies (S2) their depositional environment point to organic build up - fore reef deposit within marine slope.

Four cycles of Bouma sequence are overlying the lowest part, these cycles deposited by turbidity current that bypasses the mid fan and deposited their bedload as a sheet sands or depositional lobes (Bouma, 2000). Each cycle commence with deposition of sandstones (facies K4), which consist of massive badly sorted sandstone were representing *Ta* interval of Bouma sequence (Bouma *et al.*, 1985). *Td* and *Te* interval can be found overlaying *Ta* interval directly, *Tb* and *Tc* interval are difficult to identify or may be eroded by the next gravity flow coming from the same distributary, eroded parts of the underlying deposit (Bouma *et al.* 1985 and Bouma, 2000). A number of layers are stack on top of one another as a sheet sand layers this part of the section representing the lowest part of the mid fan and the upper part of the outer fan (Figure 7). (Reading, 1986; Bouma, 2000; Kirschner and Bouma, 2000 and Nichols, 2009).

Seventeen meter of shale beds overlying the package of sandstones layers, it is silty rich partly laminated shale beds (Facies K5) embrace thin siltstones - fine sandstones characterized by parallel lamination and highest porosity, which is generally well cemented, the sand to shale ratio of the shale beds is low, and the carbonate material of the sandstone beds were ranging between 32 - 44% of the total content, it make the thin siltstones-fine sandstones hard and easy recognized in the field. A block of coral bearing limestone bed of 0.8m. thick and 4m. width (Figure 8) is embedded within the shale facies. It is sliding from the upper part of the shelf as a result of active turbidity current and storms originate from the tectonic earthquake. The shales – sandstones layers deposited in the mid fan as levee and over bank deposit (Figure 7). (Darling and Sneider, 1992; Basu and Bouma, 2000; Bouma, 2000; Steling *et al.*, 2000 and Nichols, 2009).

The upper succession of Kolosh Formation in Bekhme section embraces poorly fossiliferous marl facies (K3-2) characterized by low P/ B ratio, benthonic forams represented *Lokhartia spp.*, *Rotalia spp.*, *eponids spp. and Quinqueloculina spp.* These genera were indicated to shallow marine environment (inner – middle shelf) (Berggren, 1974; Petters,1978 and Leckie and Olson, 2003), The present of *lokharitia* point to regional regression during late Paleocene

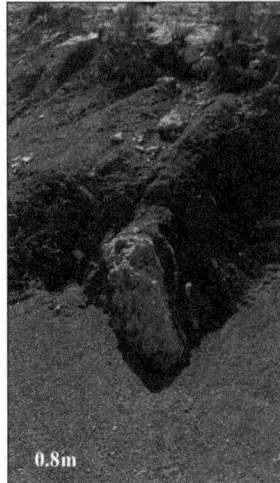

Fig. 8. Sliding coral bearing limestone bed in the upper part of shale facies.

(Berggren, 1974). Marl facies embedded a layer of conglomerate facies (K6) contains intrabasional (preigenetic) clasts. It is interpreted to represent the distal toe deposit of flows debouching from a main feeder canyon at the foot of the slope (Browne *et al.*, 2000). The sedimentological evidences of the upper succession indicated to the upper fan sediment (Figure 7).

The Kolosh formation in Aqra section is 40m thick, The variable thicknesses of the two sections are due to the bathymetry of the Kolosh basin which is a result of the inherited anticlines of Cretaceous age. Aqra section consists of alternated thick marl facies (K3-2), it is very poorly fossils, representing by few bad preserved benthonic forams only, and pelloidal lime packstone facies (K2)), the succession is pale grey color. The sedimentological and biological attributes indicated to semirestricted platform (Flugel, 2004) and represented upper fan deposit.

10. References

Abawi, T. S. and Abdo, G. S., (2001) Biostratigraphy of Aaliji Formation (Paleocene – Eocene) in Jumbor well 18- Northren Iraq. Rafidain Journal of Science, Vol. 12, No. 1, pp. 60 – 69.

Abdel-kireem, M. R., (1986) Contribution to the stratigraphy of the upper Cretaceous and the lower Tertiary of the Sulaimaniah – Dokan region, Northern Iraq. N. Jb. Geol. Paleont. Abh., Vol. 172, No. 1, pp. 121 - 139.

Al-Mutwali, M. M., (2001) Palaeocene – Early Eocene benthonic foraminiferal biostratigraphy and paleoecology of Kolosh Formation Shaqlawa area, Northeast Iraq. Iraqi Journal of Earth Science, Vol. 1, No. 2, pp. 12 – 24.

Al-Mutwali, M. M. and Al-Wazan, A. M., (2010) Planktonic foraminiferal biostratigraphy of Kolosh Formation (Paleocene) in Dohuk area North Iraq. Iraqi Journal of Earth Science, Vol. 10, No. 1, pp. 1-22.

Al-Omari, F. J., Al-Radwani, M. A. and Al-Mutwali, M. M., (1993) Biostratigraphy of Kolosh Formation at Shaqlawa area, North eastern Iraq.Journal of the Geological Society of Iraq, Vol. 21, No. 2, pp. 91-104.

Al-Omari, F. S., (1995) Globorotalia uncinata Zone (Lower - Middle Paleocene), Rafidain Journal of Sinece, Vol. 6, No. 1, pp. 64 – 76.

Al-Qayim, B. and Salman, L., (1986) Lithofacies analysis of Kolosh Formation, Shaqlawa area, North Iraq, Journal of the Geological Society of Iraq, Vol. 19, No. 3, pp. 107 – 117.

Arenillas, I., Arz, J. A. and Molina, E., (2004) A new high - resolution planktonic foraminiferal zonation and subzonation for the Lower Danian. Lethaia, Vol. 37, pp. 79 - 95.

Arenillas, I., Arz, J. A. and Molina, E. and et al., (2006) Chicxulub impact event is Cretaceous / Paleocene boundary in age: new micropaleontological evidence. Earth and Planetary Science Letters, Vol. 249, pp. 241 - 257.

Basu, D., and A. H. Bouma, (2000), Thin-bedded turbidites of the Tanqua Karoo: physical and depositional characteristics, in A. H. Bouma and C. G. Stone, eds., Fine-grained turbidite systems, AAPG Memoir 72/SEPM Special Publication 68, pp. 263–278.

Berggren, W. A., (1974) Paleocene benthonic foraminiferal biostratigraphy, biogeography and paleoecology of Libya and Mali. Micropaleontology, Vol. 20, No. 4, pp. 449 – 465.

Berggren, W. A., Kent, D. V., Swisher, C. C., and Aubry, M. P., (1995) A revised Cenozoic geochronology and chronostratigraphy, In: Berggren, W., Kent, D., Aubry, M.-P., and Hardenbol, J. (eds.), geochronology, time scales and global stratigraphic correlations, Society for Sedimentary Geology, Tusla, Okla., Special Publication, Vol. 54, pp. 129 - 212.

Berggren, W. A. and Pearson, P., (2005) A revised tropical to subtropical Paleogene planktonic foraminiferal zonation. Journal of Foraminiferal Research, Vol. 35, No. 4, pp. 279 - 298.

Bilotte, M., Bruxelles, L., Canerot, J., Laumonier, B., Coincon, R. S., (2007) Comment to " Latest - Cretaceous / Paleocene Karsts with marine infillings from Languedoc (South of France); paleogeographic, hydrogeologic and geodynamic implication by P. J. Combes et al.". Geodinamica Acta, Vol. 20, No. 6, pp. 403 - 413.

Blow, W. H., (1979) The Cainozoic Globigerinida, Laiden Brill, The Netherlands, Vol. I, II, III, 1413 P.

Boggs, S. Jr., (2006) Principles of sedimentology and stratigraphy.4th ed., Pearson Prentice Hall, Upper Saddle River, New Jersey, 662 P.

Bolli, H.M., (1966) Zonation of the Cretaceous to Pliocene marine sediments based on planktonic foraminifera. Boletin Informativo Asociacion Venezolana de Geologia, Mineria y Petroleo. Vol. 9, No. 1,pp.3 - 32.

Bouma, A. H., Normark, W. R. and Barnes, N. E., (1985) Submarine fans and related turbidite systems: New York, Springer-Verlag, 351 P.

Bouma, A. H., (2000) Fine – grained, mud – rich turbidite systems: model and comparisons with coarse – grained, sand – rich systems, In: Bouma, A. H. and Stone, C. G., (eds.) Fine – grained turbidite systems, American Association of Petroleum Geologists, Memoir 72/ SEPM Special Publication No. 68, Tulsa, Oklahoma, U.S.A., pp. 9 - 20.

Browne, G. H., Slatt, R. M. and King, P. R., (2000) Contrasting styles of basin-floor fan and slope fan deposition: Mount Messenger Formation, New Zealand. in A. H. Bouma and C. G. Stone, eds., Fine-grained turbidite systems, AAPG Memoir 72/SEPM Special Publication 68, pp. 143–152.

Chacon, B. and Martin-Chivelet, J., (2005) Majer palaeoenvironmental changes in the Campanian to Palaeocene sequence of Caravaca (Subbetic zone, Spain). Journal of Iberian Geology, Vol. 31, No. 2, pp. 299 - 310.

Darling, H. L., and Sneider, R. M., (1992) Production of low resistivity, low contrast reservoirs offshore Gulf of Mexico Basin: Gulf Coast Association of Geological Societies Transactions, Vol. 42, pp. 73–88.

Dunnington, H. V., (1958) Generation, migration, accumulation and dissipation of oil in Northern Iraq. Reprint in GeoArebia (2005), Vol. 10, No. 2, pp. 39 - 84

Emery, D. and Myers, K. J., (2006) Sequence Stratigraphy. Blackwell Science Ltd., 297 P.

Flugel, E. (2004) Microfacies of carbonate rocks, analysis, interpretation and application, Springer, Berlin, 976 P.

Folk, R. L., (1974) Petrology of sedimentary rocks. Hemphill Pub. Comp., Texas, 128 P.

Gibson, T. G., (1989) Planktonic benthonic foraminiferal ratios: modern patterns and tertiary applicability, Marine Micropaleontology, Vol. 15, pp. 29 – 52.

Gradstein, F. M., Ogg, J. G., Smith, A. G., Bleeker, W. and Lourens, L. J., (2004) A new Geologic time scale, with special reference to Precambrian and Neogene. Episodes, Articles, Vol. 27, No. 2, pp. 83 – 100.

Ingersoll, R. V., Bdullard, T. F., Ford, R. L., Grimm, J. P., Pickle, J. D. and Sares, S. W., (1984) The effect of grain size on detrital modes : A test of Gazzi – Dickinson point - counting method. Journal of Sedimentary Petrology, Vol. 45, pp. 103 – 116.

Jassim, S. Z. and Buday, T., (2006) Middle Paleocene – Eocene megasequence, In: Jassim, S. Z. and Golf, J., C.,(2006) Geology of Iraq. Published By Dolen, Prague and Moravian Museum, Brno., pp. 155 - 168.

Jassim, S. Z., & Goff, J., C., (2006) Geology of Iraq. Published By Dolen, Prague and Moravian Museum, Brno., pp. 155 - 168.

Kassab, I., Al-Omari, F. S. and Al-Safawee, N. M., (1986) The Cretaceous – Tertiary boundary in Iraq (Represented by the subsurface section of Sasan Well No. 1.N. W. Iraq). Journal of the Geological Society of Iraq, Vol. 19, No. 2, pp. 129 - 167.

Keller, G., Li, L. and Macleod, N., (1995) The Cretaceous / Tertiary boundary stratotype section at El Kef, Tunisia: how catastrophic was the mass extinction ?. Palaeogeography, Palaeoclimatology, Palaeoecology, Vol. 119, pp. 221 - 254.

Keller, G., Adatte, T., Tantawy, A. A., Berner, Z., Stinnesbeck, W., Stueben, D. and Leanza, H. A., (2007) High stress late Maastrichtian – early Danian palaeoenvironment in the Neuquen Basin, Argentina. Cretaceous Research, Vol. 28, pp. 939 - 960.

Keller, G., Khosla, S. C., Sharma, R., Khosla, A., Bajpai, S., and Adatte, T., (2009) Early Danian planktic foraminifera from Cretaceous-Tertiary intertrappean Beds at Jhilmili, Chihindwara district, Madhya Pradesh, India. Journal of Foraminiferal Research, Vol. 39, No. 1, pp. 40 - 55.

Kirschner, R. H. and Bouma, A. H., (2000) Characteristics of a distributary channel – levee – overbank system, Tanqua Karoo, In: Bouma, A. H. and Stone, C. G., (eds.) Fine – grained turbidite systems, American Association of Petroleum Geologists, Memoir 72/ SEPM Special Publication No. 68, Tulsa, Oklahoma, U.S.A., pp. 233 – 244

Leckie, R. M. and Olson, H. C., (2003) Foraminifera as proxies for Sea – level change on siliciclastic margins. SEPM (Society for Sedimentary Geology), Special Publication, No. 75, pp. 5 - 19.

Li, L. and Keller, G., (1998) Maastrichtian climate, productivity and faunal turnover in planktic foraminifera in south Atlantic DSDP Sites 525 and 21. Marine Micropaleontology, Vol. 33, pp. 55 - 86.

Molina, E., Alegret, L., Arenillas, I. and Arz, J. A., (2005) The Cretaceous/ Paleogene boundary at the Agost section revisited: paleoenvironmental reconstruction and mass extinction pattern. Journal of Iberian Geology, Vol. 31, No. 1, pp. 135 - 148.

Munim, A., (1976) Upper Cretaceous and lower Tertiary foraminifera of North Iraq, Dohuk area, Ustredaui, Ustav Geol. Praha. S. O. M. Unpublished Report, Baghdad. pp. 1 - 157.

Nichols, G., (2004) Sedimentary and stratigraphy, Blackwell Publishing Company, UK, 355 P.

Nichols, G., (2009) Sedimentology and stratigraphy, 2nd ed., Wiley- Blackwell Publishing Company, UK, 419 P.

Numan, N. M., (1997) A plate tectonic scenario for the Phanerozoic succession in Iraq, Iraqi Geological Journal, Vol. 30, No. 2, pp. 85 – 110.

Olsson, R. K., Hemleben, C., and Berggren, W. A., (1999) Atlas of Paleocene planktonic foraminifera, Smithsonian Contributions to Paleobiology, No. 85, 252 P.

Petters, S. W. (1978) Maastrichtian – Paleocene foraminifera from NW Nigeria and their paleogeography. Acta Palaeontologica polonica, Vol. 23, No.2, pp. 131 – 154.

Pettijhon, F. J., Potter, P. E. and Siever, R., (1973) Sand and sandstone. Spring – Verlag, New York, 618 P.

Pettijhon, F. J. (1975) Sedimentary rocks, (3rd ed). Harper and Row, New York, 628 P.

Phillip, J. M., Roger, J., Vaslet, D., Cecca, F., Gredin, S. and Memesh, A. M. S., (2002) Sequence Stratigraphy, Biostratigraphy and Paleontology of the (Maastrichtian – Paleocene) Aruma Formation in outcrop in Saudi Arabia. GeoArabia, Vol. 7, No. 4, pp. 699- 718.

Randazzo, A. F. and Zachos, L. G., (1984) Classification and description of dolomite fabrics of rocks from the Floridian aquifer. Sedimentary Geology, Vol. 37, pp. 151 – 162.

Reading, H. G., (1986) Sedimentary environments and facies. Blackwell Scientific Pub. 616 P.

Sharbazheri, Kh. M., Ghafor, I. M. and Muhammed, Q. A., (2009) Biostratigraphy of the Cretaceous/ Tertiary boundary in the Sirwan Valley (Sulaimani Region, Kurdistan) NE-Iraq. Geologica Carpathica, Vol. 60, Issue 5, pp. 381 – 396.

Sharland, P. R., Archer, R., Casey, D. M., and et al., (2001) Arabian plate sequence Stratigraphy, GeoArabia, Special Publication 2, Gulf PetroLink, Manama, Bahrain. 371 P.

Schlager, W., (1992) Sedimentology and sequence stratigraphy of reefs and carbonate platforms. Continuing education course notes, American Association of Petroleum Geologists, Vol. 34, 71 P.

Stelting, C. E., Bouma, A. H. and Stone, C. G., (2000) Fine - grained turbidite systems: Overview, In: Bouma, A. H. and Stone, C. G., (eds.) Fine – grained turbidite systems, American Association of Petroleum Geologists, Memoir 72/ SEPM Special Publication No. 68, Tulsa, Oklahoma, U.S.A., pp. 1- 7.

Toumarkine, M. and Luterbacher, H. P., (1985) Paleocene and Eocene planktic foraminifera. In: Bolli, H. M., Saunders, J. B. and Perch-Nielsen, K. (eds), Plankton Stratigraphy. Cambridge Earth Science Series, pp. 87 - 154.

Van Bellen, R. C.; Dunnington, H. V.; Wetzel, R. and Morton, D. M., (1959) Iraq. lexique stratigraphique international Center National de la Recherche Scientifique, III, Asie, Fasc. 10a,Paris, 333 P.

Van Vliet, A., (1978) Early Tertiary deepwater fans of Guipuzcoa, Northern Spain, In: Stanley, D. J. and Kelling, G.,(eds) Sedimentation in submarine canyons, fans, and trenches. Dowden, Hutchinson & Ross, Inc. U.S.A., pp. 190 - 209.

Wilson, J. L., (1975) Carbonate facies in geologic history, Springer – Verlag, Berlin, 471 P.

Yan, Z., Wang, Z., Wang, T., Yan, Q., Xiao, W. and Li, J., (2006) Provenance and tectonic setting of clastic deposits in the Devonian Xicheng basin, Qinling orogen. central China, Journal of Sedimentary Research, Vol. 76, pp. 557 – 574.

Section 4

Tectonostratigraphy

Tektono-Stratigraphy as a Reflection of Accretion Tectonics Processes (on an Example of the Nadankhada-Bikin Terrane of the Sikhote-Alin Jurassic Accretionary Prism, Russia Far East)

Igor V. Kemkin
Far East Geological Institute,
Far Eastern Branch of Russian Academy of Sciences,
Russia

1. Introduction

The Sikhote-Alin folded belt is one of the Asia eastern margin regions, where terranes of ancient accretionary prisms are widely distributed. Within this region there are fragments, as a minimum, of three such different-aged prisms - Jurassic, Early Cretaceous and Middle Cretaceous (fig. 1). Accretionary prisms are complicate-dislocated sedimentary complexes that are formed at the basis of continental or island-arc slopes as a result of accretion of fragments of the sedimentary cover and morphologically positive structures of an oceanic plate during its subduction. The subduction of an oceanic plate under a continent or an island arc is commonly accompanied by intense deformation of the sedimentary cover deposits. At first, in the frontal part of the slope basis the upper part of a sedimentary cover (trench turbidite) under the action of "bulldozer effect" (off-scraping) repeatedly imbricate, forming a series of tectonic slices of terrigenous composition (fig. 2). Secondly, pelagic and hemipeladic sediments located below of an imbrication zone, plunging into a subduction zone rumple under the action of simple shear (effect of drag folding) into the different-amplitude and disharmonious overturned folds, which axial planes are inclined towards a trench. Crumpling of sediments continues until ultimate strength limit of rocks is reached and the ruptures are appearing. Further along these ruptures numerous underplating and doubling of a primary cut-section of an oceanic plate cover occur that results in formation of imbricate-underthrusted structure of the accreted rocks (Hashimoto & Kimura, 1999; Kimura, 1997; Kimura & Mukai, 1991; Moore & Byrne, 1987; Seely et al., 1974; Sokolov, 1992, 1997; Sokolov et al., 2001; etc.). For this reason the accretionary prisms are a tectonic package of repeatedly alternating tectonic slices consisting of marine (pelagic and hemipelagic deposits and also fragments of seamountns), oceanic-margin (turbidites), and chaotic (melange and olistostromes) formations. Structure of ancient accretionary prisms is even more complex because of a numerous post-accretion deformations such as thrusts and strike-slip faults.

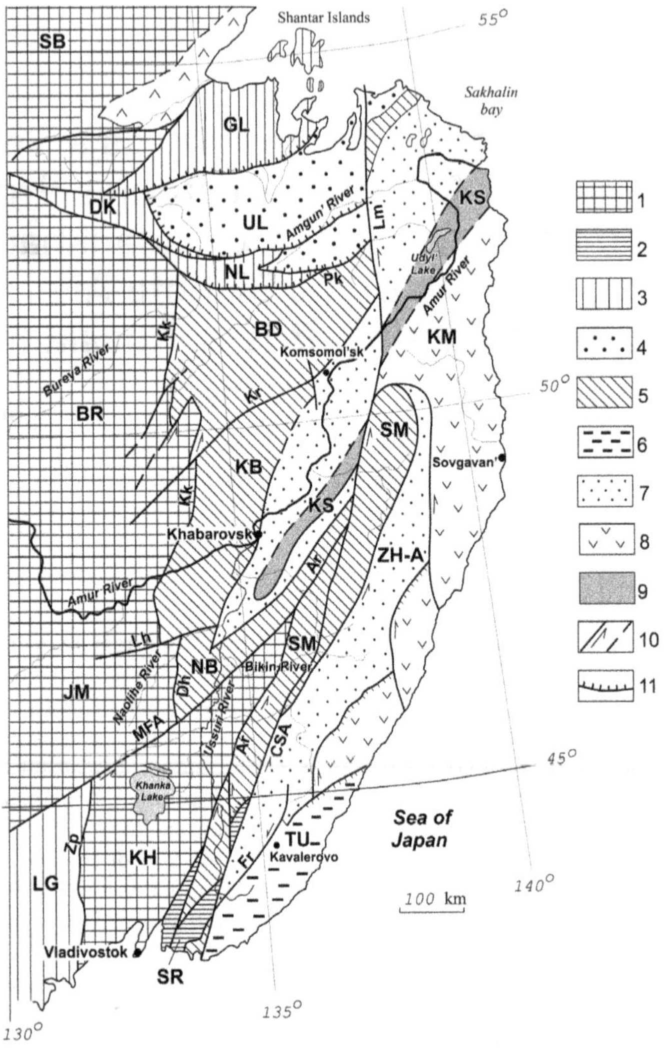

1 – Early Paleozoic continental blocks: Bureya (BR)–Jiamusi (JM)–Khanka (KH) superterrane and Siberian craton (SB); 2 – Fragment of an Early Paleozoic continental margin - Sergeevka terrane (SR); 3 – Permian-Triassic accretionary prisms – Dzhagdy-Kerbin (DK), Nilan (NL), Galam (GL), Laoelin-Gradek (LG) terranes; 4 – Jurassic turbidite basin - Ulban and Un'ya-Bom terranes (UL); 5 – Jurassic accretionary prism – Samarka (SM), Nadankhada-Bikin (NB), Khabarovsk (KH) and Badzhal (BD) terranes; 6 – Tithonian-Hauterivian accretionary prism - Taukha terrane (TU); 7 – Early Cretaceous turbidite basin - Zhuravlevka-Amur terrane (ZH-A); 8 – Hauterivian-Albian island arc – Kema (KM) terrane; 9 – Hauterivian-Albian accretionary prism - Kiselevka-Manoma (KS) terrane; 10 – left-lateral strike-slip faults; 15 – thrusts. Faults: Lm- Limourchansky, Pk – Paoukansky, Kk – Koukansky, Kr – Koursky, MFA - Misha-Fushung-Alchan, Lh – Lyaolihe, Dh – Dahezhen, Ar - Arsen'evsky, CSA – Central Sikhote-Alin, Fr – Fourmanovsky, Zp – Zapadno-Primorsky.

Fig. 1. The tectonic scheme of Sikhote-Alin region and adjoining areas (After Khanchuk, 2006 with additions)

It is obvious, that the absence of normal sedimentary contacts between various lithological groups of rocks in such tectono-sedimentary complexes complicates the decision of a number of geological tasks such as study of stratigraphic sequence of deposits, clearing-up structure of an area of accreted formations distribution, its geological evolution, etc. Nevertheless, in some individual tectonic slices, the fragments of primary consecutive section of pelagic to hemipelagic or hemipelagic to oceanic-margin formations are observed. Numerous results of microfaunistic researches of such fragments (Kemkin, 1996; Kemkin & Golozubov, 1996; Kemkin & Kemkina, 1999; Kemkin & Khanchuk, 1992; Kemkin & Khanchuk, 1993; Kemkin & Philippov, 2002; Kemkin & Rudenko, 1998; Kemkin & Taketani, 2008; Kemkin et al., 1999; Kirillova, 2002; Philippov et al., 2000; Philippov et al., 2001; Philippov & Kemkin, 2007; Zyabrev, 1998; Zyabrev & Matsuoka, 1999, etc.) have allowed us to reconstruct primary cut-sections of accreted paleooceanic deposits. Their lowermost part is composed by pelagic cherts which are gradually replaced by hemipelagic cherty-clayey formations, and, further, by terrigenous rocks. Such sequences of deposits named by Oceanic Plate Stratigraphy sequences (Berger & Winterer, 1974; Isozaki et al., 1990; Wakita & Metcalfe, 2005, etc.) reflect the history of sedimentary process on an oceanic plate during its drift from the spreading zone to the subduction zone. Each lithological group of these sequences is very informative. For example, cherts characterize a history and features of pelagic sedimentation. Hemipelagic deposits (siliceous mudstones, mudstones and aleuroargillite) fix the moment of the approach of a foremost site of an oceanic plate to convergent border. The terrigenous rocks, which accumulation occurred in a trench, are the indicator of the time of beginning of immersing of an oceanic plate into subduction zone and, accordingly, of subsequent accretion of fragments of its sedimentary cover.

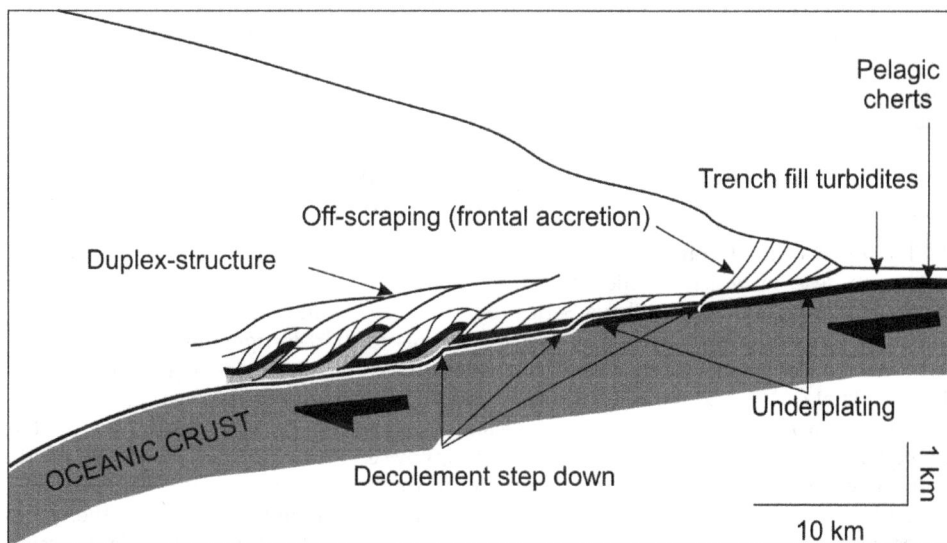

Fig. 2. Model of an accretionary prism structure (After Hashimoto & Kimura, 1999)

Knowing the age of rocks of such Oceanic Plate Stratigraphy sequences in various tectonic slices of the accretionary prism, it is possible to specify the time when individual oceanic fragments were accreted. These date give us the base to subdivide the accretionary prism to several tectono-stratigraphic units responding to certain stages of the prism formation. Subsequent mutual correlation and comparison of the allocated tectono-stratigraphic units allows to reconstruct a succession of the accretion of an oceanic plate fragments and to specify the prism structure as a whole.

Thus, a study of the accretionary prisms is very important as for specification of geological structure and evolution of regions composed of such prisms, as for elucidation of specific features of the accretion process in different areas of a convergent boundary, as well as for correlation of geological events at a junction zone of lithospheric plates, and for reconstruction of a geodynamic evolution of continental margins along which the accretionary prisms were formed.

One of these ancient prisms is a Jurassic accretionary prism, which was being formed during more than 70 m.y. by the consecutive accretion of different-aged and different-facies formations of Paleo-Pacific under the Paleo-Asian continent east margin. In the Sikhote-Alin region, the Jurassic prism is represented by four terranes: Samarka, Nadankhada-Bikin, Khabarovsk and Badzhal.

This article presents results of lithological-biostratigraphic studying of chert-terrigenous formations of the Nadankhada-Bikin terrane of the Sikhote-Alin Jurassic accretionary prism, which allow allocating in terrane structure three different-aged tectono-stratigraphic complexes reflecting successive stages of its formation.

2. Regional tectonic position and the previous researches

The Nadankhada-Bikin terrane is located in the lower reaches of the Ussuri River, in the area between the Black River and the Naolihe River mouths (fig. 3). It extends for about 350 km in a northeast direction along the northwest edge of the Bureya-Jiamusi-Khanka superterrane and is about 60 km in width. The terrane is separated from the Bureya-Jiamusi-Khanka superterrane by Dahezhen fault in the west and by Misha-Fushung-Alchan fault in the south. The Lyaolihe and the Arsen'evsky faults separate it from the Khabarovsk and the Samarka terranes, respectively. The Ussuri River valley divides this terrane into two parts: Nadankhada, located in China, and Bikin, located in Russia.

The Bikin part of the terrane is composed mainly (fig. 4 A) of cherts and terrigenous rocks and much less of volcanogenic formations that had formerly been referred to the Carboniferous-Permian Samurskaya Series (Bersenev, 1969), and later, after the findings of Triassic conodonts in the cherts, to the Triassic-Jurassic chert-terrigenous suite and Early Cretaceous volcanogenic-terrigenous Koultoukha suite (Liht, 1997). A structure of this area of the Sikhote-Alin was considered as a multiple sedimentary alternation of cherts, terrigenous rocks and rare layers of volcanites. However, it was established by the subsequent researches (Phillippov, 1990; Phillippov & Kemkin, 2004) that the cherts compose generally the stratum-like bodies occurring among terrigenous deposits and limited by faults. The results of microfaunistic study have shown that the age of the cherts ranges from Middle Triassic to Early Jurassic (in some cut-sections up to Middle Jurassic), and the age of the terrigenous rocks is Middle-Late Jurassic (up to Early Berriasian in some cut-sections). These data convincingly testify that repeated alternation of Triassic cherts and Jurassic terrigenous rocks can not be sedimentary. It is result of tectonic recurrence. As a whole, the structure of the Bikin part of the terrane represents a package of repeated alternation of tectonic slices of cherts and terrigenous rocks.

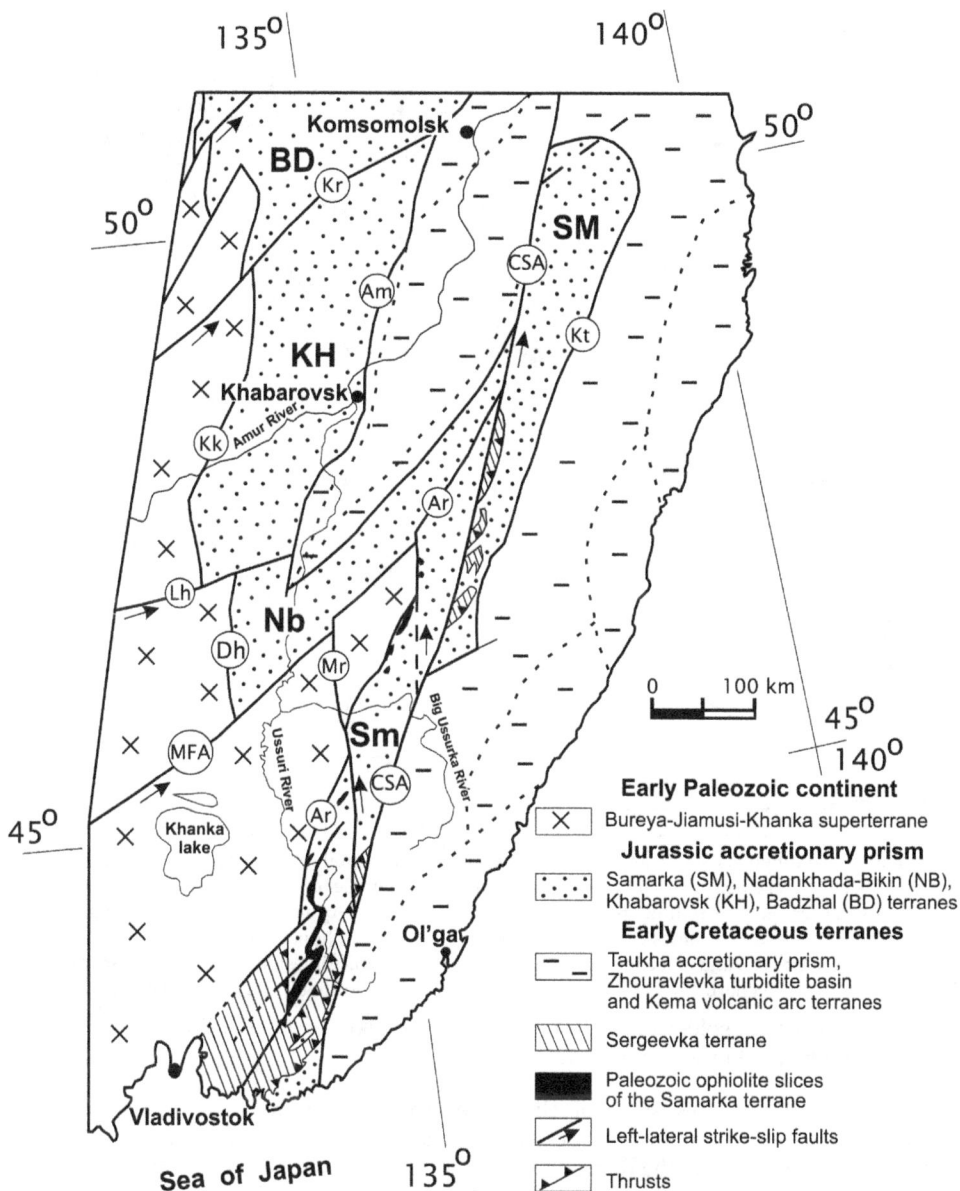

Fig. 3. Terranes of Jurassic accretionary prism (After Kemkin & Philippov, 2001 with additions)

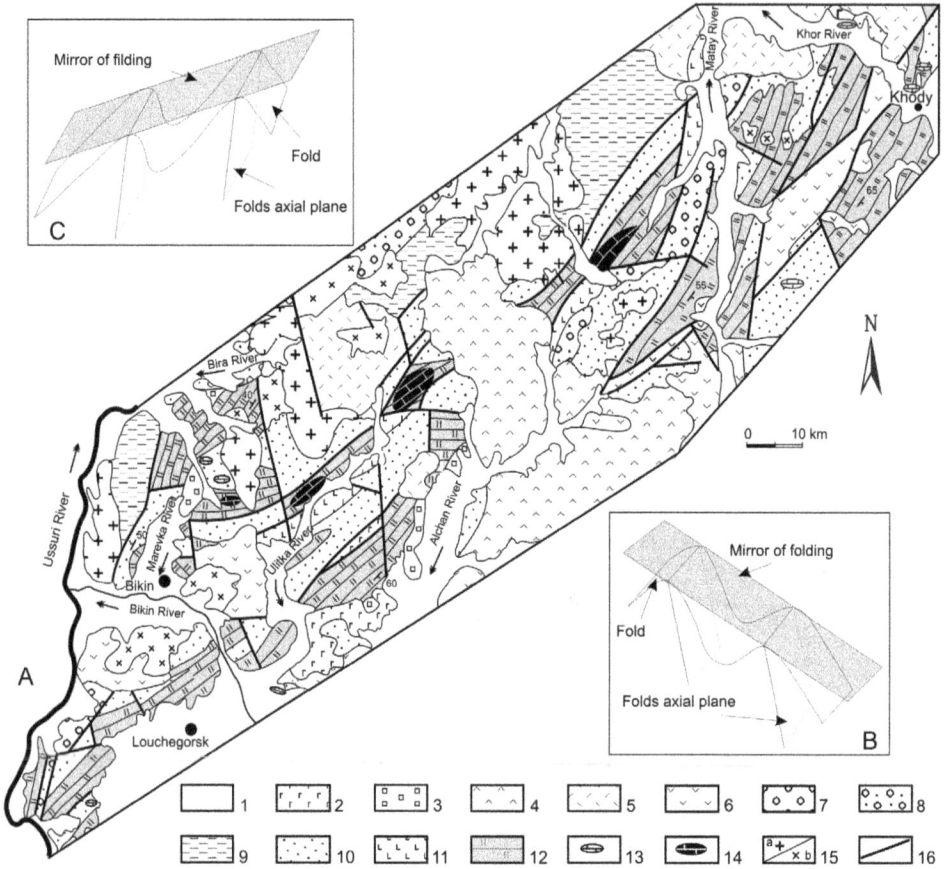

1 - Quaternary sediments; 2 - Neogene basalt; 3 - Cenozoic continental terrigenous rocks; 4 - Paleogene andesite, 5 - Late Cretaceous rhyolite and rhyolitic tuff, 6 - Albian-Cenomanian basalt and andesite; 7 - 9 - shallow-water terrigenous sediments of different ages: Albian (7), Aptian-Albian (8), Berriasian (?)-Valanginian (9); 10 - 12 - volcanic-sedimentary rocks of Jurassic accretionary prism, dominated by clastic (10), volcanic (11), and siliceous (12) rocks; 13 - Late Paleozoic limestone blocks; 14 - Late Triassic limestone interbeds in chert; 15 - Early Cretaceous granite (a) and granodiorite (b); 16 - faults.

Fig. 4. The scheme of geological structure of the terrane Bikin part (After Philippov, 1990 and Pfilippov & Kemkin, 2004 with additions)

The rocks composing this stack of slices are crumpled in the compressed different-amplitude and asymmetric folds of northeast (on some sites submeridional) strike. In the central and eastern parts of the Bikin area the axial planes of folds have northwest vergence with a mirror of folding inclined on the southeast (fig. 4 B). In the western part, on the contrary, the folds axial planes are inclined on the southeast and the mirror of folding is gently immersing to the northwest (fig. 4 C).

The inversion of the folds axial planes in the central and eastern parts is caused by the flexure-like bending of the Jurassic prism deposits during the left-lateral motion of the Bureya-Jiamusi-Khanka superterrane block along the Misha-Fushung-Alchan fault (Khanchuk, 1994).

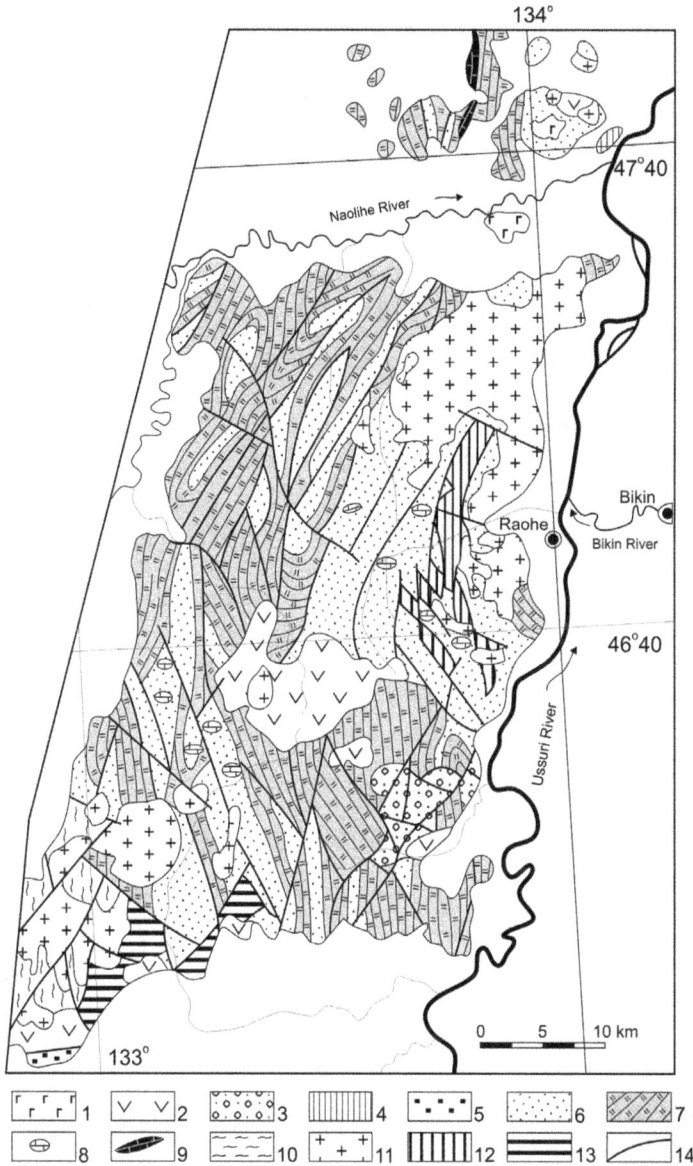

1 - Neogene basalt; 2 - Late Cretaceous rhyolite and rhyolitic tuff, 3 - 5 - shallow-water terrigenous
sediments of different ages: Aptian-Albian (3), Tithonian-Berriassian (4), Late Triassic (5); 6 - 7 -
sedimentary rocks of Jurassic accretionary prism, dominated by clastic (6) and siliceous (7) rocks; 8 -
Late Paleozoic limestone blocks; 9 - Late Triassic limestone interbeds in chert; 10 - metamorphic rocks;11
- Early Cretaceous granite; 12 - Early Mesozoic (?) ophiolite (Zhaohe complex); 13 - Middle Paleozoic
ophiolite (Dahezhen complex); 14 - faults.

Fig. 5. The scheme of geological structure of the terrane Nadankhada part (After Shao et al.,
1992 with additions)

The similar situation takes place in the Nadankhada part of the terrane (so-called Nadankhada ridge), which occupies a territory between the Naolihe, Tsikhulinhe and Ussuri rivers (fig. 5). Up to middle 1980, this territory was considered to be composed of mainly Upper Paleozoic volcanogenic-chert-terrigenous formations (Li et al., 1979). Later, during the joint Chinese-Japanese researches, it has been established that the cherts contain Triassic and the siliceous shales contain Middle Jurassic radiolarians (Kojima & Mizutani, 1987; Mizutani et al., 1986).

According to the subsequent works (Mizutani et al., 1989, 1990; Shao et al., 1990, 1992 etc.), geological structure of the Nadankhada ridge was defined as a complex alternation of tectonic slices of terrigenous rocks and cherts crumpled in the asymmetric and overturned folds of mostly northeast strike. In the southwestern part of the Nadankhada area, the northeast orientation of the folds axes is gradually changing to the submeridional and further to the northwestern and sublatitudinal one (fig. 5).

3. Structure of the Nadankhada-Bikin terrane chert-terrigenous sequences and age of the chert-terrigenous formations

Deformations behavior of the Nadankhada-Bikin terrane deposits and spatial orientations of the major structural elements of rocks occurrence (i.e. a general dipping of layers, a vergence of the folds axial planes, a direction of the folding mirror inclination) are such that the tectonic slices of the lower structural level of the terrane are cropped out in its central and

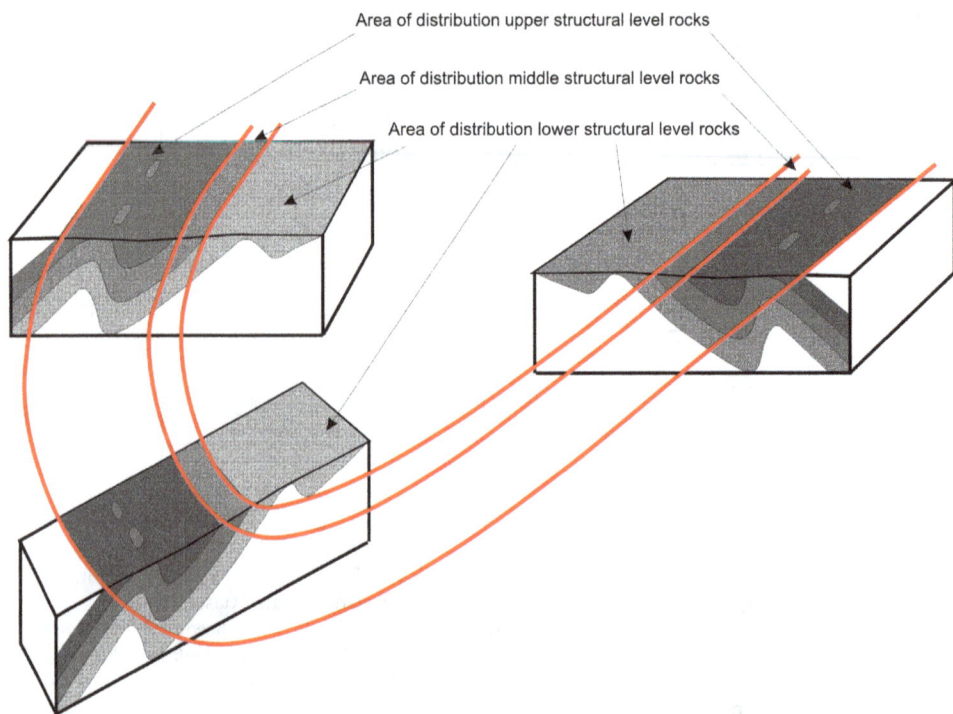

Fig. 6. Block-diagram illustrating distribution of rocks of various structural levels of the Nadankhada-Bikin terrane folded formations on a denuded earth surface

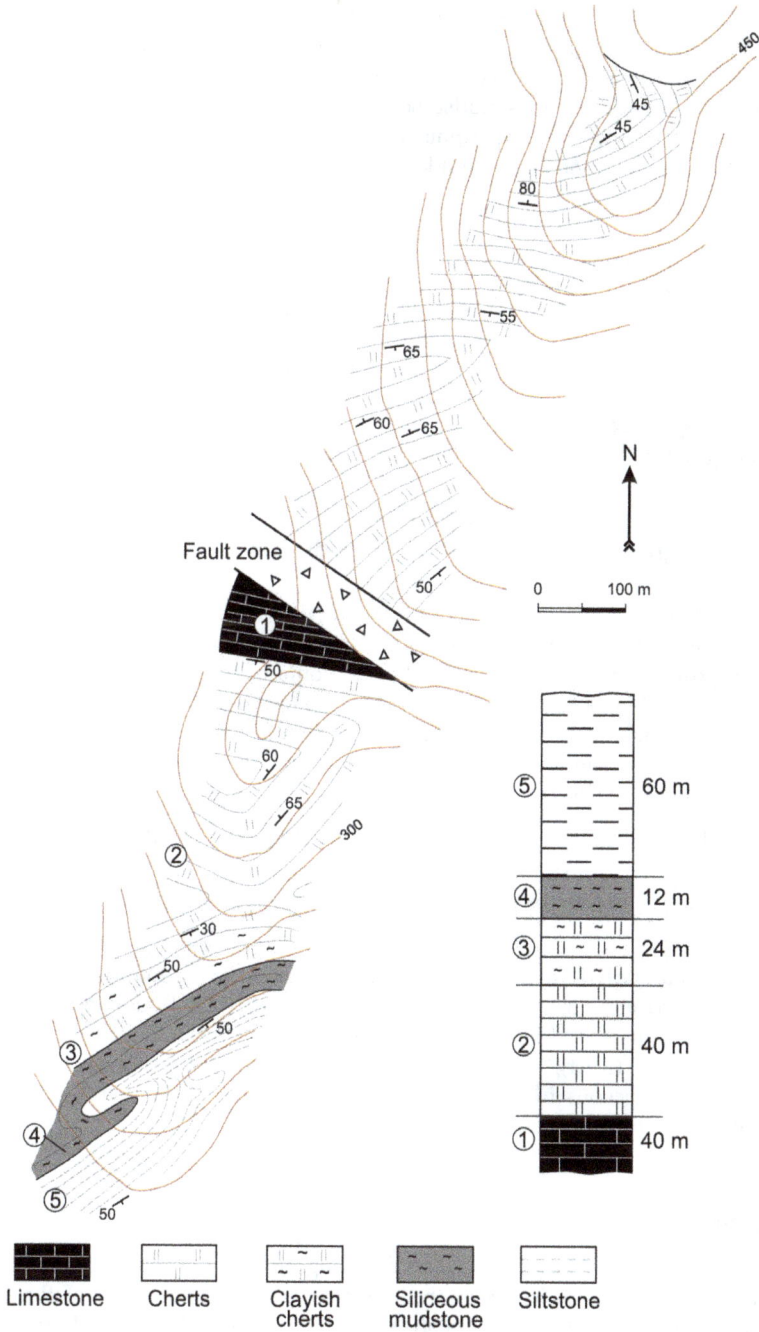

Fig. 7. Example of structure of the chert-terrigenous sequence fragment on the right bank of
the Ulitka River (After Philippov & Kemkin, 2004)

northeastern parts, and those of the upper level - in the southeastern, southwestern and western parts of the terrane (fig. 6). In some tectonic slices, fragments of primary cut-section of the paleooceanic plate sedimentary cover are observed, in which cherts are sequentially and gradually replaced by the siliceous mudstones and then by the terrigenous rocks (fig. 7). The results of lithological and biostratigraphic researches of such chert-terrigenous sequences fragments belonging to various structural levels of the terrane show some differences in their composition, structure and age.

3.1 Lower structural level

The most complete fragment of cut-section of the chert-terrigenous sequence has been investigated in the interstream area between the Ulitka and Zolotoy Klyuch Rivers (Phillippov & Kemkin, 2004). Here are observed (from the bottom to top):

1. Grey, light grey and pink-grey bedded cherts	60 m
2. Grey and light grey limestones and siliceous limestones	40 m
3. Grey and light grey bedded cherts	40 m
4. Grey and greenish-grey massive clayish cherts	24 m
5. Light greenish-grey slightly layered siliceous mudstones	12 m
6. Dark grey layered mudstones and silty mudstones	60 m
7. Dark grey siltstones with rare layers of fine-grained sandstones	20 m
8. Alternation of light grey fine-grained sandstones and dark grey siltstones, (rhythms from 20 to 40 cm). Sandstones contain layers of alkaline basalts and hyaloclastites (10-20 m, rarely up to 100 m). In the upper part of this pack, sandstones are replaced by chaotic formations represented by aleuro-psammites containing different-sized lumps, blocks and fragments of cherts and sandstones.	more than 400 m

According to the radiolarian and conodont fauna (Phillippov, 1990) the age of the cherts ranges from Middle Triassic (Anisian) to Middle Jurassic (Bathonian-Callovian). The limestones embedded in the middle part of the chert cut-section contain Late Carnian-Early Norian conodonts.

Oxfordian-Tithonian radiolarians *Archaeodictyomitra minoensis* (Mizutani), *Archaeodictyomitra* ex gr. *apiarium* (Rust), *Cinguloturris cylindra* Kemkin et Rudenko *Podobursa* sp., *Pseudodictyomitra primitiva* Matsuoka & Yao, *Pseudodictyomitra carpatica* (Lozyniak), *Spongocapsula perampla* (Rust), *Stichocapsa* ex gr. *cribata* Hinde, *Xitus* sp. and others have been extracted from the siliceous mudstones and silty mudstones (fig. 8), and Late Tithonian-Early Berriasian Buchia *Buchia* cf. *unshensis* (Pavl.), *B.* cf. *okensis* (Pavl.) has been found in the siltstones (Phillippov & Kemkin, 2004).

3.2 Middle structural level

Fragments of the chert-terrigenous sequence belonging to the middle structural level are exposed in the numerous outcrops in the interstream areas and banks of the Peshkova, Perepelinaya, Cheremshanka, Kamenushka, and Kedrovka Rivers (all are the right tributaries of the Ussuri River), as well in the road quarries. The reconstructed cut section looks as following:

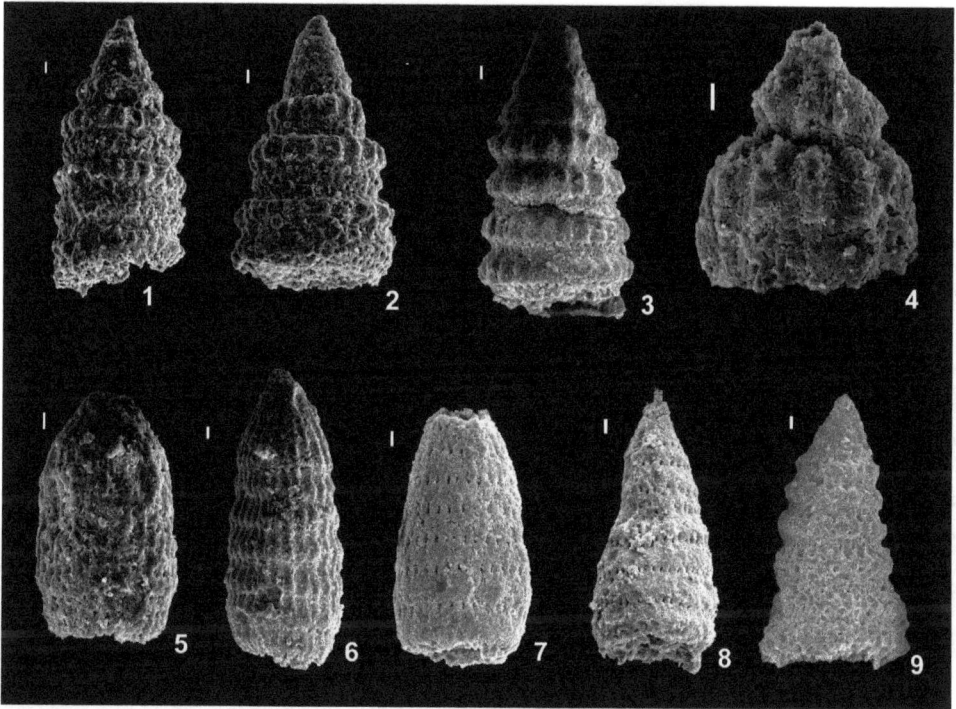

1-2. *Cinguloturris cylindra* Kemkin et Rudenko (sample Sv-3), 3. *Pseudodictyomitra carpatica* (Lozyniak)
(sample Sv-3), 4. *Eucyrtidiellum pyramis* (Aita) (sample Sv-3), 5. *Archaeodictyomitra* ex gr. *apiarium* (Rust)
(sample Sv-3), 6. *Archaeodictyomitra minoensis* (Mizutani) (sample Sv-3), 7 - 8. *Pseudodictyomitra* ex gr.
nuda Shaaf (sample Sv-3), 9. *Wrangellium* cf. *puga* (Schaaf) (sample Sv-5). All markers – 10 µк.

Fig. 8. Some Upper Jurassic radiolarians from silty mudstones of Lower structural level of
the Nadankhada-Bikin terrane Bikin part

1. Alternating grey and dark grey cherts (each bed 1-7 cm thick) and yellowish-grey siliceous mudstones (each bed 0.5-5 cm thick)	30 m
2. Grey bedded cherts	66 m
3. Grey and dark grey clayish cherts alternating in the upper part of the layer with thin layers of greenish-grey siliceous mudstones	22 m
4. Greenish-grey siliceous mudstones	8 m
5. Dark grey and black mudstones and silty mudstones	40 m
6. Dark grey siltstones with thin rare layers of fine-grained sandstones in the upper part of the layer	90 m

Based on numerous data of radiolarian analysis, the age of the cherts is established as
Middle Triassic (Anisian)-early Middle Jurassic (Aalenian).

The siliceous mudstones contain abundant radiolarians (fig. 9) *Archaeodictyomitra* ex gr.
rigida Pessagno, *Sethocapsa* cf. *funatoensis* Aita, *Stichocapsa* cf. *robusta* Matsuoka, *Xitus* sp.,
Stichocapsa cf. *convexa* Yao, *Protunuma turbo* Matsuoka, *Stichocapsa decora* Rust, *Dictyomitrella*
cf. *kamoensis* Mizutani et Kido, *Parvicingula* cf. *dhimenaensis* s.l. Baumgartner, *Sethocapsa* sp.,

Transhsuum cf. *brevicostatum* gr. (Ozvoldova), *Transhsuum* cf. *maxwelli* gr. (Pessagno), *Archicapsa* cf. *pachyderma* Tan, *Pantanellium* sp., *Tricolocapsa* sp., *Unuma* sp., *Eucyrtidiellum unumaensis* Yao, *Parahsuum officerence* (Pessagno et Whalen), *Cyrtocapsa mastoidea* Yao, *Stichocapsa japonica* Yao, *Unuma latusicostatus* (Aita), *Parvicingula* cf. *nanoconica* Hori et Otsuka, *Tricolocapsa* cf. *fusiformis* Yao, *Tricolocapsa* cf. *plicarum* Yao, *Podobursa* sp., *Mesosaturnalis* sp., *Hsuum* cf. *mirabundum* Pessagno et Whalen, *Unuma typicus* Ichikawa et Yao, *Napora saginata* Takemura, *Napora* sp., that indicate Bajocian age of host rocks.

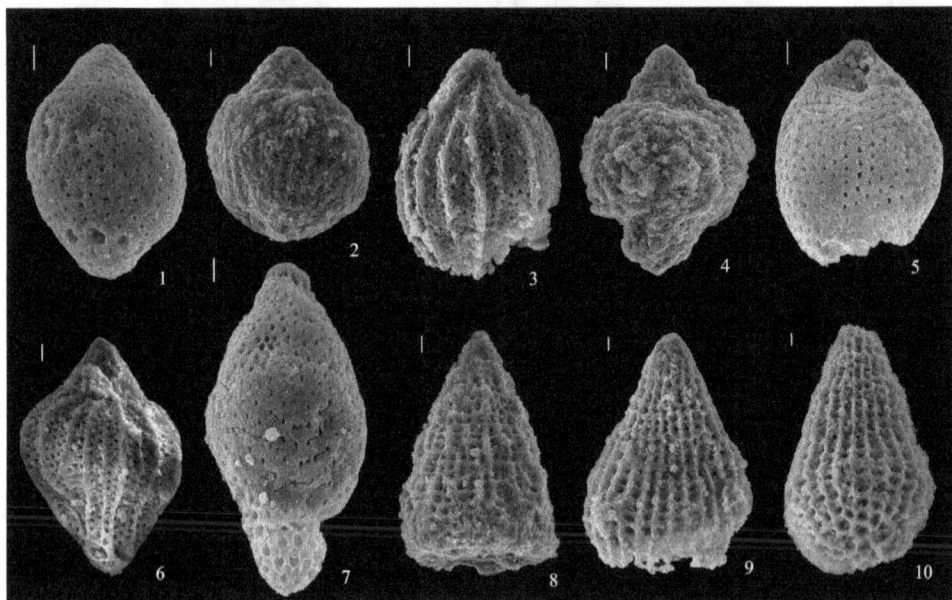

1. *Tricolocapsa fusiformis* Yao (sample Khor-3), 2. *Tricolocapsa* cf. *plicarum* Yao (sample Khor-3), 3. *Protunuma turbo* Matsuoka (sample Khor-2), 4. *Unuma latusicostatus* (Aita) (sample Khor-2), 5. *Stichocapsa* cf. *robusta* Matsuoka (sample Khor-A-436), 6. *Unuma typicus* Ichikawa et Yao (sample Khor A-436), 7. *Cyrtocapsa mastoidea* Yao (sample Khor A-436), 8. *Hsuum* cf. *belliatulum* Pessagno and Whalen (sample Khor A02P-3/5), 9 - 10. *Stichocapsa* cf. *decora* Rust (9 - sample Khor A02P-1/6, 10 - sample Khor Al-9). All markers – 10 µк.

Fig. 9. Some Middle Jurassic radiolarians from siliceous mudstones, mudstones and siltstones of Middle structural level of the Nadankhada-Bikin terrane Bikin part

Bathonian radiolarians *Stichocapsa decora* Rust, *Archicapsa pachyderma* Tan, *Tricolocapsa fusiformis* Yao, *Hsuum belliatulum* Pessagno et Whalen and others have been extracted from the silty mudstones, and Bathonian–Callovian radiolarians *Stylocapsa lacrimalis* Matsuoka, *Gongylothorax sakawaensis* Matsuoka, *Stichocapsa robusta* Matsuoka, *Stichocapsa naradaniensis* Matsuoka, *Stylocapsa hemicostata* Matsuoka, *Tricolocapsa plicarum* Yao and others has been found in the siltstones.

It should be noted, that the same age radiolarians (Middle Jurassic) have been found (Kojima & Mizutani, 1987) in siliceous mudstones (fig. 10) on a left bank of the Ussuri River, in China (20 km to the west from Shichang village). But originally they were considered as Bathonian–Callovian in age. However, according to the specified data on a time intervals of the Jurassic radiolarian species distribution (Baumgartner et al., 1995), their age is Bajocian. These data indicate that fragments of the middle structural level are also extended in the Nadankhada part of the given terrane.

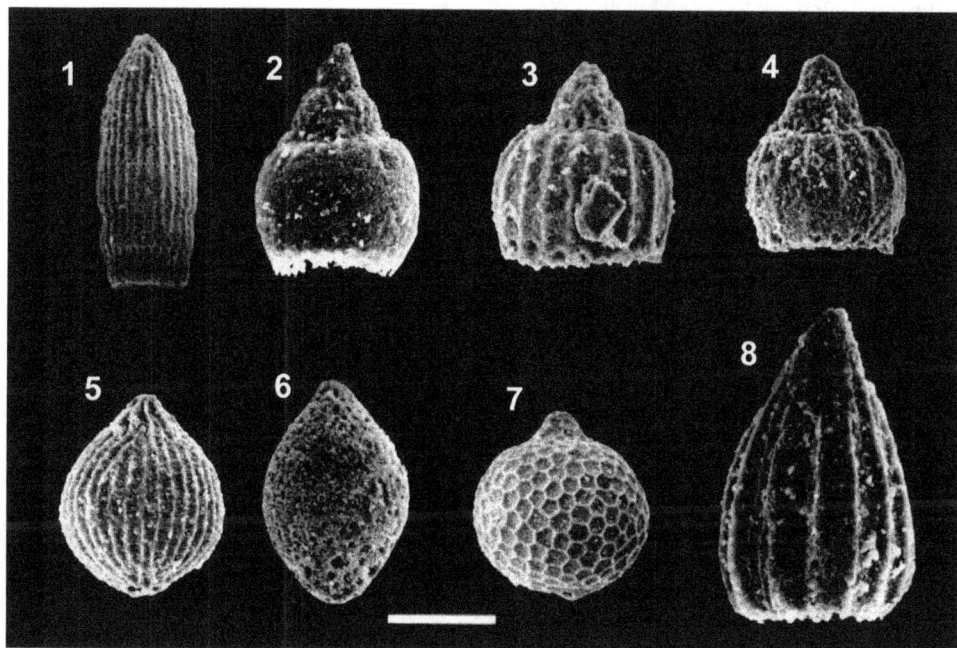

1. *Archaeodictyomitra* sp. (sample 40040/1331) scale bar = 73 µк, 2. *Eucyrtidiellum unumaensis* Yao (sample 42015/1331) scale bar = 54 µк, 3. *Eucyrtidiellum ptyctum* (Riedel et Sanfilippo) (sample 40060/1331) scale bar = 38 µк, 4. *Eucyrtidiellum* sp. (sample 40037/1331) scale bar = 49 µк, 5. *Tricolocapsa plicarum* Yao (sample 42211/1331) scale bar = 80 µк, 6. *Tricolocapsa fusiformis* Yao (sample 39862/1320) scale bar = 54 µк, 7. *Tricolocapsa ruesti* Tan Sin Hok (sample 40054/1331) scale bar = 73 µк, 8. *Thanarla* sp. (sample 42016/1331) scale bar = 54 µк.

Fig. 10. Some Middle Jurassic radiolarians from siliceous mudstones of the Nadankhada-Bikin terrane Nadankhada part (After Kojima & Mizutani, 1987)

3.3 Upper structural level

A fragment of cut-section with gradual cherts-to-terrigenous rocks transition, belonging to the upper structural level, was investigated in detail in the Nadankhada part of the terrane (Yang & Mizutani, 1991; Yang et al., 1993). This cut-section locates about 6 km to the north from the Hongqiling Nongchang village, representing over 40 m exposure of cherts, clayish cherts and siliceous mudstones in which five consecutive radiolarian assemblages have been established (Yang et al., 1993). Four of them, extracted out of the cherts and clayish cherts, characterize their host rocks as being Middle Carnian–Early Pliensbachian in age. The fifth assemblage is related to the siliceous mudstones with Late Pliensbachian radiolarians. The terrigenous rocks associated with these cherty formations are Middle Jurassic in age. The silty mudstones and siltstones contain, correspondingly, Aalenian and Bathonian radiolarians (Yang et al., 1993). In the area of the terrigenous rocks distribution, the chaotic formations representing different-sized lumps, blocks and fragments of Carboniferous-Permian limestones, Triassic cherts, basalts, gabbro and serpentinous ultramafic rocks among the schistose aleuro-psammites are observed. Middle Jurassic radiolarians have been extracted from them too (Yang et al., 1993).

Fragments of the similar chert-terrigenous sequence have been also investigated in the Bikin part of the terrane, in the numerous bank outcrops of the middle reaches of the Khor River and its inflows (Phillippov, 1990). According to the microfaunistic data, the following cut-section was reconstructed: the lower part of this sequence (about 90 m thick) is composed of cherts that are changed by the clayish cherts up the sequence. The age of these rocks,

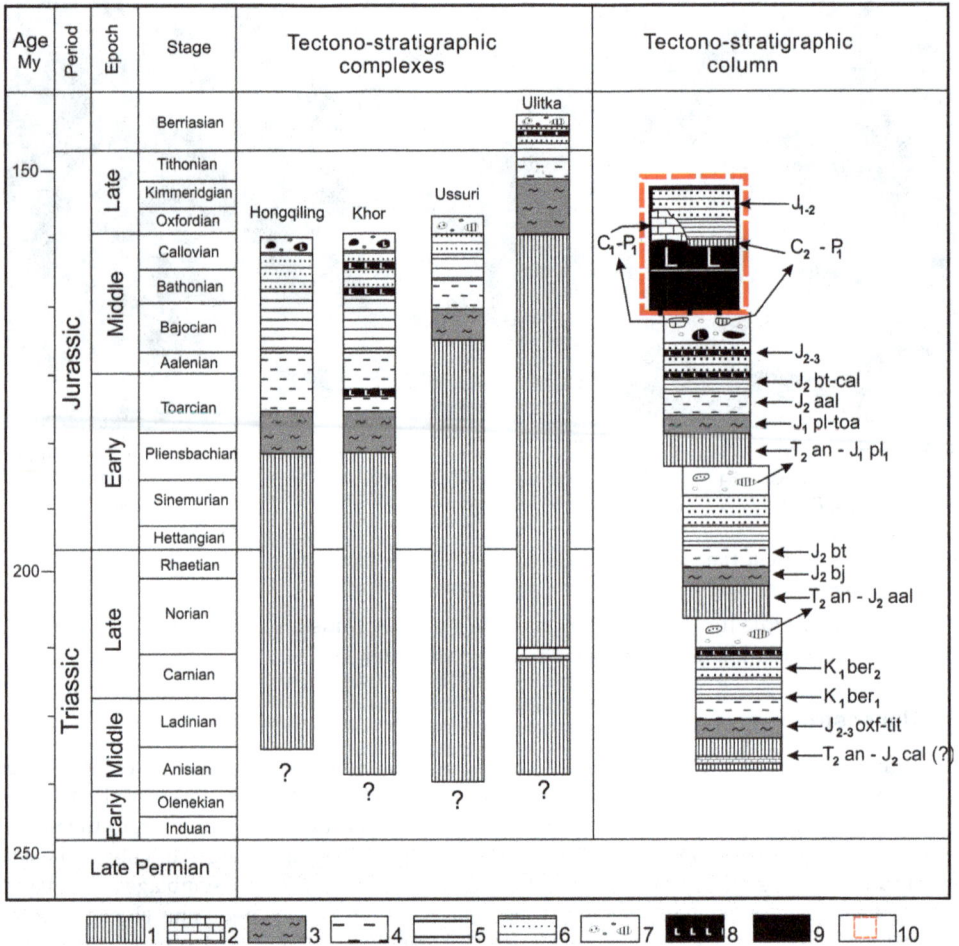

1 – cherts; 2 – limestones; 3 - cherty mudstones; 4 – mudstones; 5 – siltstones; 6 – siltstones-sandstones alternation; 7 – subduction melange; 8 – basalts; 9 – ultramaphic rocks; 10 - supposed tektono-stratigraphic unit.

Fig. 11. Tektono-stratigraphic complexes and column of the Nadankhada-Bikin terrane

according to the conodont fauna (Klets, 1995), ranges from Anisian to Rhaetian, however, taking into account that they are gradually replaced by the siliceous mudstones containing Pliensbachian radiolarians, their upper age border is likely to be early Pliensbachian. The sequence upper part is composed of terrigenous rocks representing rhythmical alternation of siltstones and sandstones. Middle Jurassic Radiolarians such as *Gongylothorax sakawaensis* Yao, *Eucyrtidiellum* sp., *Diacantocapsa normalis* Yao, *Protunuma turbo* Matsuoka, *Tricolocapsa tetragona* Matsuoka, *Tricolocapsa* sp., *Stichocapsa* sp. and others have been determined in the siltstones layers. Upwards of the terrigenous rocks there is a section of chaotic formations (up to 200 m thick) with rock blocks and fragments represented by Carboniferous-Permian limestones, Triassic cherts, basalts and sandstones. Besides that there are also 10-40 m thick layers of basic volcanites, such as hyaloclastites, tuffs, basalts lavas and picritebasalts among the terrigenous rocks.

4. Discussion and conclusions

Biostratigraphic researches of the above mentioned cut-sections have revealed that the chert-terrigenous deposits of the different structural levels of the Nadankhada-Bikin terrane are slightly differing in their ages (fig. 11).

In particular, the age of the cherts-to-terrigenous rocks transitive siliceous mudstones varies from Pliensbachian to Oxfordian-Tithonian. It means that the chert-terrigenous rocks composing this terrane are fragments of the sedimentary cover of the different-aged sites of the paleooceanic plate (fig. 12) or, in other words, they are the fragments of three different-aged Oceanic Plate Stratigraphy Sequences, which are characterized by the different time of accretion (the accretion starting time is correlated with the age of terrigenous rocks, accumulation of which occur in a trench).

Hence, in the terrane structure, it is possible to allocate as a minimum three successive tectono-stratigraphic complexes (fig. 13) representing the primary cut-sections of the sedimentary cover from the different-aged sites (i.e. areas at different distances from the sea-floor spreding center) of the Paleo-Pacific Ocean and reflecting a process of their consecutive accretion. Let's name them Ulitka (Triassic-Early Cretaceous), Ussuri (Triassic – Late Jurassic) and Khor-Hongqiling (Triassic - Middle Jurassic) Formations.

It should be noted that the age of rocks of the allocated Formations, as well the time of their accretion are gradually rejuvenate from the upper structural level to the lower one. As a whole, the Nadankhada-Bikin terrane structure is characterized by the inverted stratification of formations that make up the terrane. Compared to the upper structural level (the Khor-Hongqiling Formation), the lower level (the Ulitka Formation) is composed of relatively younger rocks, thus the rocks of the middle level (the Ussuri Formation) are of intermediate age. At the same time, a primary stratigraphic succession of deposits within each complex is normal (from older to younger bottoms up). Such structure of the Nadankhada-Bikin terrane completely corresponds to a structure of modern accretionary prisms forming at the basis of internal slopes of the modern convergent margins trenches, and is a result of a successive subduction and partial accretion of the oceanic plate sites of different age and lithology. During the subduction the most remote from the spreading center sites of the oceanic plate, its oldest part, are accreted first, with the following underplating of its younger fragments in future. As a result, a package of tectono-sedimentary complexes is formed.

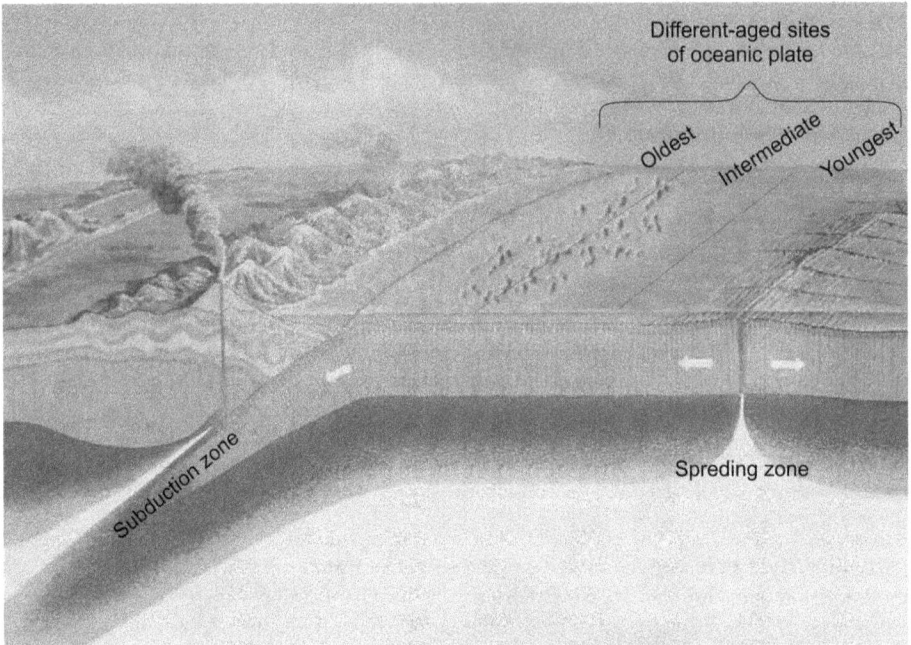

Fig. 12. The scheme illustrating spreding-subduction model of the Plate Tectonics and different-aged sites of drifting oceanic plate

The true structure of the Nadankhada-Bikin terrane thus appears to be a regular recurrence of the strongly deformed fragments of the primary sedimentary cover of the paleooceanic plate sites situating at a different distance from the spreading center, rather than a chaotic repeated alternation of tectonic slices of different age and lithology.

Based on lithological and age characteristics of the rocks, the allocated tectono-stratigraphic complexes of the Nadankhada-Bikin terrane confidently correlate with the corresponding structural complexes of the Samarka terrane which is the most thoroughly investigated unit of the Sikhote-Alin Jurassic prism. For example, the Ulitka Formation that composes the lower structural level of the Nadankhada-Bikin terrane can be compared with the Katen Formation that is also the lowermost tectono-stratigraphic unit of the Samarka terrane. The Ussuri Formation, composing the terrane middle structural level, is correlated with the Breevka Formation, and the Khor-Hongqiling Formation is comparable with the Amba-Matay Formation which also contains chaotic formations with exotic blocks and fragments of Paleozoic limestone, cherts, basalts, and gabbro within.

It should be added that in the southwestern part of the Nadankhada ridge (in China), structurally above the slices of the chert-terrigenous rocks, there are tectonic slices of ophiolite (Dahezhen Formation). These widely distributed ophiolitic slices are made up of basalts in association with Carboniferous - Permian limestones, serpentinites and gabbro (Mizutani et al., 1989; Shao et al., 1992 etc.). Tectonic slices of same composition and age compose the uppermost structural level of the Sebuchar Formation of the Samarka terrane (Kemkin, 2006). These data suggest as a minimum one more tectono-stratigraphic complex in the Nadankhada-Bikin terrane structure (see fig. 10, it is allocated by a dashed line).

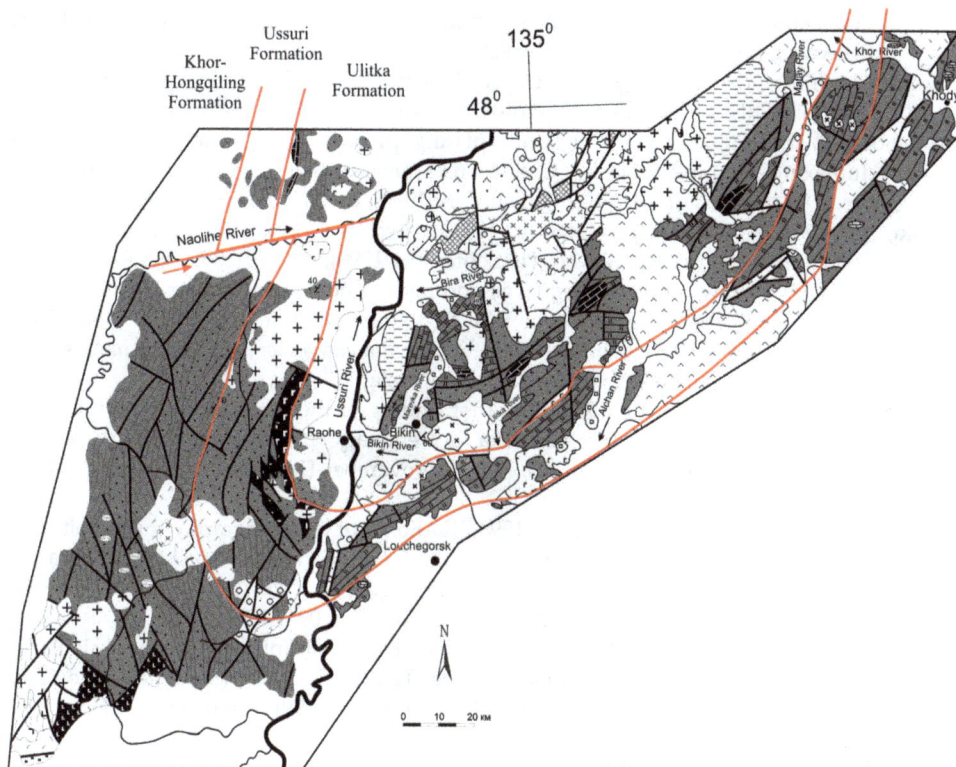

See legend on a Fig. 4 and 5.

Fig. 13. Different-aged tectono-stratigraphic complexes of the Nadankhada-Bikin terrane

The comparative analysis of even-aged Formations of the Jurassic accretionary prism different terranes indicates that during the Jurassic the geodynamic mode along the Paleo-Asian continent eastern margin was invariable.

5. Acknowledgments

This work was done with partial financial support from the Grants of the President of Russian Federation (Grant No NSH – 5162.2010.5) and the Far Eastern Branch of Russian Academy of Sciences (Grant No 09-III-A-08-403).

6. References

Baumgartner, P.O., O'Dogherty, L., Gorican, S., Urquhart, E., Pillevuit, A. & De Wever, P. (1995). Middle Jurassic to Lower Cretaceous Radiolaria of Tethys: Occurrences, Systematics, Biochronology. *Memoires de Geologie*, Universite de Lausanne, No. 23, 1172 pp, ISSN 1015-3578

Berger, W.H. & Winterer, E.L. (1974). Plate stratigraphy and fluctuating carbonate line, In: *Pelagic sediments: On land and under the sea*, Hsu, K.J., Jehkyns, H. (Eds.), Special

Publication of the International Association of Sedimentologists. No. 1, pp. 11-48, Blackwell, ISBN 780632001675, Zurich

Bersenev, I.I. (Ed.). (1969). *Geology of the USSR. Primorye Region. Part I. Geological description.* "Nedra", ISBN 7-73-10648-1, Moscow (in Russian)

Hashimoto, Y. & Kimura, G. (1999). Underplating process from melange formation to duplexing: Example from the Cretaceous Shimanto Belt, Kii Peninsula, Southwest Japan. *Tectonics*, Vol. 18, No. 1, (1 February 1999), pp. 92-107, ISSN 0278-7407

Isozaki, Y., Maruyama, S. & Furuoka, F. (1990). Accreted oceanic materials in Japan. *Tectonophysics*, Vol. 181, No. 1/2 (10 September 1990), pp. 179-205, ISSN 0040-1951

Kemkin, I.V. (1996). New data on the *geology* and age of the Koreiskaya River area (South Sikhote-Alin). *The Island Arc*, Vol.5, No. 2, (June 1996), pp. 130-139, ISSN 1440-1738

Kemkin, I.V. (2006). *Geodynamic evolution of the Sikhote-Alin and Sea of Japan region in Mesozoic*, "Nauka", ISBN 5-02-034259-9, Moscow (in Russian)

Kemkin, I.V. & Golozubov, V.V. (1996). The first finding of the Early Jurassic radiolaria in cherty allochtnons of the Samarka accretionary prism (South Sikhote-Alin). *Tikhookeanskaya Geologiya*, Vol. 15, No. 6, (November-December 1996), pp. 103-109, ISSN 0207-4028 (in Russian)

Kemkin, I.V., Kametaka, M. & Kojima, S. (1999). Radiolarian biostratigraphy for transitional facies of chert-clastic sequence of the Taukha terrane in the Koreyskaya River area, Southern Sikhote-Alin, Russia. *The Journal of Earth and Planetary Sciences*. Nagoya University, Vol. 46, (December 1996), pp. 29-47, ISSN 0919-875X

Kemkin, I.V. & Kemkina, R.A. (1999). Radiolarian biostratigraphy of the Jurassic-Early Cretaceous chert-clastic sequence in the Taukha Terrane (South Sikhote-Alin, Russia). *Geodiversitas*, Vol. 21, No. 4, (24 December 1999), pp. 675-685, ISSN 1280-9659.

Kemkin, I.V. & Khanchuk, A.I. (1992). New data on age of the paraautochthone of Samarka accretionary complex (South Sikhote-Alin). *Doklady of Russian Academy of Sciences*, Vol. 324, No. 4, (June 1992), pp. 847-851, ISSN 0869-5652 (in Russian)

Kemkin, I.V. & Khanchuk, A.I. (1993). First data about Early Cretaceous accretionary complex in Chernaya River area (Soust Sikhote-Alin). *Tikhookeanskaya geologiya*, Vol. 12, No 1, (January-February 1993), pp. 140-143, ISSN 0207-4028 (in Russian)

Kemkin, I.V. & Philippov, A.N. (2001). Structure and genesis of lower structural unit of the Samarka Jurassic accretionary prism (Sikhote-Alin, Russia). *Geodiversitas*, Vol. 23, No. 3, (28 September 2001), pp. 323-339, ISSN 1280-9659

Kemkin, I.V. & Philippov, A.N. (2002). The structure and formation of the Samarka accretionary prism in Southern Sikhote-Alin. *Geotektonika*, Vol. 36, No. 5, (September-October 2002), pp. 79-88, ISSN 0016853X (in Russian)

Kemkin, I.V. & Rudenko, V.S. (1998). New data on chert age of the Samarka accretionary prism, southern Sikhote-Alin. *Tikhookeanskaya geologiya*, Vol. 17, No. 4, (July-August 1998), pp. 22-31, ISSN 0207-4028 (in Russian)

Kemkin, I.V. & Taketani, Y. (2008). Structure and age of lower structural unit of the Taukha terrane of Late Jurassic - Early Cretaceous accretionary prism, Southern Sikhote-Alin. *The Island Arc*, Vol. 17, No. 4, (December 2008), pp. 517-530, ISSN 1440-1738

Khanchuk, A.I. (1994). Tectonics of Russian Southeast. *Chishitsu News*, No. 480, (August 1994), pp. 19-22, ISSN 0009-4854

Khanchuk, A.I. (Ed.). (2006). *Geodynamics, magmatism and metallogeny of East of Russia. Book 1.* Dal'nauka, ISBN 5-8044-0634-5, Vladivostok (in Russian)

Kimura, G. (1997). Cretaceous episodic growth of the Japanese Islands. *The Island Arc,* Vol. 6, No. 1, (March 1997), pp. 52-68, ISSN 1440-1738

Kimura, G. & Mukai, A. (1991). Underplated units in an accretionary complex: melange of the Shimanto belt of eastern Shikoku, Southwest Japan. *Tectonics,* Vol. 10, No. 1, (February 1991), pp. 31-50, ISSN 0278-7407

Kirillova, G.L. (2002). Structure of the Jurassic accretionary prism in the Amur region: aspects of nonlinear geodynamics. *Doklady of Russian Academy of Sciences,* Vol. 386. No. 4, (October 2002), pp. 515-518, ISSN 0869-5652 (in Russian).

Klets, T.V. (1995). *Biostratigraphy and Triassic conodonts of the Middle Sikhote-Alin,* Novosibirsk State University, ISBN 5-230-13602-2, Novosibirsk (in Russian)

Kojima, S. & Mizutani, S. (1987). Triassic and Jurassic Radiolaria from the Nadanhada Range, northeast China. *Transactions and Proceedings of the Palaeontological Society of Japan,* New series, No. 148, (30 December 1987), pp. 256-275, ISSN 0031-0204

Li, W.K., Han, J.X., Zhang, S.X. & Meng, F.Y. (1979). The main characterstics of the Upper Paleozoic stratigraphy at the North Nadanhada Range, Heilongjiang Province, China. *Bulletin of Chinese Academy of Geological Sciences,* Vol. 1, (January 1979), pp. 104-120, ISSN 1004-1931

Liht, F.R. (1997). Sedimentological features of Cretaceous sedimentary basins of western Sihote-Alinja. *Tikhookeanskaya geologiya,* Vol. 16, No. 6, (November-December 2001), pp. 92-101, ISSN 0207-4028 (in Russian)

Mizutani, S., Kojima, S., Shao, J.A. & Zhang, Q.Y. (1986). Mesozoic radiolarians from the Nadanhada area, northeast China. *Proceedings of the Japan Academy, Ser. B.,* Vol. 62, No. 9, (12 November 1986), pp. 337-340, ISSN 0386-2208

Mizutani, S., Shao, J.A. & Zhang, Q.Y. (1989). The Nadanhada terrane in relation to Mesozoic tectonics on continental margins of East Asia. *Acta Geologica Sinica,* Vol. 63, No. 3, (September 1989), pp. 204-216, ISSN 0001-5717

Mizutani, S., Shao, J.A. & Zhang, Q.Y. (1990). The Nadanhada Terrene in relation to Mesozoic Tectonics on Continental Margins of East Asia. *Acta Geologica Sinica,* Vol. 3, No. 1, (March 1990), pp. 15-29, ISSN 1000-9515

Moore, J.C. & Byrne, T. (1987). Thickening of fault zones: A mechanism of melange formation in accreting sediments. *Geology,* Vol. 15, (November 1987), pp.1040-1043, ISSN 0091-7613

Philippov, A.N. (1990). *Formation analysis of Mesozoic strata in Western Sikhote-Alin.* Far Eastern Branch of USSR Academy of Sciences, ISBN 5-7442-0056-8, Vladivostok (in Russian).

Philippov, A.N., Buriy, G.I. & Rudenko, V.S. (2001). Stratigraphic sequence of sedimentary deposits Samarka terrane (Central Sikhote-Alin): record of paleooceanic sedimentation. *Tikhookeanskaya geologiya,* Vol. 20, No. 3, (May-June 2001), pp. 29-49, ISSN 0207-4028 (in Russian)

Philippov, A.N. & Kemkin, I.V. (2004). «Koultoukha suite» - the tectono-stratigraphic complex of the Jurassic accretionary prism of the Western Sikhote-Alin. *Tikhookeanskaya geologiya,* Vol. 23, No. 4, (July-August 2004), pp. 43-53, ISSN 0207-4028 (in Russian)

Philippov, A.N. & Kemkin, I.V. (2007). Siliceous-argillaceous deposits of the Jurassic accretionary prism of the Khekhtsir Range, Sikhote-Alin: Stratigraphy and Genesis. *Tikhookeanskaya geologiya*, Vol. 26, No 1, (January-February 2007), pp. 51-69, ISSN 0207-4028 (in Russian)

Philippov, A.N., Kemkin, I.V. & Panasenko, E.S. (2000). Early Jurassic hemipelagic deposits of the Samarka terrane (Central Sikhote-Alin): sequence, composition and sedimentary environments. *Tikhookeanskaya geologiya*, Vol. 19, No 4, (July-August 2000), pp. 83-96, ISSN 0207-4028 (in Russian)

Seely, D.R., Vail, P.R. & Walton, G.G. (1974). Trench slope model, In: *The geology of continental margins*, Burk, C.A., & Drake, C.L. (Eds.), pp. 249-260, Springer-Verlag, ISBN 354006866X, 9783540068662, Berlin – Heidelberg - New York

Shao, J.A., Tang, K.D., Wang, C.Y., Zang, Q.Y. & Zhang, Y. P. (1992). Structural features and evolution of the Nadanhada terrane. *Science in China*. Ser. B, Vol. 35, No. 5, (May 1992), pp. 621-630, ISSN 1674-7291

Shao, J.A., Wang, C.Y., Tang, K.D. & Zhang, Q.Y. (1990). Relationship of the stratigraphy and terrane of the Nadanhada Range. *Journal of Stratigraphy*, Vol. 14, No. 4, (December 1990), pp. 286-291, ISSN 0253-4959 (in Chinese with English abstract)

Sokolov, S.D. (1992). *Accretionary tectonics of the Koryak-Chukotka segment of the Pacific Belt*. Transactions of the Geological Institute RAS, Issue 479, "Nauka", ISBN 5-02-002235-7, Moscow (in Russian)

Sokolov, S.D. (1997). Continental accretion, terranes, and nonlinear effects in geodynamics of Northeastern Russia, In: *Tectonic and Geodynamic Phenomena*, Transactions of the Geological Institute RAS, Issue 505, Perfil'ev, A.S. & Raznitsyn, Yu.N. (Eds.), pp. 42-69, "Nauka", ISBN 5-02-003693-5, Moscow (in Russian)

Sokolov, S.D., Bondarenko, G.E., Morozov, O.L., Aleksyutin, M.V., Palandzhyan, S.A. & Khudoley, A.K. (2001). Features of structure of the paleoaccretionary prism of the Taigonos Peninsula, Northeastern Russia. *Doklady of Russian Academy of Sciences*, Vol. 377, No. 6, (Aprile 2001), pp. 807-811, ISSN 0869-5652 (in Russian)

Wakita, K. & Metcalfe, I. (2005).Ocean plate stratigraphy in East and Southeast Asia. *Journal of Asian Earth Sciences*, Vol. 24, No. 6, (March 2005), pp. 679-702, ISSN 1367-9120

Yang, Q. & Mizutani, S. (1991). Radiolaria from the Nadanhada terrane, northeast China. *The Journal of the Earth Science*. Nagoya University, Vol. 38, (December 1991), pp. 49-78, ISSN 0022-0442

Yang, Q., Mizutani, S. & Nagai, H. (1993). Biostratigraphic correlation between the Nadanhada terrane of NE China and the Mino terrane of Central Japan. *The Journal of the Earth and Planetary Science*. Nagoya University, Vol. 40, (December 1993), pp. 27-43, ISSN 0919-875X

Zyabrev, S.V. (1998). The stratigraphic record of the chert-terrigenous complex of the Khekhtsir Range and kinematics of asymmetric folds as indicators of subduction-related accretion. *Tikhookeanskaya geologiya*, Vol. 17, No 1, (January-February 1998), pp. 76-84, ISSN 0207-4028 (in Russian)

Zyabrev, S.V. & Matsuoka, A. (1999). Late Jurassic (Tithonian) radiolarians from a clastic unit of the Khabarovsk complex (Russian Far East): Significance for subduction accretion timing and terrane correlation. *The Island Arc*, Vol. 8, No. 1, (March 1999), pp. 30-37, ISSN 1440-1738

Sedimentary Tectonics and Stratigraphy: The Early Mesozoic Record in Central to Northeastern Mexico

José Rafael Barboza-Gudiño
Universidad Autónoma de San Luis Potosí,
México

1. Introduction

The stratigraphy has traditionally been conceived as a geological science based on the description and subdivision of rock successions in some kind of units, following the International Stratigraphic Guide or the North American Stratigraphic Code and accord to lithostratigraphic criteria. Diversification or development of modern analytical techniques and emerged new concepts and models of sedimentary, volcanic, plutonic or metamorphic settings, associated to specific geotectonic regimes, offers new possible approaches for stratigraphic subdivision. Accord to these new criteria, revision of numerous sequences is required in order to establish a direct relationship between subdivision and events or environments in space and time, prevailing genetic or provenance criteria over the simple lithological or chronological criteria. Tectonic events are typically recorded in sediments being deposited at the same time, in such a way; each stratigraphic unit can be related to an specific geologic process or tectonic setting.

The composition of a sedimentary rock provides us information that allows interpreting several kinds of relations, like source land composition or provenance. By siliciclastic rocks, their mineral and clastic components established through petrographic studies, are characteristic of specific tectonic settings, and in this way usefully for stratigraphic subdivision, representing a signature for each unit. The heavy minerals in a siliciclastic rock are also a useful tool for the characterization of a sedimentary rock and their possible sourceland. Geochemical studies are also useful to complement stratigraphic subdivision and classification constraining provenance and a geotectonic regime or stage related to each stratigraphyc unit. Finally geochronology and especially the detrital zircon geochronology of sillisiclastic rocks is today a poderous tool to determine a maximal deposition age of a layer but also a provenance through the several clusters of individual zircon ages established by the U-Pb method. The Goal of this chapter is to present some criteria for the stratigraphic subdivision and establishment of stratigraphic units utilizing the modern, petrographic, geochemical and geochronologic thechnicks through a case study of the Early Mesozoic succession outcropping in central to northeastern Mexico (Figure 1).

After Laurentia-Gondwana collision during Late Paleozoic time, the actual central Mexico was occupied by a remnant basin at the westernmost culmination of the Ouachita-Marathon belt. This basin was bounded by a subduction zone to the east, where their oceanic floor subducted eastward below Pangea and by a transform margin to the north in southern Laurentia. The subduction process caused deformation and metamorphism during the Late Paleozoic, producing the Granjeno schist, exposed in some places in the Sierra Madre Oriental. Towards the latest Permian-Early Triassic, tokes place magmatic activity, related to a low stress stage of the same subduction process, producing the east Mexico Permo-Triassic Magmatic arc. Such ancient active margin, was probably abandoned since the Early Triassic, and a new subduction zone evolved to the west towards the end of the Triassic time,

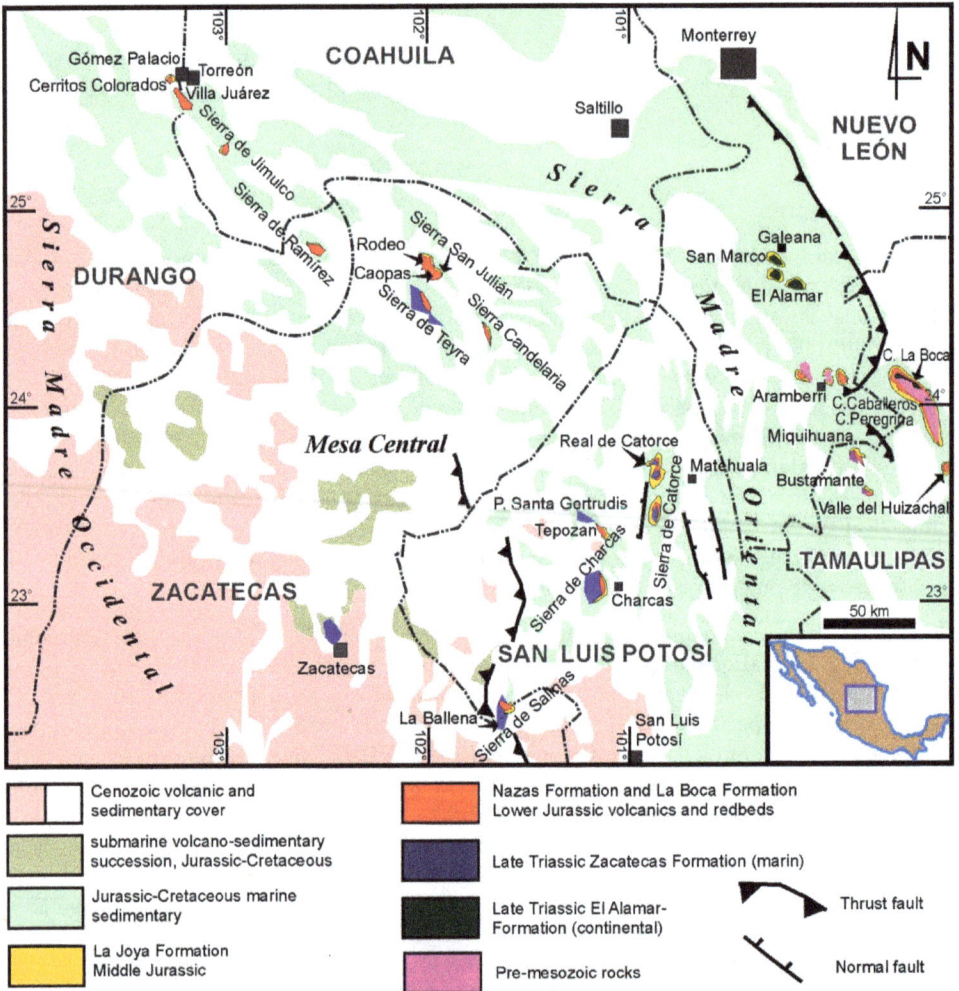

Fig. 1. Pre-Late Jurassic localities in central to northeastern Mexico modified after Barboza-Gudiño et al. (2008). Shown are Late Triassic exposures of the marine and continental post-Triassic units, exposures of pre-Mesozoic crystalline rocks, in some cases interpreted as areas of no deposition during the Triassic, and the main exposed volcanic centers of the Early Jurassic volcanic arc

producing new deformation in the Mesa Central and subsequent a new continental volcanic arc during the Early Jurassic. The earliest deposits in the pacific sub-basin located today in central Mexico, are represented by Triassic and Jurassic successions whose subdivision has been able to establish on the basis of sedimentary tectonics criteria, assigning a specific geotectonic setting or a distinctive provenance to each defined stratigraphic unit.

2. Research methods

The petrography and specially the point counting of mineral and clastic components from a sedimentary rock, is the traditional technique to determine a possible provenance in terms of their typical mineral and clastic components. Each sediment, accord to their mineralogical and clastic content, can be related to a kind of source area, like plutonic or volcanic, felsic or mafic, recycled sedimentary or metamorphic, and also related to an specific tectonic setting or tectonic regime, like stable continental craton, recycled orogen, volcanic arc and block faulted continental basement, including continental rifts. Triangular Quartz-feldspar-lithics (Q-F-L) diagrams after Dickinson (1985) are the most conventional plots to assign in an empirical way a tectonic setting of provenance for each studied rock. Figure 2 shows several triangular diagrams used for different approaches in provenance studies and based on the

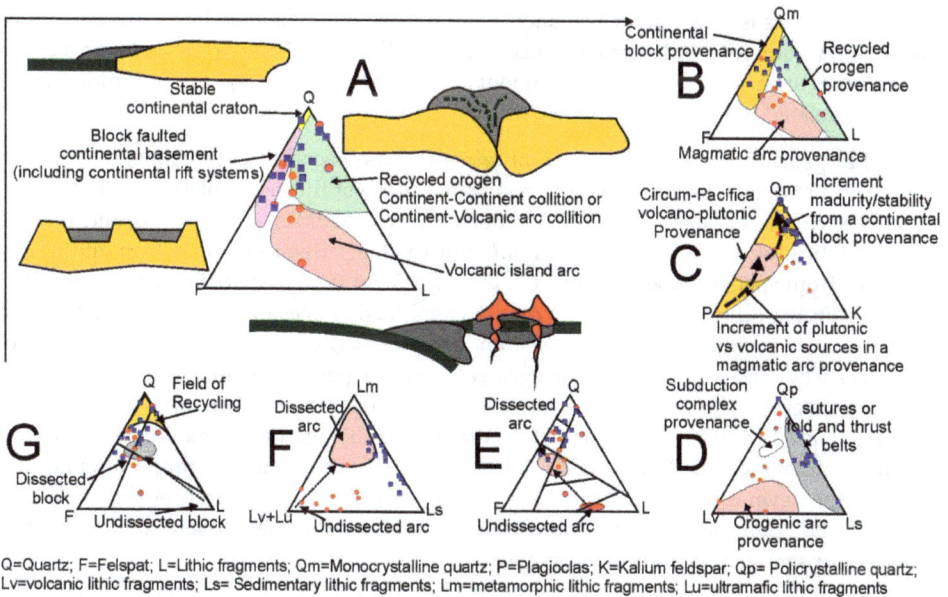

Q=Quartz; F=Felspat; L=Lithic fragments; Qm=Monocrystalline quartz; P=Plagioclas; K=Kalium feldspar; Qp= Policrystalline quartz; Lv=volcanic lithic fragments; Ls= Sedimentary lithic fragments; Lm=metamorphic lithic fragments; Lu=ultramafic lithic fragments

Fig. 2. triangular diagrams for discrimination of the tectonic regimes of provenance: Q-F-L (A), Qm-F-L (B), Qm-P-K (C) and Qp-Lv-Ls-diagrams (D), modified after Dickinson et al. (1983), Dickinson (1985); diagrams Q-F-L and Lm-LV+Lu-Ls after Marsaglia & Ingersoll (1992), modified by Garzanti et al. (2007) for discrimination of dissected/undissected magmatic arc settings (E, F) and diagram Q-F-L after Garzanti et al. (2007) for discrimination of dissected/undissected blocks. Ploted are in all diagrams, results of point counting (after Barboza-Gudiño et al., 2010) in samples from the Triassic El Alamar Formation from northeastern Mexico (quadrangles), as well as samples from the Early Jurassic La Boca Formation (circles). See discussion in text

abundances of distinct clastic components. The heavy minerals content of a siliciclastic sediment or sedimentary rock is also diagnostic for their provenance and in this way also a tool for stratigraphic correlation and subdivision, paleogeographic reconstruction and also to interpret flow patterns for a deposit.

Geochemical approaches complement provenance and stratigraphic studies, supporting interpretation of tectonic regimes related to source areas as well as correlation and subdivision of sedimentary and volcano-sedimentary successions. McLennan et al. (1993) recognized five provenance components or terrane types on the basis of whole-rock chemical and Nd-isotopic composition, incluiding: 1) old upper continental crust, 2) recycled sedimentary rocks, 3) young undifferentiated magmatic arc, 4) young differentiated magmatic arc and 5) other exotic components, like ophiolites. Mayor elements, trace elements, including rare earth elements (REE) as well as several isotopes can be indicative of sedimentary processes and sedimentary-tectonic regimes. For such geochemical studies shale are very appropriate but also fine grained sandstone because their heavy mineral content that reflect several sedimentary processes or information related to the source areas, through their REE and other trace elements or isotopic relations. Nd isotopic composition can be indicative of intracrustal igneous differentiation processes, enrichment on large ion-lithofiles elements is linked to provenance composition, and alkali-alkaline earth depletions is related to weathering and alteration processes. Finally, Zr and Hf enrichments are a direct consequence of heavy mineral enrichment on sediment and high Cr abundances is commonly related to ultramafic sources.

The U-Pb geochronology from detrital zircons has become a fundamental tool in stratigraphic studies. The most exciting results are those of the laser-ablation multicollector inductively coupled plasma mass spectrometry (LAMC-ICPMS). It is a rapid and relative cheap analytical technique that allows in an hour ca 40 age determinations from individual zircons, or from specific zones in a crystal, thanks a micron-scale spatial resolution through beam sizes in the range of 10 to 50 microns. Crystallization ages of igneous rocks, maximal deposition ages of siliciclastic sediments, and provenance analysis are the main kinds of data offered by this technique. U-Pb ages can also be performed in other minerals like sphene, to determine cooling ages in magmatic rocks, or monazite, for metamorphism age determinations. A detrital zircon analysis is based on the age determination from ca. 100 individual zircons by the LAMC-ICPMS technique. The obtained zircon age populations, represent the several blocks or source areas for the clastic components of the rock, and theoretical the youngest zircon or cluster can be considered as the maximal deposition age of the rock or sampled layer from the sequence.

For each analysis, the errors in determining $^{206}Pb/$ ^{238}U and $^{206}Pb/$ ^{204}Pb result in a measurement error of ~1-2% (at 2-sigma level) in the $^{206}Pb/$ ^{238}U age. The errors in measurement of $^{206}Pb/$ ^{207}Pb and $^{206}Pb/$ ^{204}Pb also result in ~1-2% (at 2-sigma level) uncertainty in age for grains that are >1.0 Ga, but are substantially larger for younger grains due to low intensity of the ^{207}Pb signal. For most analyses, the cross-over in precision of $^{206}Pb/$ ^{238}U and $^{206}Pb/$ ^{207}Pb ages occurs at 0.8-1.0 Ga. The detrital zircon geochronology allows the comparison of the results from the analyzed samples, in order to compare in addition to their maximal age of deposition, all provenance similarities as a signature that can also be used in some cases as a criterion for stratigraphic subdivision and correlation.

For a detrital zircon analysis, consisting usually of the age determination of 100 individual zircons, a sample of approximately 10-12 kg of fine to middle grained sandstone is prepared. The mineral separation involves crushing of the sample to gravel size on a jaw crusher and posterior pulverizing of this material on a roller mill to reduce it to a fine sand size. The gravity separation consists of a first step on a Wilfley table that separate denser grains from light grains by passing of the sample and water flow over the table that vibrates synchronically. After this first gravity separation, several magnetic minerals and magnetite or iron particles can be removed with a hand magnet, thereafter the next recommended step for separation of the zircons, is by using of methylene iodide (MI), also known as diiodomethane. In all the cases under an operating fume hood at laboratory and with properly protection for eyes and all the skin because it is a potentially hazardous chemical. Finally a Franz magnetic separator allows separation of heavy minerals according to their magnetic susceptibility. Next step before the analysis is the zircon crystals mounting in epoxy, finnaly the mounted crystals are polished to half-thickness.

By the analysis on a LAMC-ICPMS system, material is ablated from a zircon surface, the laser operates at a wavelength of 193 nm with spot sizes between 10 and 75 microns, the ablated particles are carried in helium gas into the multicollector-ICPMS. Sample preparation and analysis techniques are described in detail by Gehrels et al. (2006). Table 1. shows unpublished data of detrital zircons, obtained from a sample collected in the Late Triassic Zacatecas Formation, at La Ballena Zacatecas, south of Salinas, San Luis Potosí. In figure 3, all zircon data are plotted as a Pb/U concordia diagram (Figure 3A) and a relative age probability curve (Figure 3B), following procedure and algorithms of Ludwig (2003). Such plots show each age and its uncertainty (measurement error) as a normal distribution, and add all ages from a sample into a single curve. The analysis was performed in the Arizona LaserChron Center, University of Arizona at Tucson, following the separation and analysis techniques described above, with a New Wave/Lambda Physic DUV193 Excimer laser (operating at a wavelength of 193 nm) using a spot diameter of 15 to 35 microns and a Multicollector ICPMS GVI Isoprobe.

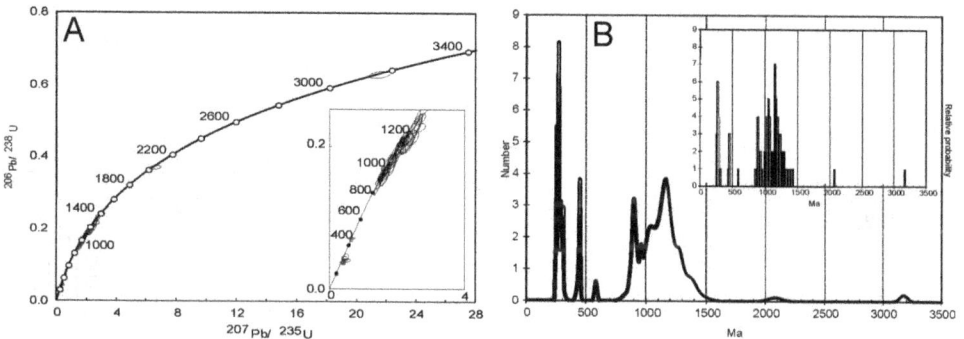

Fig. 3. Plots of zircon ages obtained from sample LB-18: A. Concordia diagram, see detail for the younger zircons, B. Relative Probability curve and histogram, the main populations are Grenvillian (900-1200 Ma), panafrican (500-600 Ma) and Early Paleozoic (400-450 Ma) as characteristic of peri-Gondwanan terranes and finally, a prominent permo-triassic cluster, typical in all samples from north-central and northeastern Mexico, showing a great influence of the East Permo-Triassic Magmatic Arc

U (ppm)	$\frac{^{206}Pb}{^{204}Pb}$	$\frac{U}{Th}$	$\frac{^{206}Pb*}{^{207}Pb*}$	± (%)	$\frac{^{207}Pb*}{^{235}U*}$	± (%)	$\frac{^{206}Pb*}{^{238}U}$	± (%)	error corr.	$\frac{^{206}Pb*}{^{238}U*}$ (Ma)	± (Ma)	$\frac{^{207}Pb*}{^{235}U}$ (Ma)	± (Ma)	$\frac{^{206}Pb*}{^{207}Pb*}$ (Ma)	± (Ma)	Best age (Ma)	± (Ma)
203	9195	1.5	16.9582	10.6	0.3155	10.8	0.0388	2.0	0.19	245.4	4.9	278.4	26.2	565.9	230.8	245.4	4.9
123	7610	1.1	16.5928	21.8	0.3319	21.9	0.0399	1.0	0.05	252.5	2.5	291.0	55.4	613.2	477.1	252.5	2.5
183	13210	1.7	18.9447	12.3	0.2941	12.3	0.0404	1.0	0.08	255.3	2.5	261.8	28.4	319.5	279.6	255.3	2.5
177	21035	2.4	17.5712	5.9	0.3263	6.1	0.0416	1.5	0.24	262.7	3.8	286.8	15.2	488.1	130.7	262.7	3.8
256	22170	1.9	18.9409	6.8	0.3068	7.2	0.0421	2.3	0.32	266.1	5.9	271.7	17.2	320.0	155.7	266.1	5.9
135	13940	1.3	19.7009	8.2	0.2950	8.3	0.0422	1.5	0.17	266.2	3.8	262.5	19.2	229.9	189.5	266.2	3.8
177	11875	1.8	19.4868	5.5	0.2992	5.9	0.0423	2.1	0.36	267.0	5.5	265.8	13.7	255.1	125.9	267.0	5.5
209	25115	2.7	18.2658	7.1	0.3215	7.1	0.0426	1.0	0.14	268.9	2.6	283.0	17.6	401.9	158.3	268.9	2.6
325	5030	2.9	13.6809	13.2	0.4348	13.5	0.0431	3.0	0.22	272.3	7.9	366.6	41.6	1016.6	268.2	272.3	7.9
282	23350	1.6	18.2460	5.3	0.3419	5.4	0.0452	1.2	0.22	285.3	3.3	298.6	14.1	404.3	118.9	285.3	3.3
82	230590	4.6	17.8439	13.5	0.3616	14.0	0.0468	3.6	0.26	294.9	10.4	313.4	37.8	454.0	301.6	294.9	10.4
153	11715	1.3	18.6630	11.0	0.3471	11.1	0.0470	1.5	0.13	295.9	4.3	302.5	29.0	353.5	249.0	295.9	4.3
108	5045	1.3	17.8777	17.7	0.3652	19.2	0.0474	7.6	0.39	298.3	22.0	316.1	52.3	449.8	395.8	298.3	22.0
118	7905	1.5	17.4578	8.0	0.3862	8.1	0.0489	1.0	0.12	307.8	3.0	331.6	22.9	502.3	177.4	307.8	3.0
429	14630	1.6	16.1814	4.9	0.5894	6.1	0.0692	3.7	0.60	431.1	15.4	470.5	23.1	667.1	105.0	431.1	15.4
299	35260	51.2	17.5955	5.1	0.5553	5.4	0.0709	1.8	0.33	441.3	7.6	448.4	19.5	485.0	112.0	441.3	7.6
138	21790	0.9	18.0138	3.8	0.5486	3.9	0.0717	1.0	0.26	446.3	4.3	444.1	14.0	432.9	83.7	446.3	4.3
116	11240	7.5	16.6156	5.5	0.5990	5.6	0.0722	1.1	0.19	449.3	4.7	476.6	21.3	610.2	118.6	449.3	4.7
355	30470	4.0	15.6006	4.0	0.8361	4.5	0.0946	2.0	0.46	582.7	11.4	617.0	20.7	744.9	84.4	582.7	11.4
686	306140	37.7	13.7685	2.1	1.4009	5.8	0.1399	5.4	0.93	844.0	43.0	889.3	34.5	1003.6	42.0	844.0	43.0
204	90590	5.0	14.2983	2.4	1.3954	2.8	0.1447	1.6	0.54	871.2	12.6	887.0	16.9	926.5	49.1	871.2	12.6
315	43815	4.3	13.8975	2.0	1.4671	2.3	0.1479	1.2	0.52	889.0	9.8	916.9	13.8	984.7	39.7	889.0	9.8
1377	131280	16.9	13.8971	2.3	1.4712	6.1	0.1483	5.7	0.92	891.4	47.1	918.6	37.0	984.7	47.4	891.4	47.1
445	87625	0.8	14.0163	1.0	1.4670	1.6	0.1491	1.2	0.78	896.1	10.4	916.9	9.6	967.3	20.4	896.1	10.4
377	56630	1.9	14.1573	1.5	1.4574	1.8	0.1496	1.1	0.59	899.0	8.9	912.9	10.8	946.9	29.7	899.0	8.9
357	110960	4.3	14.0480	1.2	1.4741	2.2	0.1502	1.9	0.84	902.1	15.7	919.8	13.3	962.7	24.1	902.1	15.7
122	51165	1.2	13.9913	3.3	1.4973	3.6	0.1519	1.4	0.40	911.8	12.1	929.3	21.8	971.0	67.1	911.8	12.1
158	140330	1.9	14.0649	1.6	1.5109	2.2	0.1541	1.6	0.69	924.1	13.3	934.8	13.7	960.2	32.9	924.1	13.3
419	85995	8.6	13.8326	1.7	1.5580	3.4	0.1563	3.0	0.87	936.2	25.7	953.7	21.1	994.2	34.6	936.2	25.7
173	56605	7.3	14.1350	1.4	1.5629	1.7	0.1602	1.0	0.59	958.0	8.9	955.6	10.5	950.1	27.9	958.0	8.9
233	46225	2.6	13.9229	2.8	1.6148	3.0	0.1631	1.1	0.36	973.8	9.9	976.0	18.8	980.9	56.8	980.9	56.8
290	69175	1.9	13.8741	3.0	1.6383	3.2	0.1649	1.2	0.36	983.7	10.6	985.1	20.3	988.1	61.3	988.1	61.3
181	72165	2.4	13.7557	1.5	1.6183	2.0	0.1615	1.3	0.66	964.9	11.7	977.3	12.4	1005.5	29.9	1005.5	29.9
395	153670	1.7	13.7092	1.7	1.6194	2.8	0.1610	2.3	0.80	962.4	20.1	977.8	17.6	1012.4	34.1	1012.4	34.1
102	45075	1.7	13.6187	2.1	1.6627	2.3	0.1642	1.0	0.44	980.2	9.1	994.4	14.5	1025.8	41.6	1025.8	41.6
169	58350	23.1	13.5581	2.5	1.6371	3.0	0.1610	1.6	0.52	962.2	13.9	984.6	18.7	1034.8	51.1	1034.8	51.1
343	81585	3.5	13.5510	2.1	1.7129	3.3	0.1683	2.7	0.80	1003.0	24.8	1013.4	21.4	1035.9	40.2	1035.9	40.2
195	143910	3.0	13.5288	1.6	1.7365	1.9	0.1704	1.2	0.60	1014.3	10.9	1022.2	12.5	1039.2	31.5	1039.2	31.5
149	153020	4.9	13.4958	3.1	1.7936	3.4	0.1756	1.4	0.42	1042.7	13.9	1043.1	22.3	1044.1	62.8	1044.1	62.8
118	34550	2.5	13.4821	2.6	1.7102	3.0	0.1672	1.4	0.46	996.8	12.6	1012.4	19.0	1046.1	53.3	1046.1	53.3
357	164815	3.7	13.4173	4.4	1.5548	6.2	0.1513	4.3	0.70	908.3	36.5	952.4	38.1	1055.9	88.7	1055.9	88.7
437	60685	1.2	13.4055	2.1	1.6421	3.9	0.1597	3.2	0.83	954.8	28.6	986.5	24.4	1057.6	42.9	1057.6	42.9
142	99225	4.4	13.3669	3.9	1.5829	5.0	0.1535	3.1	0.62	920.3	26.4	963.5	31.1	1063.4	79.3	1063.4	79.3
308	75850	1.7	13.3639	2.5	1.7659	8.1	0.1712	7.7	0.95	1018.5	72.8	1033.0	52.8	1063.9	50.7	1063.9	50.7
596	110020	22.1	13.2798	2.3	1.6626	5.0	0.1601	4.4	0.89	957.5	39.5	994.4	31.6	1076.6	45.6	1076.6	45.6
77	22050	2.7	13.2578	4.9	1.7902	5.0	0.1721	1.0	0.20	1023.8	9.5	1041.9	32.5	1079.9	98.0	1079.9	98.0
632	131425	2.9	13.1722	3.1	1.6320	7.3	0.1559	6.7	0.91	934.0	57.9	982.6	46.3	1092.9	62.1	1092.9	62.1
240	62105	1.3	13.1405	1.3	1.8856	1.7	0.1797	1.0	0.60	1065.4	9.8	1076.0	11.0	1097.7	26.4	1097.7	26.4
245	63750	3.3	13.0188	1.2	1.9102	2.3	0.1804	1.9	0.86	1069.0	19.1	1084.7	15.1	1116.3	23.4	1116.3	23.4

U	206Pb	U/Th	206Pb*	±	207Pb*	±	206Pb*	±	error	206Pb*	±	207Pb*	±	206Pb*	±	Best age	±
67	14780	3.4	12.9971	3.6	1.8563	4.2	0.1750	2.2	0.51	1039.5	20.6	1065.7	27.8	1119.6	72.1	1119.6	72.1
273	131670	6.1	12.8997	3.0	1.8371	4.9	0.1719	3.8	0.79	1022.4	36.2	1058.8	32.0	1134.6	59.5	1134.6	59.5
242	166460	1.4	12.8364	1.2	1.9616	1.8	0.1826	1.4	0.76	1081.3	13.9	1102.4	12.3	1144.4	23.5	1144.4	23.5
132	44335	3.4	12.8266	2.8	2.0791	3.0	0.1934	1.0	0.34	1139.8	10.4	1141.9	20.3	1145.9	55.3	1145.9	55.3
129	55470	3.2	12.8114	1.6	2.1384	2.1	0.1987	1.4	0.66	1168.3	15.2	1161.3	14.9	1148.3	32.0	1148.3	32.0
268	62045	3.4	12.7971	1.1	2.1083	1.6	0.1957	1.2	0.74	1152.1	12.8	1151.5	11.3	1150.5	21.9	1150.5	21.9
204	92235	2.5	12.7927	1.5	2.0901	2.0	0.1939	1.3	0.66	1142.6	14.0	1145.6	13.9	1151.2	30.2	1151.2	30.2
471	162380	4.8	12.7808	1.7	1.9166	3.4	0.1777	3.0	0.87	1054.2	29.0	1086.9	22.9	1153.0	33.6	1153.0	33.6
94	39370	1.2	12.7387	2.8	2.0581	3.0	0.1901	1.0	0.34	1122.2	10.3	1135.0	20.3	1159.6	55.6	1159.6	55.6
92	31350	0.8	12.7128	2.3	2.0509	3.7	0.1891	2.9	0.78	1116.5	29.3	1132.6	24.9	1163.6	45.0	1163.6	45.0
351	119845	1.1	12.6918	1.9	2.1203	2.4	0.1952	1.6	0.65	1149.3	16.5	1155.4	16.7	1166.9	36.7	1166.9	36.7
241	72590	2.5	12.6819	1.0	2.1751	1.4	0.2001	1.0	0.70	1175.7	10.7	1173.1	9.9	1168.4	20.0	1168.4	20.0
220	148075	4.8	12.6763	2.1	2.0923	4.8	0.1924	4.3	0.90	1134.1	44.9	1146.3	33.0	1169.3	41.6	1169.3	41.6
772	121430	4.2	12.6491	2.6	1.8997	5.4	0.1743	4.8	0.87	1035.7	45.5	1081.0	36.2	1173.5	52.3	1173.5	52.3
70	9760	1.8	12.6067	2.6	1.7891	3.2	0.1636	1.8	0.57	976.6	16.4	1041.5	20.7	1180.2	51.6	1180.2	51.6
185	50300	3.0	12.5611	1.8	2.1231	2.7	0.1934	2.0	0.75	1139.9	21.0	1156.3	18.6	1187.3	35.4	1187.3	35.4
538	181480	2.3	12.5352	1.7	2.0457	2.4	0.1860	1.6	0.69	1099.6	16.6	1130.9	16.3	1191.4	34.3	1191.4	34.3
184	64655	1.8	12.5331	1.6	2.1877	1.9	0.1989	1.0	0.54	1169.2	10.7	1177.1	13.0	1191.8	31.0	1191.8	31.0
216	29240	2.9	12.4548	1.6	1.8255	4.6	0.1649	4.3	0.94	984.0	39.4	1054.7	30.1	1204.1	30.5	1204.1	30.5
648	64865	3.0	12.3798	3.8	2.0693	5.0	0.1858	3.3	0.66	1098.5	33.7	1138.7	34.5	1216.0	74.2	1216.0	74.2
153	106115	2.2	12.3252	3.2	2.1904	3.7	0.1958	1.8	0.49	1152.7	19.0	1178.0	25.8	1224.7	63.5	1224.7	63.5
260	72780	2.3	12.2890	2.3	2.3024	2.5	0.2052	1.0	0.39	1203.3	11.0	1213.0	18.0	1230.5	45.7	1230.5	45.7
451	296585	23.9	12.2503	3.3	2.2196	3.7	0.1972	1.6	0.43	1160.3	17.1	1187.2	26.0	1236.7	65.5	1236.7	65.5
323	93820	2.1	12.1228	3.7	2.2935	4.6	0.2017	2.7	0.59	1184.2	29.2	1210.3	32.2	1257.1	71.8	1257.1	71.8
539	226165	4.5	12.0447	1.2	2.3000	1.6	0.2009	1.0	0.64	1180.3	10.8	1212.3	11.0	1269.8	23.2	1269.8	23.2
313	160110	8.0	12.0110	1.3	2.6135	3.0	0.2277	2.7	0.89	1322.3	31.8	1304.4	21.9	1275.2	26.1	1275.2	26.1
195	52755	1.8	11.9707	2.0	2.3492	2.7	0.2040	1.9	0.69	1196.6	20.2	1227.3	19.2	1281.8	38.2	1281.8	38.2
366	100525	3.8	11.9134	5.9	2.2525	7.2	0.1946	4.1	0.58	1146.4	43.5	1197.6	50.5	1291.1	114.1	1291.1	114.1
205	102540	2.9	11.7945	2.1	2.5124	3.4	0.2149	2.6	0.77	1255.0	29.8	1275.6	24.5	1310.6	41.5	1310.6	41.5
205	102540	2.9	11.7945	2.1	2.5124	3.4	0.2149	2.6	0.77	1255.0	29.8	1275.6	24.5	1310.6	41.5	1310.6	41.5
248	109930	1.7	11.7158	2.9	2.4425	8.9	0.2075	8.4	0.94	1215.7	93.0	1255.2	64.1	1323.6	56.8	1323.6	56.8
370	100165	1.7	11.4239	2.6	2.4966	7.4	0.2069	7.0	0.93	1212.0	76.8	1271.0	54.0	1372.3	50.8	1372.3	50.8
134	85690	2.8	11.3613	1.8	2.7651	2.3	0.2278	1.4	0.63	1323.2	17.0	1346.2	16.8	1382.9	33.6	1382.9	33.6
408	41555	3.2	11.1950	4.2	2.4862	5.2	0.2019	3.2	0.61	1185.3	34.4	1268.0	37.9	1411.1	79.5	1411.1	79.5
667	158940	8.2	11.1108	3.9	2.6701	4.3	0.2152	1.7	0.40	1256.3	19.6	1320.2	31.8	1425.6	75.3	1425.6	75.3
257	85185	2.1	7.7723	3.4	6.5784	3.6	0.3708	1.0	0.28	2033.3	17.4	2056.5	31.5	2079.8	60.4	2079.8	60.4
157	350120	2.5	4.0278	2.1	21.5463	2.3	0.6294	1.0	0.43	3147.3	24.9	3163.5	22.7	3173.8	33.6	3173.8	33.6

Table 1. U-Pb detrital zircon data obtained from 86 zircons, Sample LB-18, Late Triassic, La Ballena, Sierra de Salinas

3. The early mesozoic record in central and northeastern Mexico

The Mesa Central in central to northeastern Mexico is a plateau situated between the Sierra Madre Oriental to the East and the Sierra Madre Occidental to the west. The Mesozoic rocks of the Mesa Central and Sierra Madre Oriental provinces, are mostly covered by Cenozoic sedimentary and volcanic successions and the more expanded Mesozoic outcrops in this region are cretaceous limestone. Triassic and Lower Jurassic rocks are subordinated and occur regularly in uplifted areas. The Late Jurassic-Cretaceous cover consisting mostly of limestone and marls in this region shows compressive deformation produced during the Laramide orogeny and is frecuently detached from the underlying Triassic-Lower Jurassic succession, mostly composed of siliciclastic and volcanic rocks. There are in the Mesa Central three widely recognized Triassic to Middle Jurassic stratigraphic units (Barboza-Gudiño et al., 1998, 1999): (1) Upper Triassic Submarine fan deposits in Central Mexico (Zacatecas Formation), (2) A Lower to Middle Jurassic volcanic succession (Nazas Formation, Pantoja-Alor, 1972), and (3) A Middle Jurassic fining upward succession of red beds that general change gradually from conglomerate or breccias on their basis to sandstone and mudstone on the top, changing transitionally into Oxfordian transgressive shallow marine limestone. In the Sierra Madre Oriental, The corresponding Late Triassic-Middle Jurassic stratigrapy consists of three units, comparable in age but not in their facies with the units exposed in the Mesa central: (1) Late Triassic rocks interpreted as a fluvial facies, exposed in southern Nuevo León and Tamaulipas, and defined as El Alamar Formation (Barboza-Gudiño et al., 2010), (2) Lower Jurassic red beds and interlayered volcanic and volcanoclastic deposits, defined as La Boca Formation (Mixon et al, 1959) or Hizachal-Formation (Imlay et al., 1948, Carrillo-Bravo, 1961), (3) Finnaly, La Joya Formation (Mixon et al, 1959), consisting of breccias or conglomerates and red sandstones, representing a widely identified erosional unconformity. Such propossed stratigraphic division is a result of field observation and description of sedimentary facies based on lithological-sedimentological studies supported by data obtained by most of the previously described analytical methods (Figure 4, 5).

3.1 Upper Triassic Submarine fan deposits in Central Mexico (Zacatecas Formation)

Upper Triassic marine rocks in the Mesa Central province, were first described by Burckhardt and Scalia (1905) at Arroyo La Pimienta in the vicinity of Zacatecas city (Figure. 1), who described a light metamorphosed succession composed of greenstone, sandstone and alternating shale or "phylithe" containing a triassic fauna which include several ammonites, and bivalves. These rocks were first named "Triásico de Zacatecas" (Gutierrez-Amador, 1908) and are currently known as the "Zacatecas Formation", (Carrillo-Bravo 1968 in Silva-Romo et al., 2000, Martínez-Pérez., 1972, Carrillo-Bravo 1982). Outcrops of comparable rocks were later reported near La Ballena Zacatecas (Cantú Chapa, 1969), Charcas and Sierra de Catorce, San Luis Potosí (Martínez-Pérez, 1972). The age of the strata exposed in this localities was supported by fauna of amonoids in La Ballena or Sierra de Salinas (Gómez-Luna et al., 1998) and amonoids (Cantú-Chapa, 1969 and Gallo-Padilla et al., 1993) and conodonta (Cuevas Pérez., 1985) in the Charcas area, remaining unknown in the Sierra de Catorce and other outcrops in northern Zacatecas and northwestern San Luis Potosí because a lack of fossils.

With minor differences between all studied localities, the Zacatecas Formation consists of a siliciclastic succession, mostly composed of interstratified sandstone, siltstone, shale and conglomeratic sandstone. At Arroyo La Pimienta, Zacatecas, the Triassic Succession consists of alternating dark gray to brown shale and thin to middle bedded sandstone, the upper part of the exposed sequence, consists of fossiliferous black shale or phylite with quartzite lenses, at this locality, Centeno-García (2005) reported ancient MORB remnants underlying the Triassic succession. In La Ballena, at the border zone between Zacatecas and San Luis

Potosí states, The Zacatecas Formation, named also locally La Ballena Formation (Centeno-García and Silva-Romo, 1997), consists of interstratified sandstone, siltstone, shale and conglomeratic sandstone. At Sierra de Charcas, west of Charcas, San Luis Potosí, in La Trinidad Anticlinorium and several minor outcrops to the north in the Santa Gertrudis area, (Figure 1), the triassic succession consists of turbiditic sandstones, conglomeratic sandstones, and greywacke, alternating with siltstone and shale. The greywacke and sandstone beds contain internally graded bedding, commonly showing partial developed Bouma sequences, load and groove casts are the most common sole marks, as well as slump deposits and wildflysh. The succession at Sierra de Catorce in northern San Luis potosí consists of finely laminated shale and intercalated thin siltstone and sandstone layers. There are no reports of Triassic fossils in this locality, and an older age has been also suggested by a possible Late Paleozoic flora (Franco-Rubio, 1999) and Late Paleozoic spores (Bacon, 1978). Finally intensely deformed and poorly understood successions consisting of probable triassic sandstone and shale containing ophiolitic rocks and older exotic blocks, are known as Taray Formation in northern Zacatecas (Córdoba-Méndez, 1964), El Chilar Complex, Queretaro, recently dated as a possible Late triassic deposit (Davila-Alcocer at al., 2008) and other outcrops in the Sierra de Guanajuato and Durango.

The most recent studies from triassic rocks in western San Luis Potosí and Zacatecas include sedimentologic, stratigraphic, geochronologic and tectonic studies (Centeno-García and Silva-Romo,1997; Silva-Romo et al., 2000; Hoppe et al., 2002, Bartolini et al., 2001, Barboza-Gudiño et al., 2010), interpreting all Triassic marine successions in the Mesa Central, as part of a submarine fan, named the "Potosí fan" by Centeno-García (2005). All facies are compatible with those of a submarine fan, including facies "A" (after Mutti and Ricci Lucchi, 1972) that represent channels, facies "B" and "C", representing channel margins and facies "D", "E" "F" and "G" (suprafan lobe, levee, and inter-channel flats), corresponding in the Charcas outcrops to a midfan or suprafan zone. The most common facies associations in Real de Catorce are lithofaces "D", "E", and "G". in La Ballena the lithofacies succession , interpreted as middle to lower fan include well developed "C" and "B" facies, "A" facies in La Ballena correspond to channel margins, suprafan lobes and channel environments, respectively.

There are no known exposures of the base of the Zacatecas Formation in the Mesa Central, but the presence of oceanic crust supposedly underlying the Zacatecas Formation at the vicinity of Zacatecas city allow to interpret a remnant basin for the latest Paleozoic and Early Mesozoic time, floured by oceanic crust at the western margin of Pangea, where some of the earliest deposits of the Potosí Fan (Zacatecas Formation), occurred during the Middle to Late Triassic time. Eastward, the first deposits of the Potosí Fan, also Middle to Late Triassic in age, rest hypothetical, over an older, Precambrian-Paleozoic crust, like the Oaxaquia block (Ortega-Gutierrez et. al., 1995), as indicated by upper crustal xenoliths contained in volcanic rocks of the same region. The thickness is also very difficult to estimate, because the strongly deformed strata and the previously mentioned lack of exposures incliuding their base. As a structural thickness of this rock body which include a considerable structural increase, can be mentioned the Tapona-1 Well drilled by PEMEX in the Sierra de la Tapona, northwest of Charcas, where the total depth of the well represent Triassic turbidites of the Zacatecas Formatio, without reaching the base (PEMEX internal report, cited in Tristán-González et al., 1995).

Fig. 4. A. Triassic turbidites of the Zacatecas Formation (Charcas), B. Conglomeratic sandstones in El Alamar Formation (San Marcos), C. Lower Jurassic greenstone, quarzite and red to yellow mudstones, Capas Cerro El Mazo (Real de Catorce), D. Spherulitic Rhyolite, Nazas Formation (Sierra de San Julián), E. Lower Jurassic red sandstone, La Boca Formation (Huizachal), F. Distorted accretionary lapilli in epiclastic layers of La Boca Formation (La Boca Canyon), G. Polymictic breccia, La Joya Formation (San Marcos), H. La Joya Formation (Js Br.) overlies Paleozoic schist and phylites (Pz sh.), by no deposition of Triassic and Jurassic red beds, the breccias change gradually upwards into shallow marine limestone of the Upper Jurassic Novillo Formation, an equivalent of the basal part of the Zuloaga Formation (Aramberri)

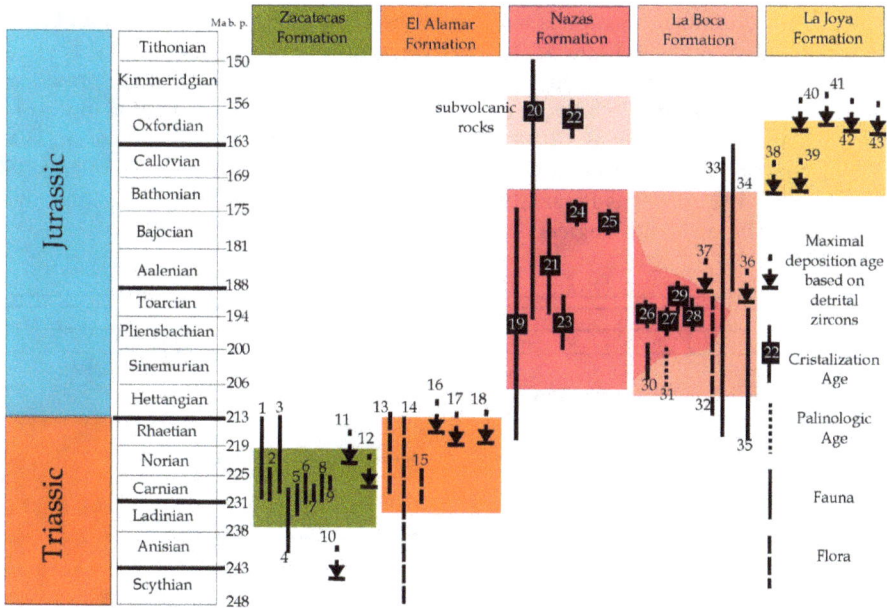

Fig. 5. Paleontologic, palinologic and isotopic ages of pre-Oxfordian units from Central and northeastern Mexico: Data and sources: 1. *Avicula hofmanni, Cassianella (Burkhardtia)* aguilerae, *Halovia austriaca* mojsisovics, *Palaeoneilo aguilerae*. (Zacatecas, Burkchardt and Scalia, 1905); 2. *Juvavites* sp., *Pleurotoma* sp.(Zacatecas, Gutierrez-A., 1908); 3. *Pseudomonotis* sp., (Zacatecas, Maldonado-K.,1948); 4. *Berichitidae Spath*, 1934 (La Ballena, Gallo-P. et al., 1993); 5. *Clionitidae* Tozer, 1994, *Trachiceratidae* Haug, 1994 (La Ballena, Gómez-L. et al., 1997); 6. *Sirenites* sp. (La Ballena, Chavez-A., 1968); 7. *Juvabites* sp. (Charcas, Cantú-Ch., 1968); 8. *Aulacoceras* sp. (Charcas, Gallo-P. et al., 1993); 9. *Neogondolella polygnatiformis, Epigondolella Primitia* (Charcas, Cuevas-P., 1985); 10. This work (La Ballena); 11. Charcas (Barboza-Gudiño et al., 2010); 12. Real de Catorce (Barboza-Gudiño et al., 2010); 13. *Podosamites* sp. (Source of Novillo C., Mixon et al, 1959); 14. *Araucarioxylon* (La Boca C., Mixon et al, 1959); 15. *Laurozamites yaqui* (reported as *"Pterophyllum fragile"* by Mixon et al, 1959), *Ctenophyllum braunianum* (reported as *"Pterophyllum inaequale"*, by Mixon et al., 1959), *"Elatocladus ex* gr. *Carolinensis* (reported as *"Cephalotaxopsis carolinensis"* by Mixon et al., 1959), (Source of Novillo C., Weber, 1997); 16. San Marcos, N.L. (Barboza-Gudiño et al., 2010); 17. Cañón de La Boca (Barboza-Gudiño et al., 2010); 18. San Marcos, N.L. (Barboza-Gudiño et al., 2010); 19-20. Rb-Sr, (w), N Zacatecas, (Fries & Rincon-O, 1965); 21.K-Ar,(h),Rodeo, (López-I., 1986); 22. U-Pb, (zr), Caopas (Jones et al, 1995); 23. 40Ar/Ar39(pl), Villa Juárez (Bartolini & Spell, 1997); 24. U-Pb (zr), Real de Catorce (Barboza-Gudiño et al., 2004); 25. U-Pb (zr) Charcas (Zavala-Monsivais et al., in press.); 26. U-Pb (zr) Aramberri (Barboza-Gudiño et al., 2008); 27. U-Pb(zr), Huizachal C. (Fastovsky et al., 2005); 28. U-Pb (zr), Huizachal (Zavala-M. et al., 2009); 29. U-Pb (zr), Aramberri (Zavala-M. et al., 2009); 30. *Bocatherium mexicanum*, Huizachal C. (Clark et al, 1994); 31. Palinomorpha, La Boca C. (Rueda-Gaxiola et al., 1993); 32. Williamsonia netzahualcoyotl (Carrillo-Bravo, 1961); 33. Pterosaurio, Huizachal Canyon., *Sphenodon* sp.. Nov., Huizachal (Reynoso-Rosales, 1992); 35. *Cynosphenodon huizachalensis*, Huizachal (Reynoso-Rosales (1996); 36-41. Huizachal (Rubio and Lawton, 2011); 42. Miquihuana (Barboza-Gudiño and Zavala-Monsivais, 2011); 43. Real de Catorce (Barboza-Gudiño and Zavala-Monsivais, 2011).

The age of the Zacatecas Formation is well established in Zacatacas, La Ballena and Charcas, through their fossil fauna (Figure 4). In The Sierra de Catorce, Sierra de Las Teyra and several other minor outcrops in the Presa de Santa Gertrudis area, aren't any byostratigraphic ages available because a Lack of fossils. For age determination in the Sierra de Catorce, Barboza-Gudiño et al. (2010) provided a maximal age of deposition of ca. 230 Ma (Figure, 5), consistent with a previously interpreted Late Triassic age (Martínez-Pérez, 1972, López-Infanzón, 1986, Cuevas-Pérez, 1985, Barboza-Gudiño et al., 1999), based only on stratigraphic position and lithological similarities.

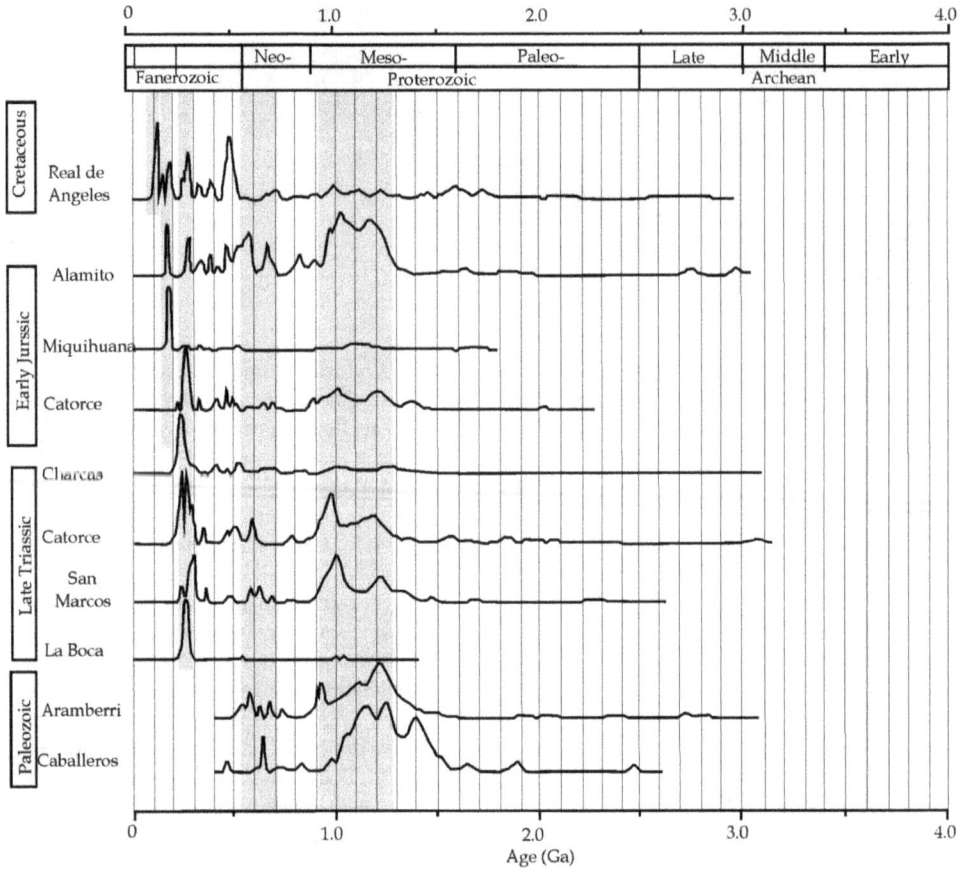

Fig. 6. Probability curves of several published detrital zircon results from Paleozoic and Mesozoic rocks from northeastern Mexico (Venegas-Rodríguez et al., 2009; Barboza-Gudiño et al., 2010; Barboza-Gudiño et al., 2011; Barboza-Gudiño & Zavala Monsiváis, 2011).

Figure 5 include results from two detrital zircon analyses from samples collected in the Zacatecas Formation from Charcas and Real de Catorce; these results were previously reported by Barboza-Gudiño et al (2010). The results are plotted as age-probability curves (Ludwig, 2003) besides results from other stratigraphic units discussed also in this chapter. A maximal depositional age between 225-230 Ma for sample collected in the Sierra de

Charcas, correspond also to the Late Triassic similar to the sample collected in the Sierra de Catorce. Both samples show notable contributions of a Permo-Triassic zircons (245-280 Ma), which correspond with the east Mexico Permo-Triassic magmatic arc, that yield K-Ar and Rb-Sr ages from 284 to 232 Ma (Torres et al., 1999, Dickinson and Lawton, 2001). Paleozoic ages between 420-467 Ma, correspond probably to Ordovician-Silurian magmatic rocks described in peri-Gondwanan terranes of Mexico like the Acatlán Complex (Miller et al., 2007) or as detrital zircon age populations present in the Granjeno Schist in northeastern Mexico (Nance et al., 2007, Barboza-Gudiño et al., 2011) or El Fuerte Formation in Sinaloa (Vega-Granillo et al., 2009). Populations corresponding to the pan-African (700-500 Ma) and Grenvillian (900-1300 Ma) events, as well as subordinate Paleoproterozoic to Archean zircons are also present in both samples.

Sediment-petrographic studies were also performed to interpret provenance by point counting in the sandstones of the Zacatecas Formation, the results, plotted in the provenance diagrams of figure 2, suggest continental block and recycled orogen provenances. Geochemical results (Barboza-Gudiño et al., 2010) are indicative of provenance from igneous rocks, accord to the Chondrite normalized REE, showing a negative Eu anomaly, formed by intracrustal differentiation including plagioclase fractionation. There is a notable similarity with LREE enrichment and a flat HREE sector in all Triassic samples. The relations $Th/Sc \approx 1$ and $Zr/Sc \approx 10$-100 are product of zircon addition, indicative of sediment recycling, as typical for trailing edge turbidites in a passive margin (McLennan et al., 1993). In addition, the initial ε_{Nd} ratios as reported by Centeno-García and Silva-Romo (1997) in sandstones collected in La Ballena and Zacatecas, are -5.2 and -5.5 respectively, and are indicative of an old upper continental crust provenance, as well as the Nd-model ages of 1.3 to 1.6 Ga in agreement with a model of source in an old continental block at the east-northeast of the region, like the Proterozoic Oaxaquia microcontinent.

3.2 Upper Triassic fluvial succession in the Sierra Madre Oriental: El Alamar formation

The Upper Tri assic succession of continental strata exposed in the Sierra Madre Oriental was defined as El Alamar Formation by Barboza-Gudiño et al. (2010). Previously Upper Triassic and Lower Jurassic rocks in northeastern Mexico were referred as the Huizachal Group (Mixon et al., 1959), consisting of the Upper Triassic to Lower Jurassic La Boca Formation and the unconformable overlying La Joya Formation of Middle to Late Jurassic age. After Barboza's definition (op.cit.) of the only Triassic El Alamar Formation, La Boca Formation in consequence consists of a red beds and interlayered volcanic and volcanoclastic succession Early to Middle Jurassic in age. A detailed description of the evolving stratigraphic nomenclature of the early Mesozoic units in the region is given by Barboza-Gudiño et al., (2010).

The name El Alamar Formation is derived from El Alamar Canyon in the Sierra de Pablillo, Nuevo León, where the proposed unit stratotype and type section are located at El Alamar Canyon, were the exposed sequence consists of more than 350 m of mostly gray and brawn-red colored conglomeratic sandstones, siltstones and mudstones. In the Huizachal-Peregrina anticlinorium in Tamaulipas incomplete sections of El Alamar Formation rests unconformably on Paleozoic metamorphic, sedimentary and magmatic rocks. The best exposures are Alamar canyon in the Sierra de Pablillo and the San Marcos area south of

Galeana, Nuevo León, along federal highway 58 (San Roberto-Linares). El Alamar Formation is the oldest unit exposed in the Galeana region and there are no exposures of their basis or any older strata.

El Alamar Formation consist of thick bedded, medium- to coarse-grained arkosic sandstone, usually containing basal conglomeratic lag horizons, changing upwards into finely laminated sandstones-siltstones and interlayered mudstones. The most abundant primary structures are trough cross-beds and channel scours, tabular burrows are common in fine-grained facies. There is a notable abundance of petrified wood (possible *Araucarioxylon* sp.), commonly associated with conglomerate and coarse grained sandstone facies, which represent channel and channel bar deposits. Several decimeter cylindrical burrow casts interpreted as possible rhizoliths or probable lungfish burrows are associated with the siltstone and mudstone facies, which represent floodplain deposits. These lithologies represent upward -fining cycles, tens of meters in thickness, and are interpreted as basal channels overlain by sand flats and overbank deposits, formed in low sinuosity streams and braided channels. Michalzik (1991) interpreted this cyclic deposition as a Donjek type fluvial system, and recognized different lithofacies after codes of Miall (1977): conglomerate facies (Gm, Gt), trough and planar cross-beds as well as horizontal-bedded coarse sandstone (St, Sp, Sh), horizontal bedded and plannar cross-bedded sandstone (Sh, Sl), laminated and massive siltstone facies (Fl, Fm), and mudstone to siltstone facies with carbonate concretions (Fm, P).

The facies associations recognized in El Alamar Formation correspond to proximal alluvial fan, braided stream and distal meandering stream deposits. The Triassic strata in Tamaulipas unconformably overlie Paleozoic strata or Precambrian-Paleozoic basement and are overlain by Jurassic redbeds of La Boca Formation, during in Nuevo Leon their base is not exposed and there is no evidence of deposition of the Lower Jurassic La Boca Formation in this area, where El Alamar Formation is in turn unconformably overlain by the Middle to Upper Jurassic La Joya Formation. The El Alamar Formation is absent in the Aramberri-Miquihuana area in southern Nuevo León and Tamaulipas where Jurassic red beds and volcanic rocks or in some places Upper Jurassic Limestones rest on Paleozoic metamorphic rocks.

Mixon et al. (1959) reported in the lower unit of their Huizachal Group or actual El Alamar Formation, a floral assemblage of Late Triassic age, reinterpreted by Weber (1997)suggesting that this flora is well indicative of the Late Triassic but more precisely of a Carnian and probably Norian age. A Late Triassic age is also in agreement with an Early Triassic maximal age of deposition (245 Ma), accord to detrital zircon geochronology. The El Alamar Formation is thus of the same age as the Late Triassic Zacatecas Formation, which represents the marine counterpart of the El Alamar fluvial system. We show below that they also have very similar provenance characteristics.

3.3 Lower to Middle Jurassic volcanic arc in central and north-central Mexico: Nazas formation

Volcanic and volcano-sedimentary successions rest unconformable on the Triassic Zacatecas Formation and are known in north-central to northeastern Mexico as the Nazas Formation (Pantoja-Alor, 1972). These volcanic rocks were assigned to the Late Triassic-Lower Jurassic volcanic arc, related to the active continental margin of western North America (Blickwede,

2001; López-Infanzón, 1986, Grajales-Nishimura et al., 1992; Jones et al., 1995; Bartolini, 1998; Bartolini et al., 2003; Barboza-Gudiño et al., 1998, 1999, 2004). The type locality of the Nazas Formation, as defined by Pantoja–Alor (1972) is the Cerritos Colorados area west of Villa Juarez, in northern Durango. The most common volcanic rocks in the type locality are rhyolitic ash flow tuffs including well preserved gray to green colored ignimbrite horizons, alternating with red-brown epiclastic materials.

In other localities of the Mesa Central, the volcanic products of the Nazas Formation are intermediate to felsic. In the Sierra de San Julián, in northern Zacatecas, Blickwede (2001) describes a 1,000 m thick volcanic succession consisting of lava flows, air-fall and ash-flow tuffs, and lahars. In the Caopas-Rodeo uplift, porphyritic rhyolite with quartz and sanidine phenocrysts of the Caopas schist and the andesitic Rodeo formation, first considered, to be pre-Jurassic because of their strongly deformed aspect (de Cserna, 1956, Córdoba-Méndez, 1964), are coeval with the Nazas Formation accord to later geochronologic studies (López-Infanzón, 1986, Jones et al., 1995, (Table 2).

In western San Luis Potosí, volcanic successions including andesitic and dacitic lava flows and volcanic breccias, as well as rhyolitic domes and ash flow tuffs, are exposed in the Sierra de Catorce, Charcas and Sierra de Salinas or La Ballena area, they rest on Triassic turbiditic layers of the Zacatecas Formation or local in Real de Catorce, on the marginal marine beds of the Lower Jurassic informal unit "Capas Cerro El Mazo" (Barboza-Gudiño et al., 2004). The Capas Cerro el Mazo unit consists of conglomeratic sandstone and medium- to coarse-grained litharenites with fragments of plants, and gray to green and red to purple siltstone and mudstone layers interfingered with volcanic greenstones at the base of the Nazas Formation, which include dacitic and rhyolitic pyroclastic and porphyritic rocks. Detrital zircon geochronology in this succession support a maximum Early Jurassic depositional age and three primary sources of detrital zircons that include Grenvillian (~900-1200 Ma) and Pan-African basement rocks (~500-700 Ma) as well as the Permo-Triassic magmatic arc (~245-280 Ma). Petrographic studies indicate a recycled orogen and continental block provenance. Barboza-Gudiño et al. (2004) reported U-Pb isotopic analyses of zircon for a rhyolite in the upper part of the succession of the Sierra de Catorce, yielded an age of 174.7 ± 1.3 Ma.

In the Sierra de Salinas, basaltic-andesitic fluidal, porphyritic lava is the dominant rock type, composed of probable hornblende phenocrysts, scarce pyroxene, olivine and abundant acicular plagioclase as the main component of the groundmass. In the Sierra de Salinas and some outcrops of the Sierra de Charcas, the lavas are brecciate at the base of the andesitic flows, arranged in a *"puzzle structure"*, like an autoclastic breccia, related to a flow front or basal breccias, engulfed by igneous material of the same composition. General, the most common rocks in the Sierra de Charcas are andesitic to rhyolitic pyroclastic products, Including breccias, lapilli tuffs, and ash flow tuffs.

The volcanic successions described here are part of the Early Jurassic volcanic arc, related to the ancient active margin of Pangea. The available ages (Tables 2) indicate that the volcanic arc was probably active for a period of 40 Ma during the Jurassic. The volcanic rocks of the described localities correlate with the Nazas Formation of northern Durango and Zacatecas and, therefore, represent a key unit for the stratigraphic subdivision, and paleogeographic and paleotectonic interpretations of north and northeastern Mexico.

Locality	State	Rock type	Method	Material dated	Age (Ma)	Source
Caopas	Zacatecas	meta-rhyolitic sub-volcanic	Rb-Sr	whole rock	195±20	Fries & Rincon-O. 1965
Caopas	Zacatecas	meta-rhyolitic sub-volcanic	Rb-Sr	whole rock	156±40	Fries & Rincon-O. 1965
Caopas	Zacatecas	meta-andesite (Rodeo Form.)	K-Ar	hornblende	183±8	López-Infanzón, 1986
Caopas	Zacatecas	meta-rhyolitic sub-volcanic	U-Pb	zircon	158±4	Jones et al., 1995
Villa Juárez	Durango	rhyolite	$^{40}Ar/^{39}Ar$	plagioclase	195.3±5.5	Bartolini & Spell, 1997
Catorce	San Luis Potosí	rhyolite	U-Pb	zircon	174.7±1.3	Barboza-Gudiño et al., 2004
Charcas	San Luis Potosí	rhyolitic ignimbrite	U-Pb	zircon coherent gr.	176.8 +4.9/-1.7	Zavala et al. In press
Huizachal	Tamaulipas	rhyolitic ash flow	U-Pb	zircon	189.0±0.2	Fastovsky et al., 2005
Aramberri	Nuevo León	rhyolitic ignimbrite	U-Pb	zircon	193.1±0.3	Barboza-Gudiño et al., 2008
Huizachal	Tamaulipas	rhyolite	U-Pb	zircon	194.1 +4.1/-4.5	Zavala-M. et al., 2009
Aramberri	Nuevo León	rhyolitic ignimbrite	U-Pb	zircon	189.5 ±3.8	Zavala-M. et al., 2009

Table 2. Selected isotopic ages of Jurassic volcanic rocks from the Nazas Formation in north-central to northeastern Mexico. For location of the areas see figure 1.

In Nuevo León state, volcanic rocks are exposed a few kilometers north of Aramberri, consisting of ignimbrites, volcanic breccias and tuffs of intermediate to felsic composition, which overlie Paleozoic schist and unconformable underlie transgressive Upper Jurassic strata. An U-Pb zircon age determination in a rhyolitic-rhyodacitic ignimbrite yields an essentially concordant age of 193.1 ± 0.3 Ma.

Lower Jurassic volcanic rocks are also exposed at Huizachal Valley, Tamaulipas (Jones et al., 1995, Fastovsky et al., 2005) and the Miquihuana-Bustamante area (Bartolini et al., 2003). In the Huizachal Valley the volcanic andesitic and rhyolitic rocks represent the basal part of the exposed Mesozoic succession, underlying and partially intruding Lower Jurassic red beds of La Boca Formation or more precisely, at this locality as in all exposures in Nuevo

León and Tamaulipas, the volcanic rocks are considered part of the La Boca Formation and occur in form of rhyolitic domes with steeply dipping flow bands resulting from magma injection, flow-like bands and lava flows or lobes with spherulitic structures and associated ash flow tuffs. Zircon grains from a pyroclastic flow at this locality were dated by U-Pb at 189 ± 0.2 Ma (Fastovsky et al., 2005) as well as new ages from Zavala-Monsivais et al. (2009).

Geochemical analysis from intermediate to felsic rocks shows a calc-alkaline character. Trace element abundances characteristic of rocks generated by subduction processes (Barboza-Gudiño et al, 2008), and generally related to the enrichment of large-ion lithophile elements (e.g., Rb, Ba, K) and light rare earth elements (e.g., La, Ce) in fluids and melts released from the subducting plate to the overlying mantle. The trace element patterns shown in the normalized multi-element diagram are a strong evidence for an origin in a continental arc setting. Rare earth element (REE) abundances are enriched in light REE relative to heavy REE and have a relatively steep slope for the LREE (La-Eu) and a flat pattern for the HREE (Gd-Lu). The most evolved samples (rhyolites), show negative Eu-anomalies, indicative of plagioclase fractionation.

The general features of the exposed volcanic sequences, the petrography of the diverse materials and the geochemical data, support the conclusion that all the studied pre-Oxfordian volcanic rocks originated in a continental arc. Our analyses provide a general idea of the compositional variations among the sequences or localities, which, however, are common in volcanic arcs composed of different volcanic centers. These variations also document the changes in composition of all volcanic products during magmatic evolution in space and time. The following observations are evidence for an origin of these rocks in a continental volcanic arc. The volcanic units are unconformably overlain by Upper Jurassic red beds of the La Joya Formation and shallow marine limestone of the Zuloaga Formation.

3.4 Lower Jurassic Fluvial, epiclastic and volcanogenic deposits (La Boca Formation)

Imlay et al. (1948) proposed the name Huizachal Formation for redbeds exposed in the Huizachal valley, 20 km southwest of Ciudad Victoria, Tamaulipas unkonformably underlying oxfordian limestone. Mixon et al. (1959) separated two units of redbeds in the Huizachal-Peregrina anticlinorium and defined La Boca Formation as the older unit and the younger, La Joya Formation, which represent an erosional unconformity as a basal conglomerate for the transgressive Callovian-Oxfordian limestones and evaporates coeval with opening of the Gulf of Mexico basin. Both units were defined by the same authors as the Huizachal group.

The La Boca Formation is well exposed in the Huizachal Peregrina Anticlinorium (Carrillo-Bravo, 1961; Rueda Gaxiola et al., 1993, 1999), the Miquihuana-Bustamante area in Tamaulipas and near Aramberri, Nuevo León. In the Huizachal-Peregrina Anticlinorium consists of ca. 1500 m of red sandstones, siltstones and mudstones, as well as interlayered polymictic matrix supported conglomerates and conglomeratic sandstones. The sandstones and siltstones are well stratiefed in medium sized beds, with internal fine lamination. Conglomeratig sandstones are thick bedded, with well developed curved cross lamination and pelitic rocks occur mostly as massive sized beds.

La Boca Formation also interfinger with volcanogenic deposits at several localities, , including Huizachal and La Boca Canyons and contains a fossil assemblage of vertebrate fauna, which allows assignment to an Early to Middle Jurassic age for red beds outcropping in the Huizachal Canyon (Clarck, et al., 1994). The volcanic activity was coeval with the Nazas arc activity and The continental redbeds assigned to the upper part of La Boca Formation were deposited probably during a period of crustal extension that followed the Early Jurassic magmatic arc activity (Fastovsky et al., 2005; Barboza-Gudiño et al., 2008), or in a back-arc setting related to the Nazas arc. As previously mentioned, the interlayered volcanic rocks yielded an age of 189.0 ± 0.2 Ma (U-Pb, zircon) in the Huizachal Valley (Fastovsky et al, 2005) and a 193 ± 0.2 Ma U-Pb zircon age (Barboza-Gudiño et al., 2008) for an ignimbrite of the Aramberri area. Volcanic arc provenances and notable content of early to middle Jurassic aged zircons, are characteristic of these units. The results of detrital zircon geochronology (Figure 5) are also in agreement with an Early Jurassic age of deposition for the succession, and provenances from Grenvillian-panafrican basement and permotriassic rocks. The La Boca Formation overlies unconformable the Triassic El Alamar Formation in La Boca canyon, a Permian turbiditic succession in the Peregrina Canyon of the Huizachal Peregrina Anticlinorium, Early Jurassic volcanic rocks in Aramberri and is in turn unconformably overlain by the callovian-oxfordian La Joya Formation.

3.5 Erosional unconformity and Middle Jurassic alluvial to lagunar deposits (La Joya Formation)

La Joya Formation was defined by Mixon et al. (1959) as Middle to Late Jurassic conglomeratic and redbeds sequence exposed in the Huizachal-Peregrina anticlinorium. They proposed the Rancho La Joya Verde in the Huizachal valley as type locality.

La Joya Formation in the Huizachal-Peregrina anticlinorium, consist of a fining upward megasequence composed of a basal polymictic breccia or conglomerate-fanglomerate facies followed by red sandstones, siltstones and mudstones. The clastic components are volcanic or plutonic, metamorphic and sedimentary rocks as well as abundant white quartz and brown-gray chert. The sandstones and siltstones are brown-red, occasionally with gray, purple and green layers. To the top changes gradually into lagoon or shallow marine deposits, containing interlayered evaporites, regularly gypsum and fine laminated limestone. The La Joya Formation changes typically in their thickness, varying from 0 to more than 200 m in the different localities from central to northeastern Mexico. La Joya Formation represents the basal strata of the marine Upper Jurassic-Cretaceous succession.

The interpretation of the different facies of La Joya Formation include several depositional environments as interpreted from Michalzik (1988), as follows: alluvial fan fanglomerates, channels and distal alluvial fan conglomerates, shallow marine carbonates and caliche crusts, finer grained alluvial plain deposits and lagoon to sabckha evaporites.

The age was established in accordance to their stratigraphic position, overlying Early Jurassic redbeds and volcanic rocks, and underlying Kallovian-Oxfordian gypsum and limestone. Detrital zircon geochronology results are also in agreement with a Middle to early Late Jurassic age for this unit, yielded maximal depositional ages between 175 and 178 Ma (Rubio-Cisneros and Lawton, 2011) and (Barboza-Gudiño and Zavala-Monsivais, 2011), considering that two of the results presented by Rubio-Cisneros and Lawton (2010) for La Boca Formation, correspond in the interpretation by this study, to La Joya Formation, following definition from Mixon et al. (1959).

4. Conclusions

Figure 7 shows the stratigraphic correlation of the Early Mesozoic units outcropping in central to Northeastern Mexico. Upper Triassic turbidities that appear in the Mesa central province and are known as Zacatecas Formation, correspond to a subsea fan known as the Potosí fan (Centeno-García, 2005). The Potosí fan was formed in a geoclinal setting at the western equatorial margin of Pangaea. Meanwhile in the Sierra Madre Oriental, an in age comparable fluvial succession, defined as El Alamar Formation (Barboza-Gudiño, et al., 2010), represents remnants of a river system known as El Alamar River, that was housed in a rift associated to the break up of Pangea. According to provenance and distribution of both, continental and marine Triassic deposits, it can be interpreted that El Alamar River fed the sedimentation in to the Potosí subsea fan.

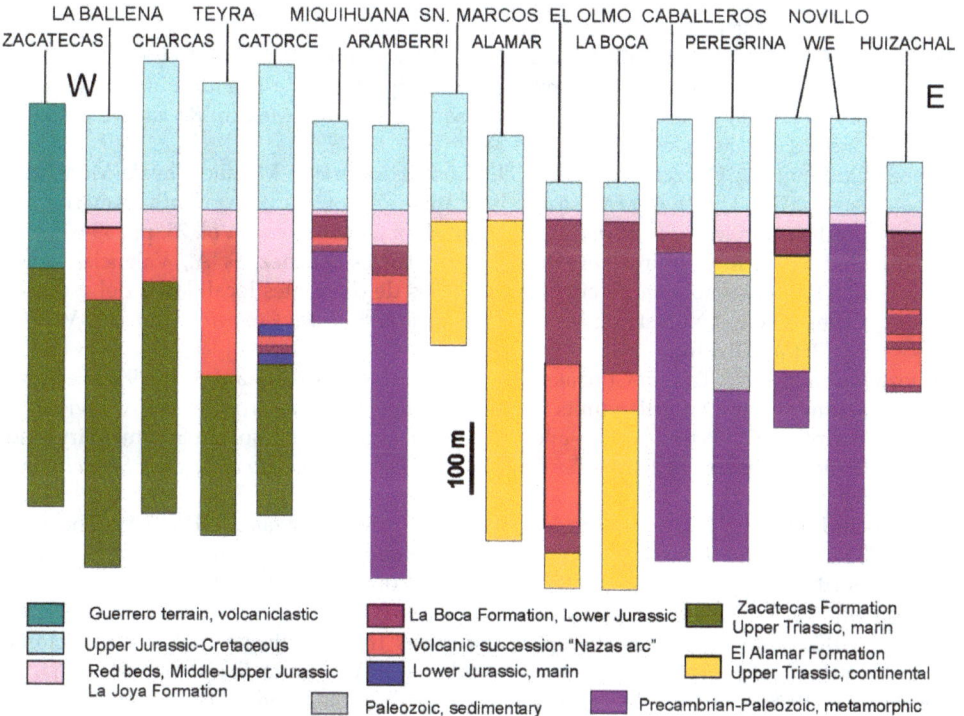

Fig. 7. Stratigraphic correlation of Early Mesozoic units from central to northeastern Mexico, for location see figure 1

During the earliest Jurassic time, continued in these region similar conditions as in the upper Triassic, where river systems drained continental blocks that had been part of Pangaea that now begining to disperse. At this time, fluvial and alluvial red beds of La Boca Formation (Mixon et al., 1959), known also as Huizachal Formation (Imlay, 1948, Carrillo-Bravo, 1961), were deposited to the mainland and towards the Pacific were deposited the marine facies of the Huayacocotla Formation, both units, contemporary and in some places interstratified with the early Jurassic volcanic arc or Nazas Formation (Pantoja-Alor, 1972).

Finally La Joya Formation (Mixon et al., 1959), initially considered part of the Huizachal Group, besides La Boca Formation, represents an erosional unconformity linked to the opening of the Golf of Mexico basin as a brake up unconformity and in this sense the basal unit of the Middle to Late Jurassic marine succession, related to a transgression coming from the east. During the Late Jurassic-earliest Cretaceous, the pacific connection from the central Mexico basin was closed through the presence of the composed Guerrero superterrane, an intraoceanic arc complex linked to subduction processes evolved to the west of the ancient early Mesozoic subduction zone. At this time, the only connection from eastern Mexico to an oceanic basin was to the Atlantic, through the new opened Gulf of Mexico.

5. References

Bacon, R. W. (1978). Geology of the northern Sierra de Catorce, San Luis Potosí, México (master thesis): *Arlington, University of Texas*, 124 p.

Barboza-Gudiño, J.R., Hoppe, M., Gómez-Anguiano M., and Martínez-Macías, P.R. (2004). Aportaciones para la interpretación estratigráfica y estructural de la porción noroccidental de la Sierra de Catorce, San Luis Potosí, México: *Revista Mexicana de Ciencias Geológicas*, Vol. 21, pp. 299-319.

Barboza-Gudiño J. R., Orozco-Esquivel M. T., Gómez-Anguiano M., and Zavala-Monsiváis A. (2008). The Early Mesozoic volcanic arc of western North America in northeastern Mexico: *Journal of South American Earth Sciences*: Vol. 25, pp. 49-63.

Barboza-Gudiño, J. R., Ramirez-Fernández, J. A., Torres-Sanchez, S. A., Valencia, V. A. (2011). Geocronología de circones detríticos de diferentes localidades del esquisto Granjeno en el Noreste de México; *Boletín de la Sociedad Geológica Mexicana*, Vol. 63 No. 2, pp. 201-216.

Barboza-Gudiño, J. R., Tristan-Gónzalez, M., and Torres-Hernández, J. R. (1999). Tectonic setting of pre-Oxfordian units from central and northeastern-Mexico: A Review, *in* Bartolini, C., Wilson, J. L., and Lawton, T. F., eds., Mesozoic Sedimentary and Tectonic History of North-Central Mexico: *Geological Society of America Special Paper* 340, pp.197-210.

Barboza-Gudiño, J.R., Torres-Hernández, J.R., and Tristán-González, M. (1998). The Late Triassic-Early Jurassic active continental margin of western North America in northeastern México: *Geofísica Internacional*, Vol. 37, pp. 283-292.

Barboza-Gudiño, J. R., Zavala-Monsivais, A., Venegas-Rodriguez, G., Barajas-Nigoche, L. D. (2010). Late Triassic stratigraphy and facies from northeastern Mexico: Tectonic setting and provenance, *Geosphere*, October 2010 Vol. 6; No. 5, pp. 621-640.

Bartolini, C. (1998). Stratigraphy, geochemistry, geochronology and tectonic setting of the Mesozoic Nazas Formation, north-central Mexico (Ph.D. thesis): *University of Texas at El Paso*, 557 p.

Bartolini, C., Spell, T. (1997). An Early Jurassic age (40Ar/39Ar) for the Nazas Formation at the Cannada Villa Juárez, northeastern Durango, México. *Geological Society of America, Abstracts with Programs*, Vol. 29, No. 2, p. 3.

Bartolini, C., Lang, H., Cantú-Chapa, A., and Barboza-Gudiño, J. R. (2001). The Triassic Zacatecas Formation in central Mexico : Paleotectonic, paleogeographic and paleobiogeographic implications, *in* Bartolini, C., Buffler, R.T., and Cantú-Chapa, A., eds., The western Gulf of Mexico Basin: Tectonics, sedimentary basins and petroleum systems: *American Association of Petroleum Geologists Memoir* 75, pp. 295-315.

Bartolini, C., Lang, H., Spell, T. (2003). Geochronology, geochemistry, and tectonic setting of the Mesozoic Nazas arc in north-central Mexico, and its continuation to north South America. In: Bartolini, C., Buffler, R.T., Blickwede, J.F. (Eds.), The Circum Golf of Mexico and the Caribbean: Hydrocarbon habitats, basin formation and plate tectonics. *American Association of Petroleum Geologists Memoir*, 79, pp. 427-461.

Blickwede, J.F. (2001). The Nazas Formation: A detailed look at the early Mesozoic convergent margin along the western rim of the Gulf of Mexico Basin. In: Bartolini C., Buffler R.T., Cantu´-Chapa, A. (Eds.), The Western Gulf of Mexico Basin: Tectonics, Sedimentary Basins, and Petroleum Systems. *American Association of Petroleum Geologists Memoir*, 75, pp. 317–342.

Burckhardt, C., and Scalia, S. (1905). La faune marine du Trias Supérieur de Zacatecas: Instituto de Geología de México Boletín 21, 44 p.

Cantú-Chapa, A. (1969). Una nueva localidad del Triásico Superior marino en México: *Instituto Mexicano del Petróleo, Revista*, Vol. 1, pp. 71-72.

Carrillo-Bravo J. (1961). Geología del Anticlinorio Huizachal-Peregrina al NW de Ciudad Victoria, Tamaulipas: *Boletín de la Asociación Mexicana de Geólogos Petroleros*, Vol. 13, pp. 1-98.

Carrillo-Bravo. J. (1982). Exploración petrolera de la Cuenca Mesozoica del Centro de México: *Boletín de la Asociación Mexicana de Geólogos Petroleros*, Vol. 34, pp. 21-46.

Centeno-García, E. (2005). Review of Upper Paleozoic and Mesozoic stratigraphy and depositional environments of central and west Mexico: Constraints on terrane analysis and paleogeography, *in* Anderson T. H., Nourse, J. A., McKee, J. W., and Steiner, M. B., *eds.*, The Mojave-Sonora Megashear hypothesis: Development, assessment and alternatives: *Geological Society of America Special Paper* 393, pp. 233-258.

Centeno-García, E. and Silva-Romo, G. (1997). Petrogenesis and tectonic evolution of central Mexico during Triassic-Jurassic time: *Revista Mexicana de Ciencias Geológicas*, Vol.14, No. 2, pp. 244-260.

Clarck, J. M., Montellano, M., Hopson, J. A., Hernández, R. R., and Fastovsky, D. E. (1994). An Early or Middle Jurassic tetrapod assemblage, from the La Boca Formation, northeastern Mexico, in Fraser, N. C., and Sues, H.-D., eds. In the Shadow of the Dinosaurs: Cambridge, *Cambridge University Press*, pp. 294-302.

Chávez-Aguirre, R. (1968). Bósquejo Geológico de la Sierra Peñón Blanco, Zacatecas, Tesis Profesional: *Facultad de Ingeniería, Universidad Nacional Autónoma de México*, 67 p.

Córdoba-Méndez, D. A. (1964). Geology of Apizolaya quadrangle (east half), northern Zacatecas, Mexico (master thesis*): The University of Texas at Austin*, 111 p.

Cuevas Pérez, E. (1985). Geologie des Alteren Mesozoikums in Zacatecas und San Luis Potosí, Mexiko (Ph.D. thesis): *Hessen Marbur/Lahn, Germany*, 182 p.

Davila-Alcocer, V. M., Centeno-García, E., Barboza-Gudiño, J.R., Valencia, V. A., Fitz-Diaz, E. (2008). Detrital zircón ages from the El Chilar accretionary complex and volcaniclastic rocks of the San Juan de La Rosa Formation, Tolimán, Queretaro, Mexico. *Geological Society of America, Abstracts with Programs*, 167-3.

de Cserna, Z. (1956). Tectónica de la Sierra Madre Oriental de México, entre Torreón y Monterrey. *XX. Congreso Geológico Internacional*, México, 87 p.

Dickinson, W. R. (1985). Interpreting provenance relations from detrital modes of sandstones, *in* Zuffa, G. G., ed. Provenance of arenites: *Reidel Publishing, Lancaster*, pp. 333-361.

Dickinson, W. R., Beard, L. S., Brakenridge, G. R., Erjavec, J. L., Ferguson, R. C., Inman, K. F., Knepp, R. A., Lindberg, F. A., and Ryberg, P. T. (1983). Provenance of North

American Phanerozoic sandstones in relation to tectonic setting, *Geological Society of America Bulletin* Vol. 94, p. 222-235.

Dickinson, W.R., and Lawton, T.F. (2001). Carboniferous to Cretaceous assembly and fragmentation of Mexico: *Geological Society of America Bulletin,* Vol. 113, pp.1142-1160.

Fastovsky, D. E., Hermes, O. D., Strater, N. H., Bowring, S. A., Clarck, J. M., Montellano, M., Hernández, R. R. (2005). Pre-Late Jurassic, fossil-bearing volcanic and sedimentary red beds of Huizachal Canyon, Tamaulipas, Mexico, *in* Anderson T. H., Nourse, J. A., McKee, J. W., and Steiner, M. B., *eds.,* The Mojave-Sonora Megashear hypothesis: Development, assessment and alternatives: *Geological Society of America Special Paper* 393, pp.233-258.

Franco-Rubio, M. (1999). Geology of the basement below the decollement surface, Sierra de Catorce, San Luis Potosí, México. in Bartolini, C., Wilson, J. L. and Lawton, T. F. (eds.), Mesozoic Sedimentary and Tectonic History of North-Central Mexico: *Geological Society of America Special Paper* 340, pp.211-227.

Fries, C., and Rincón-Orta, C. (1965). Nuevas aportaciones geocronológicas y técnicas empleadas en el laboratorio de geocronometría; *Boletín del Instituto de Geología, Universidad Nacional Autónoma de México,* Vol. 73, pp. 57-133.

Gallo-Padilla, I., Gómez-Luna., M. E., Contreras, B., and Cedillo, P. E. (1993). Hallazgos Paleontológicos del Triásico marino en la región central de México: *Revista de la Sociedad Mexicana de Paleontología,* Vol. 6, pp.1-9.

Garzanti, E., Doglioni, C., Vezzoli, G., Andò, S. (2007). Orogenic belts and orogenic sediment provenance. *The Journal of Geology,* Vol. 115, pp. 315-334.

Gehrels, G. E., Valencia, V.A, and Pullen A. (2006). Detrital zircon geochronology by Laser Ablation Multicollector ICPMS at the Arizona LaserChron Center, in Olszewski, T., ed., Geochronology: Emerging opportunities, Paleontology Society Short Course: Philadelphia, PA, *Paleontological Society Papers,* Vol. 12, pp. 67-76.

Gómez-Luna, M. E., Cedillo-Pardo. E., Contreras, B., Gallo-Padilla, I., and Martínez-Cortés, C. A. (1998). El Triásico marino de la Mesa Central de México: Implicaciones Paleogeográficas: *II convención sobre la evolución geológica de México y de recursos asociados, Simposia y Coloquio, extended abstracts,* pp. 67-71.

Grajales-Nishimura, J.M., Terrell, D., Damon, P. (1992). Evidencias de la prolongación del arco volcánico cordillerano del Triásico tardío-Jurásico en Chihuahua, Durango y Coahuila. *Boletín de la Asociación Mexicana de Geólogos Petroleros,* Vol. 42, pp. 1–18.

Gutierrez-Amador, M. (1908). Las capas cárnicas de Zacatecas: *Boletín de la Sociedad Geológica Mexicana,* Vol. 4, pp. 29-35.

Hoppe, M., Barboza-Gudiño, J. R., and Schulz, H. M. (2002). Late Triassic submarine fan in northwestern San Luis Potosí, México –lithology, facies and diagenesis, *Neues Jahrbuch Geologie und Paläontologie,* Vol. 2002, No.12, pp.705-724.

Imlay, R. W., Cepeda, E., Álvarez, M., Jr., and Diaz, T. (1948). Stratigraphic relations of certain Jurassic formations in eastern Mexico: *American Association of Petroleum Geologists, Bulletin,* Vol. 32, pp.1750-1761.

Jones, N.W., J.W. McKee, T.H. Anderson, and L.T. Silver. (1995). Jurasic Volcanic rocks in northeastern Mexico: A possible remnant of a Cordilleran magmatic arc. In C. Jacques-Ayala, C. González-León, and J. Roldán-Quintana, eds., Studies on the Mesozoic of Sonora and adjacent areas: *Geological Society of America Special Paper* 301, pp. 179-190.

López-Infanzón, M. (1986). Estudio petrogenético de las rocas ígneas en las Formaciones Huizachal y Nazas: *Boletín de la Sociedad Geológica Mexicana,* Vol. 47, No. 2, pp.1-42.

Ludwig, K.R. (2003). Isoplot 3.00: *Berkeley Geochronology Center Special Publication* Vol. 4, 70 p.

Marsaglia, K. M., e Ingersoll, R. V. (1992). Compositional trends in arc-related deep-marine sand and sandstone: A reassessment of magmatic-arc provenance, *Geological Society of America Bulletin*, Vol. 104, pp. 1637-1649.

Maldonado-Koerdell, M. (1948). Nuevos datos geológicos y paleontológicos sobre el Triásico de Zacatecas: *Anales de la Escuela Nacional de Ciencias Biológicas*, Vol. 5, pp. 291-306.

Martínez-Pérez, J. (1972). Exploración geológica del área El Estribo-San Francisco, San Luis Potosí: *Boletín de la Asociación Mexicana de Geólogos Petroleros*, Vol. 24, No. 7-9, pp. 327-402.

McLennan, S. M., Hemming, S., McDaniel, D. K., and Hanson, G. N. (1993). Geochemical approaches to sedimentation, provenance, and tectonics, *in* Johnsson, M. J., and Basu, A., eds., Processes Controlling the Composition of Clastic Sediments: *Geological Society of America Special Paper* 284., pp. 21-40.

Miall, A. D. (1977). A review of the braided-river depositional environment: *Earth Sciences Review*, Vol. 13., pp.1-62.

Michalzik, D. (1991). Facies sequence of Triassic-Jurassic red beds in the Sierra Madre Oriental (NE Mexico)and its relation to the early opening of the Gulf of Mexico. *Sedimentary Geology*, Vol. 71, pp. 243-259.

Miller, B. V., Dostal, J., Keppie, J. D., Nance R. D., Ortega-Rivera, and Lee, J. K. W. (2007). Ordovician calc-alkaline granitoids in the Acatlán Komplex, southern Mexico: Geochemical and geochronologic data and implications for the tectonics of the Gondwanan margin of the Rheic Ocean, *in* Linneman, U., Nance, R. D., Kraft, P., and Zulauf, G., eds. The evolution of the Rheic Ocean: From Avalonian-Cadomian active margin to Alleghenian-Variscan collition: *Geological Society of America, Special Paper* 423, pp. 465-475.

Mixon, R. B., Murray, G. E., and Diaz, T. G. (1959). Age and correlation of Huizachal Group (Mesozoic), state of Tamaulipas, Mexico: American Association of Petroleum Geologists Bulletin, v. 43, p.757-771.

Mutti, E., and Ricci-Lucchy, F., 1972, Le torbiditi dell'apennino settentrionale; introfuzione all'analisi di facies. Memoir Society Geology Italy 11, 161-199 (1978, English translation in International Geology Review, v. 20, p.125-166).

Nance, R.D., Fernández-Suárez, J., Keppie, J.D., Storey, C., Jeffries, T.E., 2007, Provenance of the Granjeno Schist, Ciudad Victoria, Mexico: Detrital zircon U-Pb age constraints and implication for the Paleozoic paleogeography of the Rheic Ocean, *en* Linnemann, U., Nance, R.D., Kraft, P., Zulauf, G. (eds.), The evolution of the Rheic Ocean: From Avalonian-Cadomian active margin to Alleghenian-Variscan collision: Geological Society of America Special Paper 423, 453-464.

Ortega-Gutiérrez, F., Ruiz, J, and Centeno-García, E. (1995). Oaxaquia, a Proterozoic microcontinent accreted to North America during the Late Paleozoic: Geology, v. 23, p.1127-1130.

Pantoja-Alor, J. (1972). La Formación Nazas del Levantamiento de Villa Juárez, Estado de Durango. Memorias de la Segunda Convención Nacional de la Sociedad Geológica Mexicana, p. 25-31.

Rueda-Gaxiola, J., Dueñas, M. A., Rodríguez, J. L., Minero M., and Uribe, G. (1993). Los Anticlinorios de Huizachal-Peregrina y de Huayacocotla: Dos partes de la fosa de Huayacocotla-El Alamar: *Boletín de la Asociación Mexicana de Geólogos Petroleros*, Vol. 43, pp. 1-29.

Rueda Gaxiola, J., López-Ocampo, E., Dueñas, M. A., Rodríguez, J. L., and Torres-Rivero, A. (1999). Palynostratigraphical method: Basis for defining stratigraphy and age of the Los San Pedros allogroup, Huizachal-Peregrina Anticlinorium, Mexico, *in* Bartolini, C., Wilson, J. L., and Lawton, T. F., eds., Mesozoic Sedimentary and tectonic history of north-central Mexico: *Geological Society of America Special Paper* 340, pp. 229-269.

Reynoso-Rosales, V. H. (1992). Descripción de los esfenodontes (Sphenodontia, Reptilia) de la fauna del cañón del Huizachal (Jurásico temprano a medio) Tamaulipas, México (MS. Tesis) Universidad Nacional Autónoma de México, Facultad de Ciencias, 86 p.

Reynoso-Rosales, V. H. (1996). A Middle Jurassic Sphenodon-like sphenodontian (Diapsida: Lepidosauria) from Huizachal Canyon, Tamaulipas; Journal of Vertebrate Paleontology, v. 16, p. 210-221.

Rubio-Cisneros, I. I., Timothy, F. Lawton. (2011). Detrital zircon U-Pb ages of sandstones in continental red beds at Valle de Huizachal, Tamaulipas, NE Mexico: Record of Early-Middle Jurassic arc volcanism and transition to crustal extension; *Geosphere*, Vol. 7, pp. 159-170.

Silva-Romo, G., Arellano-Gil, J., Mendoza-Rosales, C., and Nieto-Obregón, J. (2000). A submarine fan in the Mesa Central, Mexico: Journal of South American Earth Sciences, v.13. p.429-442.

Tristán-González, M., Torrez-Hernández, J. R., Mata-Segura, J. L. (1995) Geología de la Hoja Presa de Santa Gertrudis, S. L. P., Instituto de Geología, Universidad Autónoma de San Luis Potosí, Folleto Técnico no. 122, 50 p.

Torres, R., Ruiz, J., Patchett, P.J., Grajales, J.M. (1999). Permo-Triassic continental arc in eastern México: Tectonic implications, for reconstructions of southern North America. In: Bartolini, C., Wilson, J.L., Lawton, T.F. (Eds.), Mesozoic Sedimentary and Tectonic History of North-Central Mexico, Geological Society of America Special Paper, 340, pp. 191-196.

Vega-Granillo, R., Salgado-Souto, S., Herrera-Urbina, S., Valencia, V., Ruiz, J., Mesa-Figueroa, D., and Talavera-Mendoza, O., (2008) U-Pb detrital zircon data of the Rio Fuerte Formation (NW Mexico): Its peri-Gondwanan provenance and exotic nature in relation to southwestern North America: *Journal of South American Earth Sciences*, Vol. 26, pp. 343-354.

Venegas-Rodriguez, G., Barboza-Gudiño, J. R., López-Doncel, R. A. (2009). Geocronología de circones detríticos en capas del Jurásico Inferior de las áreas de la Sierra de Catorce y El Alamito en el estado de San Luis Potosí: Revista Mexicana de Ciencias Geológicas 26 (2), p. 466-481.

Weber, R., (1997). How old is the Triassic Flora of Sonora and Tamaulipas and news on Leonardian floras in Puebla and Hidalgo, México: Revista Mexicana de Ciencias Geológicas, Vol. 14 No. 2, 225-243.

Zavala-Monsiváis, A., Barboza-Gudiño, J. R., Valencia, V. A., Rodríguez-Hernández, S. E., García Arreola, M. E. (2009). Las sucesiónes volcánicas pre-Cretácicas en el noreste de Méxcio; Unión Geofísica Mexicana, GEOS, 29 (1), resúmenes de la reunión anual nov. 2009, Puerto Vallarta Jalisco., p. 53.

Zavala-Monsiváis, A., Barboza-Gudiño, J. R., Velasco-Tapia, F., García-Arreola, M. E. (in Press). Edad y correlación de la sucesión volcánica Jurásica expuesta al poniente de Charcas, San Luis Potosí; *Boletín de la Sociedad Geológica Mexicana*.

Permissions

The contributors of this book come from diverse backgrounds, making this book a truly international effort. This book will bring forth new frontiers with its revolutionizing research information and detailed analysis of the nascent developments around the world.

We would like to thank Assist. Prof. Dr. Ömer Elitok, for lending his expertise to make the book truly unique. He has played a crucial role in the development of this book. Without his invaluable contribution this book wouldn't have been possible. He has made vital efforts to compile up to date information on the varied aspects of this subject to make this book a valuable addition to the collection of many professionals and students.

This book was conceptualized with the vision of imparting up-to-date information and advanced data in this field. To ensure the same, a matchless editorial board was set up. Every individual on the board went through rigorous rounds of assessment to prove their worth. After which they invested a large part of their time researching and compiling the most relevant data for our readers. Conferences and sessions were held from time to time between the editorial board and the contributing authors to present the data in the most comprehensible form. The editorial team has worked tirelessly to provide valuable and valid information to help people across the globe.

Every chapter published in this book has been scrutinized by our experts. Their significance has been extensively debated. The topics covered herein carry significant findings which will fuel the growth of the discipline. They may even be implemented as practical applications or may be referred to as a beginning point for another development. Chapters in this book were first published by InTech; hereby published with permission under the Creative Commons Attribution License or equivalent.

The editorial board has been involved in producing this book since its inception. They have spent rigorous hours researching and exploring the diverse topics which have resulted in the successful publishing of this book. They have passed on their knowledge of decades through this book. To expedite this challenging task, the publisher supported the team at every step. A small team of assistant editors was also appointed to further simplify the editing procedure and attain best results for the readers.

Our editorial team has been hand-picked from every corner of the world. Their multi-ethnicity adds dynamic inputs to the discussions which result in innovative outcomes. These outcomes are then further discussed with the researchers and contributors who give their valuable feedback and opinion regarding the same. The feedback is then collaborated with the researches and they are edited in a comprehensive manner to aid the understanding of the subject.

Apart from the editorial board, the designing team has also invested a significant amount of their time in understanding the subject and creating the most relevant covers. They scrutinized every image to scout for the most suitable representation of the subject and create an appropriate cover for the book.

The publishing team has been involved in this book since its early stages. They were actively engaged in every process, be it collecting the data, connecting with the contributors or procuring relevant information. The team has been an ardent support to the editorial, designing and production team. Their endless efforts to recruit the best for this project, has resulted in the accomplishment of this book. They are a veteran in the field of academics and their pool of knowledge is as vast as their experience in printing. Their expertise and guidance has proved useful at every step. Their uncompromising quality standards have made this book an exceptional effort. Their encouragement from time to time has been an inspiration for everyone.

The publisher and the editorial board hope that this book will prove to be a valuable piece of knowledge for researchers, students, practitioners and scholars across the globe.

List of Contributors

Gemma Aiello, Laura Giordano, Ennio Marsella and Salvatore Passaro
Istituto per l'Ambiente Marino Costiero (IAMC), Consiglio Nazionale delle Ricerche (CNR), Napoli, Italy

Roberto Balia
University of Cagliari, Dipartimento di Ingegneria del Territorio, Italy

M.E. Weber, N. Tougiannidis and W. Ricken
University of Cologne, Germany

C. Rolf
Leibniz Institute for Applied Geosciences, Germany

I. Oikonomopoulos and P. Antoniadis
National Technical University of Athens, Greece

Giovanni Leucci
Institute for Archaeological and Monumental Heritage, National Council of Research – (CNR-IBAM), Italy

Victor Manuel Bravo Cuevas and Katia A. González Rodríguez
Area Académica de Biología, Instituto de Ciencias Básicas e Ingeniería, México

Rocío Baños Rodríguez and Citlalli Hernández Guerrero
Licenciatura en Biología, Instituto de Ciencias Básicas e Ingeniería, Universidad Autónoma del Estado de Hidalgo, México

Olga Vasilyeva
Institute of Geology and Geochemistry Ural Branch RAS, Russia

Vladimir Musatov
Lower Volga Institute of Geology and Geophysics, Russia

Donata Violanti
Earth Science Department, Turin University, Italy

Adnan M. Hassan Kermandji
Department of Biology and Ecology, Faculty of Nature and Life, University of Mentoury-Constantine, Algeria

Nabil Y. Al-Banna, Majid M. Al-Mutwali and Zaid A. Malak
Mosul University, Iraq

Igor V. Kemkin
Far East Geological Institute, Far Eastern Branch of Russian Academy of Sciences, Russia

José Rafael Barboza-Gudiño
Universidad Autónoma de San Luis Potosí, México

www.ingramcontent.com/pod-product-compliance
Lightning Source LLC
Chambersburg PA
CBHW070735190326
41458CB00004B/1181